The Chora of Metaponto 3
Archaeological Field Survey
Bradano to Basento

Vol. III

The Chora of Metaponto 3
Archaeological Field Survey
Bradano to Basento

Vol. III

Edited by Joseph Coleman Carter and Alberto Prieto

Institute of Classical Archaeology
Packard Humanities Institute
University of Texas Press, Austin

Printed in China

First edition, 2011

Requests for permission to reproduce material from this work should be sent to:
Permissions
University of Texas Press
P.O. Box 7819
Austin, TX 78713-7819
www.utexas.edu/utpress/about/bpermission.html

♾ The paper used in this book meets the minimum requirements of
ANSI/NISO Z39.48-1992 (R1997) (Permanence of Paper).

The chora of Metaponto 3 / edited by Joseph Coleman Carter and Alberto Prieto. — 1st ed.
p. cm.
Includes bibliographical references and index.
ISBN 978-0-292-72678-9 (cloth : alk. paper)
1. Metapontum (Extinct city) 2. Metapontum (Extinct city)—Rural life and customs. 3. Agricultural colonies—Italy—Metapontum (Extinct city). 4. Greeks—Italy—Metapontum (Extinct city). 5. Excavations (Archaeology)—Italy—Metapontum (Extinct city). 6. Metaponto Region (Italy)—Antiquities. I. Carter, Joseph Coleman. II. Prieto, Alberto. III. Title: Chora of Metaponto three.
DG70.M52C56 2011
937'.773—dc22

2011004131
978-0-292-73481-4 (E-book)

For reasons of economy and speed, this volume has been printed from camera-ready copy furnished by ICA, which assumes full responsibility for its content.

To the memory of Dinu Adamesteanu,
to Antonio De Siena,
to students of Metaponto, past, present, and future.

Special Acknowledgment

These volumes and others in this series could not have been conceived, much less carried through to publication, without the generous financial support of the Packard Humanities Institute.

The Director and staff of the Institute of Classical Archaeology and its collaborators express their profound gratitude.

Contents

Volume I

Volume II

Volume III

Volume IV Atlas

VI. Division Lines: New Data and Analysis

28
The 1999 Pizzica Excavations: The Division Lines Reconsidered

Joseph Coleman Carter

No aspect of the chora of Metaponto—except possibly the remarkable density of the rural sites themselves, especially during the time of the Greek occupation—has stimulated more discussion than the so-called "division lines." First identified just over forty years ago in pioneering aerial photographic studies[1] and surface surveys,[2] these linear features (hereafter referred to as division lines or simply as lines) consist of longitudinal and transverse lines that intersect at less than right angles (80°). Since that discovery, the date, function, and even the very existence of the lines have been topics of debate (see Ch. 1). An additional issue of speculation has been their relationship to the analogous orthogonal street plan of the ancient city,[3] also discovered by aerial photography and investigated through large-scale excavations beginning in 1964. The main body of remotely sensed evidence on which to base arguments, however, long continued to be the aerial photos.

Over the last quarter century there have been significant changes in data collection and analysis. The following advances have provided a new basis for reexamining the division line question:

1. A careful restudy of the early aerial photographs by Alberto Prieto (Ch. 31).[4] This review includes the photos taken during WWII and immediately after, especially those from the Gruppo Aereo Italiano (GAI) mission of 1954.
2. An intensive field survey of the chora, conducted between 1981 and 2008, that has pinpointed the surface evidence for past human activity in the chora, including visible traces of the longitudinal and transverse lines. This data has substantially augmented the information of previous surveys.[5]
3. Detailed geological and geomorphological studies of the chora (Chs. 1–2).
4. The excavations of a number of the survey sites that have exposed extensive longitudinal and transverse lines in several areas of the chora.[6]
5. The archaeological applications of highly precise digital tools, including the Global Positioning System (GPS) and total station (laser theodolite), which have greatly reduced the margin of error in the analyses of spatial relationships.
6. The development of the geographic information system (GIS) of the chora as a whole (Ch. 4), which makes use of highly detailed databases containing information on all the features of the chora, including the site, the lines, and the pottery found in association with them.

The importance of the GIS cannot be overemphasized. It makes it possible to explore the spatial and chronological relationships of the ancient features of the chora that have been discovered through survey and excavation, all with a degree of accuracy unimaginable a decade ago. The GIS also relates these items to the physical aspects of the chora, the topography, and the geomorphology (see Chs. 1–2). Now that the system has been implemented, these relationships can be investigated in the GIS with a great range and variety of statistical tools. The GIS investigation by itself has not resolved the complex question of the division lines, but, combined with excavations, it has made the picture much clearer.

The 1999 Pizzica Excavations

The situation on the ground in 1999 (and several years thereafter) was ideal for the investigation of the lines. Archaeological excavation on a scale never before known in the chora was triggered by non-archaeological excavation that would have a major economic impact on the region. This excavation was a particularly welcome event for those concerned with the preservation of the cultural heritage, as it was the

1 Schmiedt and Chevallier 1959; Adamesteanu 1968, 1973.
2 Uggeri 1969; Adamesteanu 1973.
3 Castagnoli 1958.
4 Prieto 2005 and Ch. 31.
5 By Uggeri 1969, the Soprintendenza (see Adamesteanu 1968, 1973, and 1974), and a French group (see Adamesteanu and Vatin 1976).
6 For a summary of excavated but unpublished sites between 1966 and 1969, see Adamesteanu 1973 and 1974. Amore 1992–1993.

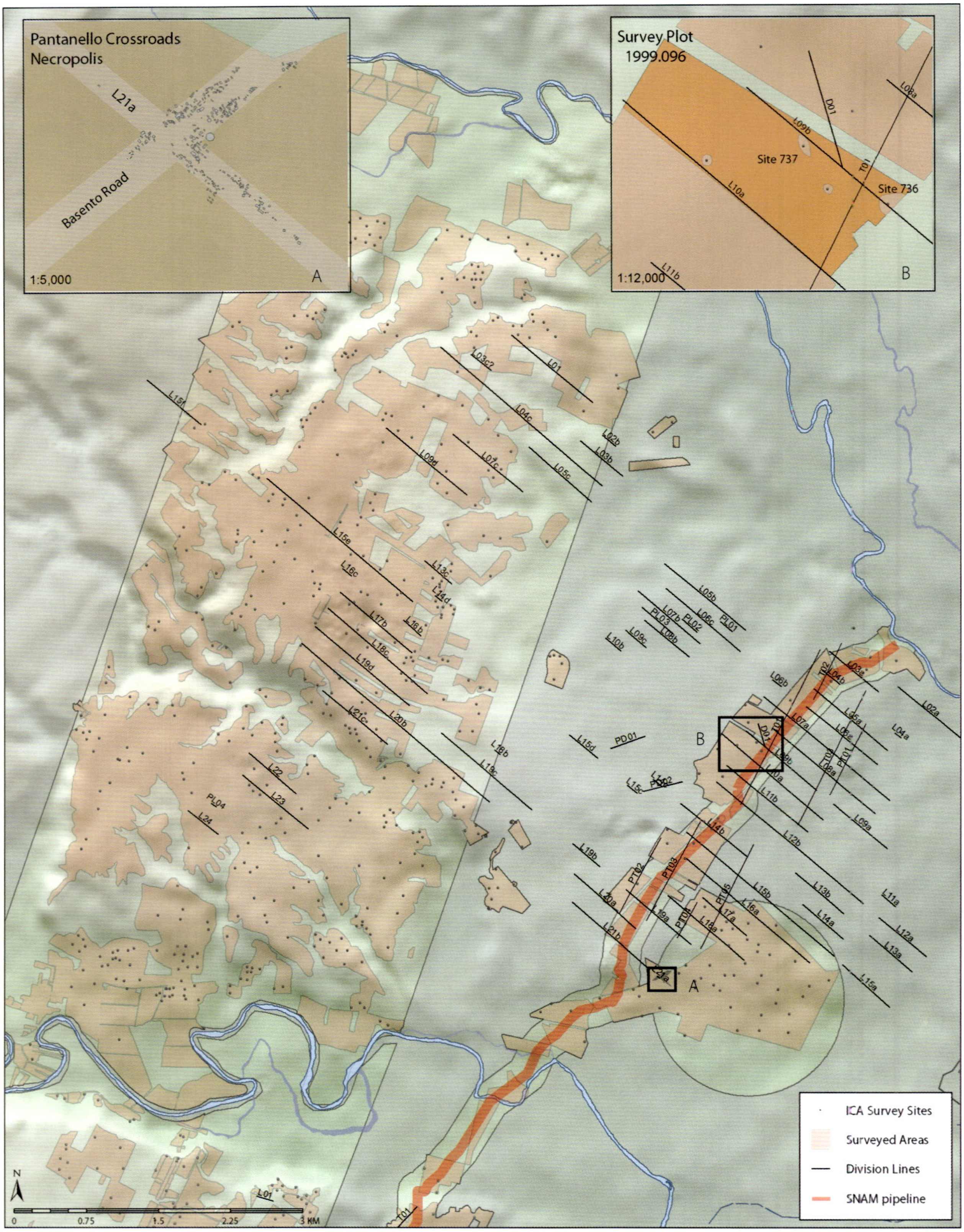

Figure 28.1 1980–1999 plan showing division lines in the area between the Bradano and Basento Rivers. The path of the SNAM pipeline is indicated in red. Inset A shows the intersection of L021 with the Basento Road, both flanked by tombs; inset B shows sites 736 and 737 (plot 1999.096) where line T01 intersects line L01, which is likewise flanked by tombs. (JT)

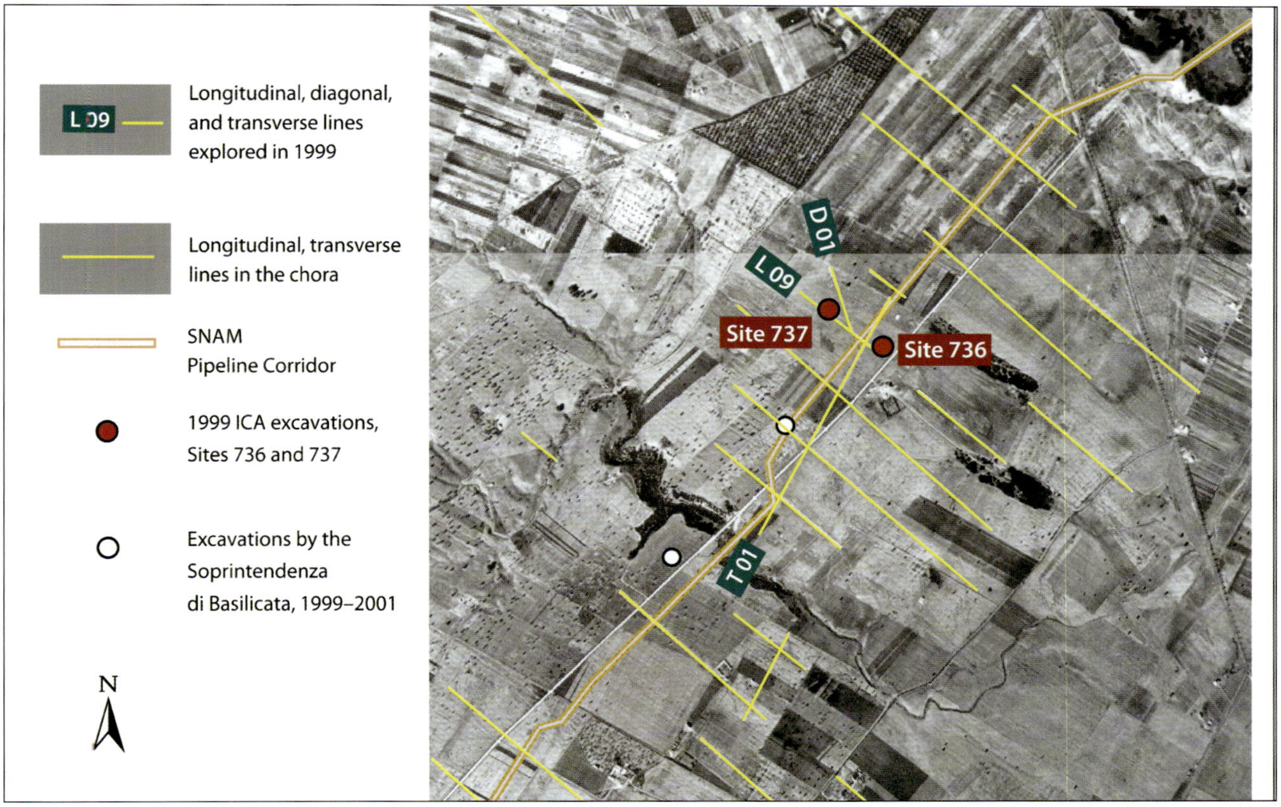

Figure 28.2 Aerial photo showing identified lines in the area of Pizzica, and location of ICA excavations in 1999. Five longitudinal divisions cross the one transverse line (T 01) preserved at Pizzica. (GAI Pos.6518/Neg.31661, reproduced with permission of the Aerofototeca, Rome, 2004)

first instance that such a high level of governmental-industrial intervention in the region also included large-scale archaeological mitigation as an integral part of the project.

The discovery of a major oil and gas deposit in the interior of Basilicata in the upper valley of the Agri River resulted in the decision by the Italian energy giant ENI and its SNAM engineering division to construct an oil conduit (*oleodotto*). The planned course ran from the source, near the town of Viggiano, down the valley of the Agri and Cavone Rivers and across the coastal plain to ENI's AGIP refinery at Taranto. The Bernalda-Brindisi gas conduit (*metanodotto*) being simultaneously constructed by SNAM followed the same course. The path of the projected pipelines led across part of the chora of Herakleia, across the whole of the chora of Metaponto (from the Cavone on the south to the Bradano River on the north) and across the southern half of the chora of Taras (from the Bradano to the outskirts of Taranto).

As planned, this path would cross the Metapontino at distance of approximately 7 km from the modern coastline (*Fig. 28.1*).[7] It would inevitably affect the archaeological sites in the chora lying in the path of the pipelines and for some distance on either side. To the untrained eye there were no visible sites; however, three decades of archaeological investigation, salvage excavations, and especially field survey had revealed an actual density of 15–20 sites per square kilometer throughout the chora. These were the sites for which there was some evidence on the surface. In contrast, sites in the valley bottoms or edges, such as the important and extensive Greek rural sanctuary at Pantanello (excavated between 1974 and 1991 by ICA), would not have been revealed by survey, because these sites were too deeply buried under the thick alluvial deposits that characterize the river valleys and coastal plain of the chora.

In spring 1999, dottoressa Maria Luisa Nava, the archaeological superintendent of Basilicata at the time, invited ICA to carry out an intensive survey along the proposed pipeline corridor. Led by Steve Thompson, the survey team discovered, as expected, a large number of sites (some 66) almost all of which either belonged to the period of the Greek polis or

[7] Also Nava 2002, 737, Fig. 7.

Figure 28.3 Looking W-NW, backhoe displacement of darker soil in plot 1999.096. (JCC)

Figure 28.4 Looking W-NW, the strip of dark soil running SE to NW that corresponds to longitudinal line L09. (JCC)

had a component dating to that time frame.[8] The surface remains indicated that some probable ancient farmhouses were in a remarkably good state of preservation (e.g., sites 728 and 730).[9] Excavations were subsequently carried out in 2000 and 2001 at many of the sites in both halves of the chora by the Soprintendenza per i Beni Archaeologici per la Basilicata, under the direction of Antonio De Siena, who was the director of the Museo Archeologico Nazionale di Metaponto at that time.

The entire 1999 survey transect measured 18 km long and 200–500 m wide, with a total area of some 5 km^2. This transect crossed the suspected positions of at least sixteen of the twenty-four longitudinal lines identified in the aerial photographs of the chora between the Bradano and Basento (*Fig. 28.2*). In some cases the survey team could easily identify the lines in the photographs with long, gully-like depressions in the landscape. The line positions were noted and dimensions taken using the GPS and total station by a joint team from ICA and the Center for Space Research (CSR) of the University of Texas at Austin.

Longitudinal line L09 in Prieto's study (Ch. 31) made a sudden and unmistakably clear physical appearance in plot 1999.096, an open trapezoidal-shaped field lying in the pipeline's path (*Fig. 28.1, inset*). The field was rapidly being prepared for the planting of grapevines in the week the team reached it. A large backhoe (*Fig. 28.3*) was systematically scooping soil—one row and one bucketful at a time—from a depth of approximately 1.2 m below the current ground level and transferring it sideways the length of the extended backhoe arm, about 3–4 m. The regularity of the procedure is important because it effectively preserved (though offset slightly) the location of the soil in the field while bringing the lower contents up to the surface. It did little to change the soil's composition or appearance, and in essence simply turned the field upside down.

What appeared to the survey team that morning in July was a clear strip of dark soil crossing the field for a distance of 360 m from SE to NW (*Fig. 28.4*), and a similar dark strip running from SW to NE intersecting the first at a distance of 100 m from the SE edge of the field. The position of these strips of darker soil corresponded precisely with the positions of longitudinal line L09 and transverse line T01 that appeared in the GAI aerial photos (*Fig. 28.5*). The angle of intersection, as in the aerial photographs, was not quite a right angle (Ch. 31). A third straight dark strip was spotted, converging diagonally with the longitudinal line at a sharp angle at a distance of 50 m to the north of the intersection. This line, labeled D01, is also present in the aerial photo (*Fig. 28.5*). The survey team's attention then focused on two restricted areas beside the longitudinal line where many fragments of tombs, tiles, limestone slabs, grave goods, and large vessels (used as tomb markers) clustered on the surface of the freshly turned soil. These areas were labeled as survey sites 736 and 737, and the individual

[8] *Ancient Territories* 1999, 12–20; Nava 2000, 680–84, and 2001.

[9] All of the sites in the path of the pipelines between the Basento and Bradano are included in this volume; those in the Basento-Cavone area will be treated in a subsequent one.

concentrations (numbering identically 17 at both sites) were recorded and the ceramic fragments collected.

We were able to determine from this initial investigation that the previously undisturbed archaeological level began about 1.2 m under the surface. The soil layer above it had been altered when the field was converted to a vineyard twenty years earlier, as we learned from the proprietor, Signor Di Donna. The 1999 earth moving operation went some 20 cm deeper than previous digging, disturbing for the first time an archaeological level that included the features identified as division lines along with tombs. This lamentable destruction of archaeological features became a serious problem—especially on the marine terraces—with the introduction of deep plowing in the 1950s and 1960s, and has worsened with each passing year. Here the backhoe destroyed grave goods and tomb constructions, and transported large, fresh fragments of stone, tile, and pottery to the surface. In the case of the linear features, it upturned dark soil that contrasted sharply with the rest of the field. The lower parts of these features, as well as a dozen mostly intact burials (six each from sites 736 and 737) lay below the longest reach of the backhoe and were spared.

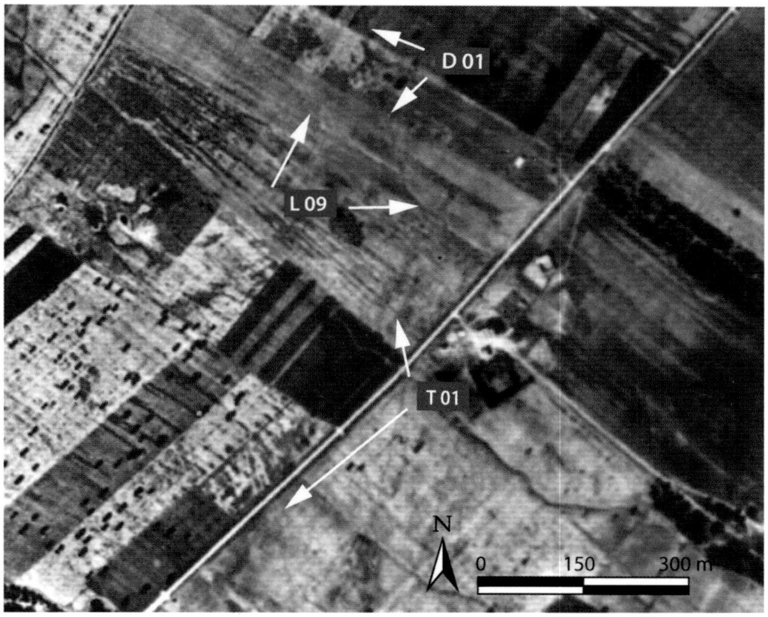

Figure 28.5 Aerial photo showing position of longitudinal line L09, transverse line T01, and diagonal line D01. (GAI Pos.6518/Neg.31661, reproduced with permission of the Aerofototeca, Rome, 2004)

A precise relationship between the division lines on the aerial photograph and the archaeological features that appeared on the ground was established for the first time in 1999, when Jessica Trelogan scanned the archive of aerial photos in the Aerofototeca in Rome. These were georeferenced to a recent satellite image of the area, which in turn was georeferenced through a GPS survey of the chora.[10] The survey was carried out by experts from the CSR, under the direction of Professor Melba Crawford. From that point onward the aerial views indicated, for first time, the precise position of sites on the ground. This combination of the aerial photographic studies and intensive field survey had clearly pinpointed a busy area of the chora, and one at risk of being seriously compromised in the immediate future by the pipeline. Permission was requested from the Soprintendenza to excavate first at sites 736 and 737, which were situated at the margins of the field where the longitudinal, transverse, and diagonal lines passed, and then in the area that marked the crossing of these lines.

A joint ICA-Soprintendenza team was formed on the spot. Within twenty-four hours De Siena obtained permission to excavate from the landowner and digging began at site 736. The backhoe that had been working in agriculture up to that moment was employed to gently remove the overburden that it had just created from an area of 15 x 15 m, which was later expanded to 20 x 26 m to encompass site 736. Our experienced foreman of twenty-five years, Giuseppe DiTaranto, closely supervised the earth-moving operation, assisted by Gaetano Petrarca and Nunzio Carella. Draftsman Saverio Careccia of the Soprintendenza recorded the burials, site plans, and sections, while the excavation was supervised by the author. The first positive results were achieved within a few hours. Two intact tombs appeared the first day: an *a cappuccina* tomb with its tile covering, and the miraculously intact plaster-lined fossa burial T 19 with its treasure of fine red-figure and black-gloss pottery (Ch. 29).

Longitudinal line L09, sites 736 and 737, trenches 6 and 7

Over the course of the following week, the excavation at site 736 was broadened to the east of the square, where further burials were encountered, and to the

[10] The procedure is described in *Ancient Territories* 1999, 18–20.

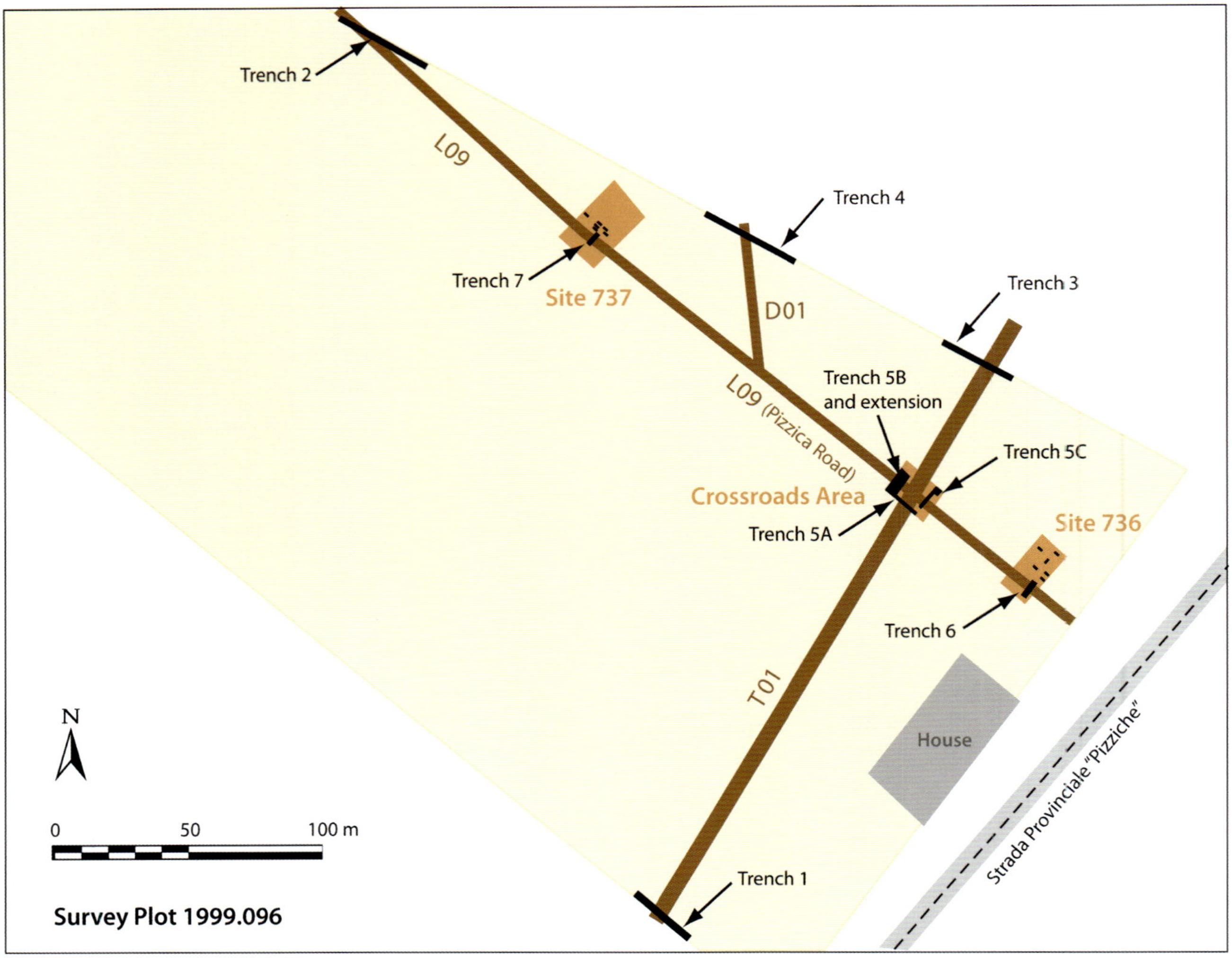

Figure 28.6 Plan of plot 1999.096 showing trenches, necropoleis, longitudinal, transverse, and diagonal lines. (DB)

south, where the lower levels of longitudinal line L09 were revealed (*Fig. 28.6*). The dark soil of L09, containing organic material as well as pottery, contrasted sharply with the earth into which it was cut, which was lighter in color and consisted of compacted sand and gravel. The distinct outline of the strip (*Fig. 28.7*) left no doubt about its orientation, which was precisely that of the light-colored linear anomaly on the aerial photographs (*Fig. 28.5*), or its width, which was a uniform 6 m (give or take a few centimeters) at the surface where the backhoe stopped. This surface was carefully cleaned, and its position recorded.

As a side note, it is worth observing that the opportunity to view the archaeological feature on the ground corresponding to a line does not come often.[11] The earliest excavation to encounter one was in 1984, at the Pantanello necropolis (*Fig. 28.1, inset A*). A largely empty strip of land approximately 9 m wide, flanked on both sides by more than 80 burials of the period 480–280 BC, was exposed for a distance of 150 m. It was found to correspond perfectly in its orientation with the system of lines known from the aerial photographs.[12] The excavators tested the hypothesis that it was a road intersecting the Basento Road (itself an indisputable road and major thoroughfare) by cutting several test trenches perpendicular to it and studying the section. The investigation appeared to confirm the hypothesis, showing that it was indeed a uniformly wide strip of level surface, 5 to 6 m wide, flanked by a drainage ditch on its south side and defined by closely spaced burials in five nuclei. The arrangement was

[11] Anecdotal accounts by Schmiedt and Chevallier 1959, Uggeri 1969, and Adamesteanu 1973, 1974 are based on visual inspection of the surface. They are valuable because they are early, in the 1950s and 1960s, before large-scale earth moving began.

[12] A second parallel but narrower strip lined by tombs was discovered by De Siena's excavation at another site in Pantanello in 1986. Carter 1998f.

Figure 28.7 Site 736, looking NW, with the dark soil of longitudinal line L09 in foreground. (JCC)

Figure 28.8 Site 736, trench 6, looking SW, with profile of longitudinal line L09 (see section drawing in Fig. 28.9)

analogous to the Basento Road which was flanked on both sides with substantial drainage ditches and a large number of burials in well-defined plots (see below, under Geomorphology and Excavations 2000–2002).[13]

At site 736, a 1-m wide trench (trench 6) was excavated down to virgin soil to obtain a stratigraphical cross section through L09 (*Fig. 28.8*). This test revealed the regular edges of the 6.6-m wide strip of gray-brown soil. The orientation and position were precisely those of the linear feature on the aerial photograph. The profile (*Fig. 28.9*), which narrowed as the trench was deepened, was not symmetrical. The southwest edge tapered down more gradually than the north edge toward the deepest part of the feature, located on its north side. The profile on both sides steps down to a relatively flat surface which measures approximately 3 m across.

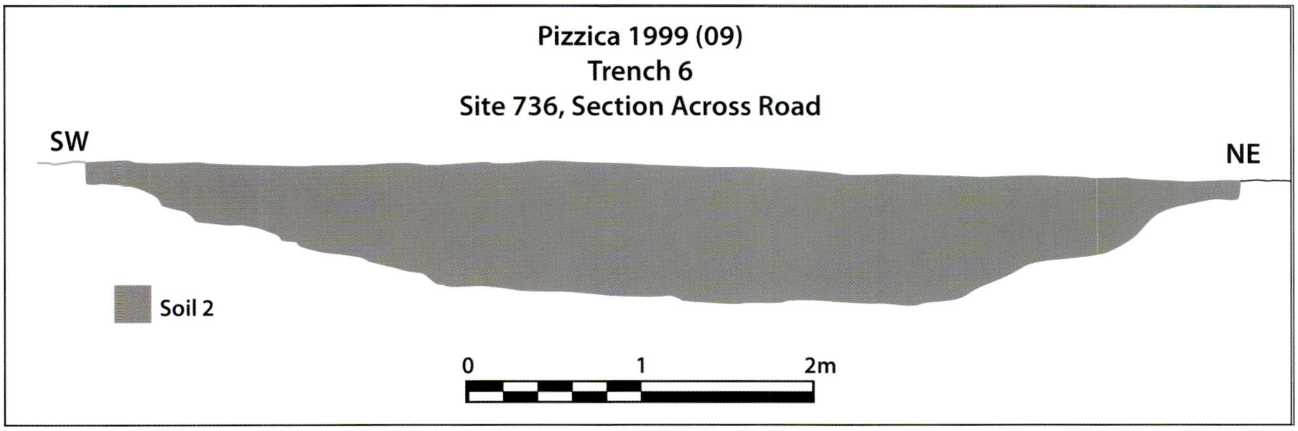

Figure 28.9 Section, site 736, trench 6, showing profile through L09.

The uniform dark soil of the fill in this incised feature contained organic material and pottery,[14] and contrasted sharply with the light brown compacted sand and gravel of the bedrock soil into which it was cut (*Fig. 28.9*). There was no indication of a road sur-

[13] Carter 1993b, 33–46. A further example of a division line road flanked on either side by a drainage ditch was discovered by De Siena in the Basento-Cavone half of the chora at San Teodoro (Nava 2003a, 670, Pl. 35.1)

[14] Soil samples were taken. The full analysis of the paleobotanical evidence will appear in another volume in this series (for a summary, see Ch. 19).

Figure 28.10 Site 737, looking N-NW showing darker soil of longitudinal line L09. (JCC)

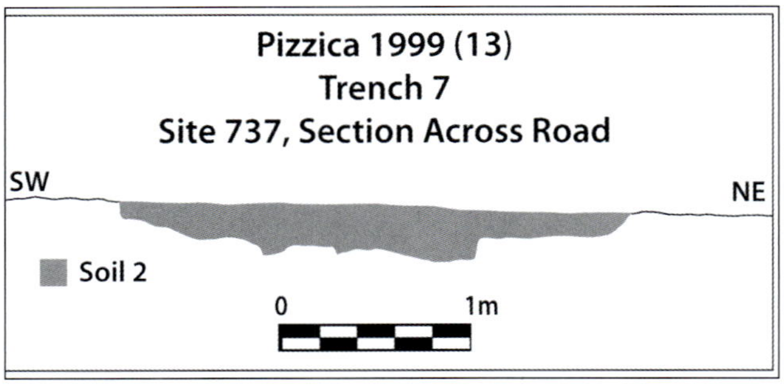

Figure 28.11 Section, site 737, trench 7, showing profile through longitudinal line L09.

face, and pottery emerged in each of the 10–15 cm thick levels as the trench was deepened.

Weighing the evidence, we concluded that it was a ditch, and this was confirmed by the many examples of similar features excavated by the Soprintendenza, 2000–2003. In these cases, there could be no doubt that they were ditches, drainage canals.[15] The ceramics provide an approximate date in the late 5th century BC for the lowest level of the fill of this truncated feature; the latest pottery from the uppermost preserved level of the fill is no later than the 4th to mid-3rd centuries BC. The pottery is described and illustrated in Chapter 29, which includes an analysis of pottery from the shallower, less well-preserved segment of L09, at site 737.

[15] Nava 2001, 940.

The strip of dark soil continued across the field in a NW direction (49.88° west of north). A decision was made to excavate approximately 200 m to the north, where the surface scatter, again consisting of fragments of large red-figure and fine black-gloss vessels, suggested that another necropolis might have flanked the line. The trench, measuring approximately 18 m x 30 m, was designated as site 737. At the lowest depth reached by the backhoe in its earlier pass through the fill, a strip of dark-colored soil appeared again. Though much attenuated, it continued across the whole length of the trench. This strip is visible in the photo just to the southwest of the excavated tombs (*Fig. 28.10*). The feature was closer to the modern surface of the field and had been more severely damaged by agricultural activity than was the case in trench 6 at the eastern edge of the field. Trench 7 confirmed this, revealing that only the lowest 20–30 cm of the feature was preserved (*Fig. 28.11*). The dark fill is only 2.5 m wide, and as in trench 6, the bottom is fairly flat. The fill, again, consisted of dark earth, but with very few fragments of pottery.

The continuation of longitudinal line L09 to the NW was confirmed by trench 2 (*Fig. 28.6*). A few fragments of pottery were recovered from the feature's fill, including: a Late Archaic (520–480 BC) fine-ware cup handle, three fine-ware body sherds, the base of a plain-ware one-handler, and a storage-ware fragment.

Crossroads area, transverse line T01, trenches 1, 3, and 5A, 5B, 5C

Trenches 1 and 3 were dug to confirm the position of transverse line T01 that appeared in the aerial photographs (*Figs. 28.2, 28.5–6*). Excavation of the feature in section revealed that a channel measuring 5.2 m across had been cut into virgin soil. The contrasting colors, differences in texture, and ceramic remains that were evident in the section through longitudinal line L09 were not so apparent here: the fill, less clearly defined, consisted of disturbed sandy gravels with less obvious organic and fewer cultural materials (see Ch. 29).

The appearance of transverse line T01 was more distinct in the trenches where it crossed longitudinal line L09. The four corners of the intersection were clearly delineated by the darker soil of the fill (*Figs. 28.12–13*). Trenches 5A, 5B, and 5C were cut to confirm the position of the intersection (*Fig. 28.14*); trenches 5B and 5C cut through the longitudinal

Figure 28.12 Crossroads area, looking SE. Trench 5B and 5B extension in foreground, trench 5A right, trench 5C in background. (SMT)

line and were dug to undisturbed virgin soil.[16] The profile of trench 5C showed that a channel filled with compact, darker soil had been cut through a gravel bank, similar to the cuts seen in trenches 6 and 7 (*Fig. 28.15*). In an interesting coincidence, a modern irrigation pipeline—since removed—had been placed in a 1-m wide trench in the center of the cut. On the east, in contrast to the gradual slopes into L09, was a vertical cut through a native conglomerate stone outcrop. The cut in the SE corner of the trench floor was the first trace of a quarrying operation, more fully revealed by the expanded excavation in trench 5B to the northwest (*Figs. 28.12, 28.14*).

As trench 5B was enlarged, regular cuts into the native conglomerate were discovered beneath the darker soil corresponding to the position of longitudinal line L09. Further excavation proved these to be the results of quarry activity: on the west side, a conglomerate slab was being prepared for extraction (*Figs. 28.12, 28.14*). It measured 0.7 x 1.6 m. The dimensions correspond nicely to conglomerate slabs used in early structures in the chora: e.g., the slabs of the temenos wall at Pantanello, the structure around the spring at San Biagio, and the foundation blocks of Oikos C in the main sanctuary of Metaponto. These projects date to the early 6th century, before the Metapontines could afford to import better quality stone. Such conglomerate slabs, however, were also

Figure 28.13 Crossroads area, looking S. Trench 5B in foreground, trench 5A to right showing section of line T01. (SMT)

employed for the construction of cist tombs dating to the mid-5th and late 4th to early 3rd centuries BC.

Further north in trench 5B another remarkable find emerged. Along with two more roughed out conglomerate blocks of similar dimensions were two wheel ruts. Occupying the precise center of the area, these were later covered by the dark earth characteristic of the fill of longitudinal line L09. The tracks were approximately 20 cm wide on the average, the distance between them ranging from 1.2 to 1.3 m.

[16] Cf. Nava 2000, Pl. 53.1, Fig. 4.

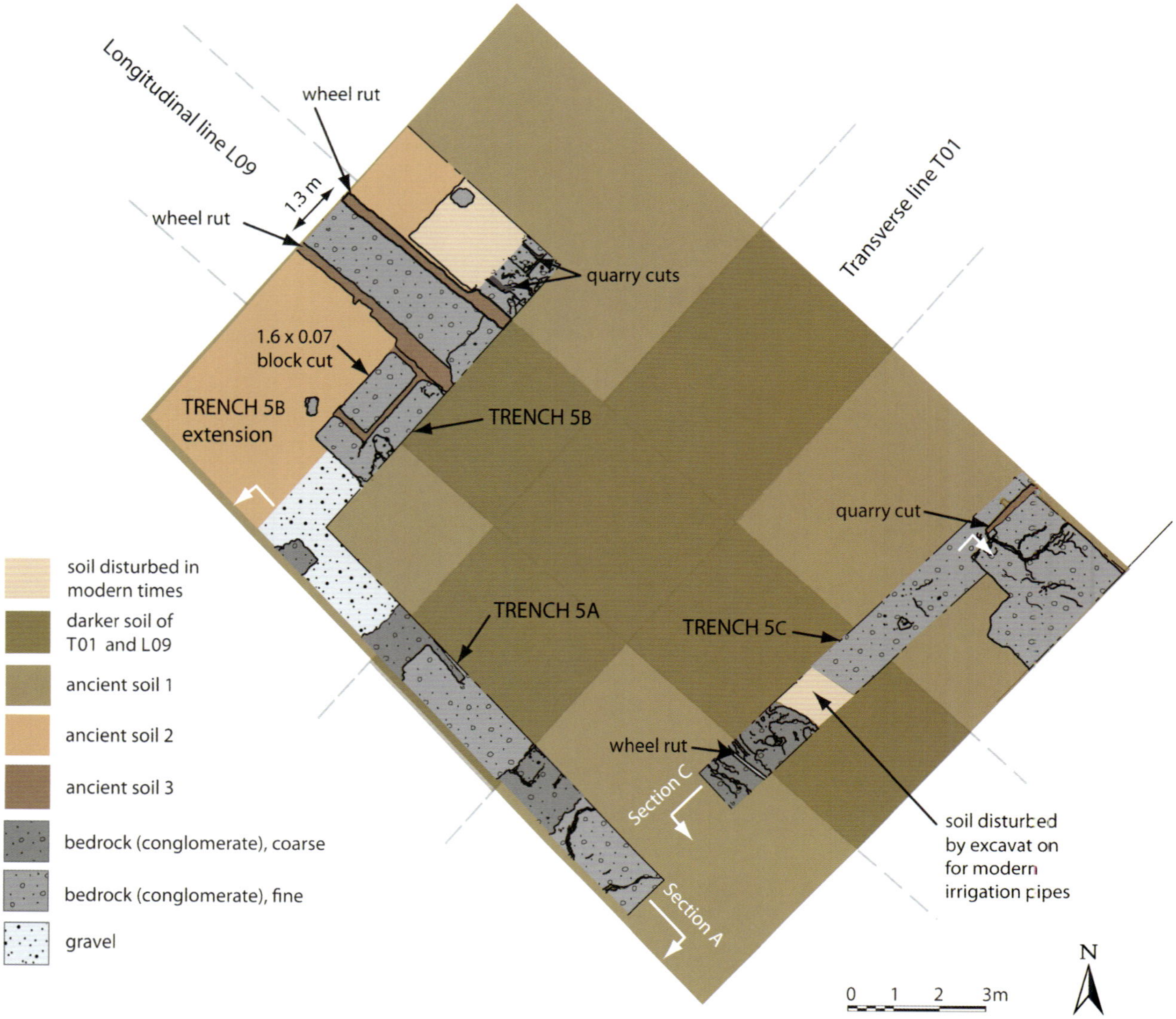

Figure 28.14 Plan, crossroads area, showing quarry activity and wheel ruts. (DB)

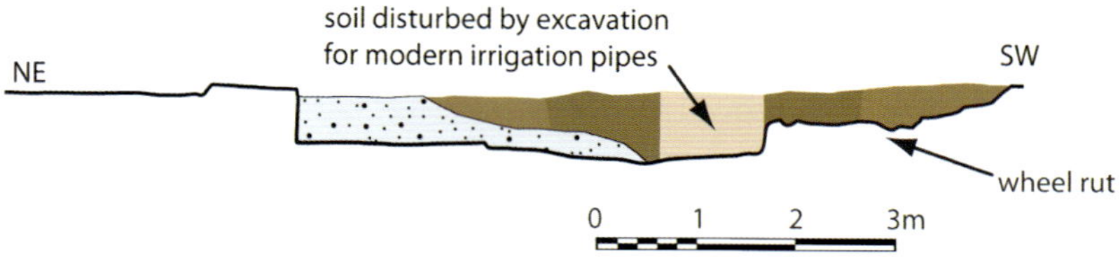

Figure 28.15 Section 5C.

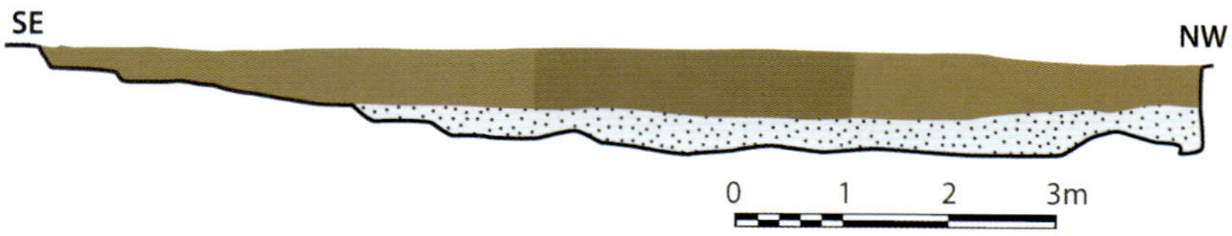

Figure 28.16 Section 5A.

This distance corresponds with the known axle width of wagons in the ancient world, particularly in ancient Italy.[17] Ancient wagon tracks are a rare discovery and almost unheard of in the countryside.[18] They are highly significant evidence for important economic activity in the chora at an early period in the colony's history.[19]

The orientation of the wheel tracks, which were uncovered for a length of slightly over 4 m, is 52° west of north, that is, very close to the orientation of the linear features determined from aerial photographs (L09, 49.88°).[20] The feature identified as a ditch or drainage canal would thus have passed right over the rutted road, which must, therefore, have antedated it. There is another example of a canal overlying a road. This was not the case at Pantanello. It should be noted, however, that these wide ruts do not appear in the conglomerate bedrock some 10 m to the south in trench 5C (*Figs. 28.14–15*), where only one narrow, slightly curving rut was found south of the axis of those in trench 5B.

It is not possible to date either the wheel ruts or the quarrying activity precisely. It would appear, however, that the road preceded the quarry cuts to the north and south of it, as the cuts seem to encroach on the tracks. The road was probably essential to the operation of the quarry.

Transverse line T01, verified by trenches 1 and 3 (*Fig. 28.6*), can also be seen in the plan and section of trench 5A (*Fig. 28.14*). As in trenches 1 and 3, the path it follows is far less obvious and more difficult to interpret than that of longitudinal line L09. The conglomerate bedrock forms a low rise about 20–30 cm high and 1 m wide in the NW corner of the trench (*Fig. 28.16*). To the south, the conglomerate floor of the trench is almost perfectly flat for a space of 4 m before it begins to rise in gentle steps. The total width to the south of the rise and the furthest extent of the step is a little over 6 m. The fill is sandy gravel, as in trenches 1 and 3, with little evidence of the presence of organic material or ceramics that characterizes longitudinal feature L09.

Diagonal line D01, trench 4

The aerial photographs show a diagonal line, D01 (*Figs. 28.2, 28.5*), and while trench 4 verified its presence, it did not contribute much more information about its possible purpose or function. This cut had steep sides and an irregular profile measuring 6 m across, corresponding to an excavated cut approximately 4 m wide (measured perpendicular to its axis). The fill, disturbed pebbles and sand, was similar to that of trenches 1 and 3 across transverse line T01.

Summary

The discovery of a series of parallel ancient man-made ditches or channels corresponding to the lines on the aerial photographs, separated like them by ca. 210 m, has fundamentally changed the way we look at the division lines and their function in the chora. These features correspond to the linear anomalies in the aerial photographs, and though they vary considerably in size and form, most if not all had the principal function of collecting water from the surface of the chora and, where possible, of directing it away from the fields towards the natural drainage of the river valleys. They were in a number of cases flanked by roads, as has become clear in the excavations at Pantanello and elsewhere. Thus they served multiple purposes, providing also a means of communication with the chora. It has been convincingly demonstrated that the *canali di bonifica* date to one period, the early first half of the 5th century BC (see Excavations 2000–2002 below).[21] The overall implications of this massive engineering project for the organization of the chora have been discussed in Ch. 23 and will be revisited below (see The Division Lines Reconsidered).

This is the place, instead, to summarize the observations made about the features discovered in plot 1999.096 of the Metaponto survey. It is clear now that the longitudinal line L09, provisionally identified as a road in July 1999,[22] was one in a series of drainage channels that were particularly numerous in the area known as Pizzica. The fill, composed of organically rich soils and artifacts, accumulated gradually in the ditch. The surface in this area is very flat, without

[17] Piggott 1983.

[18] Gissing 1996, 40.

[19] According to Adamesteanu 1980, 64, *puddinga* was used in structures at the end of the 7th/early 6th c. and again in the 4th c. BC, and occasionally for *argoi lithoi* in the 6th and 5th c. BC. Mertens 1998, 123. Nava 2000, 688.

[20] See Ch. 31.

[21] Nava 2000, 685–68; 2001, 946–50; 2002, 740–42; and 2003a, 664–67, punctuated reports on the discoveries made by Antonio De Siena and the team of the Soprintendenza Archeologica della Basilicata.

[22] *Ancient Territories* 1999.

much slope, and there was little evidence, apart from the pottery of aquatic plants, of flowing water, such as gravel and sand lenses. The ceramics which filled the ditch were deposited at various times beginning, it would seem, in the late 5th or 4th century BC, and reflect the wider history of the settlement of the chora, with pottery of the Late Archaic and Classical periods mixed in random order in the fill. From at least the mid-5th century through the 4th and early 3rd century, the small necropoleis at sites 736 and 737 flanked the ditch. Given the relatively large quantity of pottery from a 1 x 6 m trench, it may be that other ancient structures beside tombs stood very close by.

The close proximity and alignment of tombs with the channels, and the roads that probably flanked them, were common features throughout the chora, as excavations in 1984–86 and in 2000–03 have demonstrated. There is a very good reason for this, as the ditches—and the roads that in some cases may have preceded them—also had the double function of dividing properties, given that the distance separating them, ca. 210 m, corresponds to a standard Greek land measure.

The necropoleis excavations

The two small necropoleis which occupy the north side of longitudinal line L09 were approximately 200 m apart, separated by transverse line T01, which intersects L09 about 50 m north of site 736 and 150 m south of site 737 (*Figs. 28.2, 28.6*).[23] The concentrations of surface pottery fragments at both sites were given numerical labels, each of which happened to run from 1 through 17 (these groups of surface finds, the dating of which agrees with the preserved burials, are documented in Ch. 29: Section A, Surface Finds; sites 736 and 737). The soil had been thoroughly churned and mixed by the backhoe as it prepared the field for the planting of a new vineyard. As the soil was slowly removed under controlled conditions from the original 15 x 20 m squares at site 736 and site 737, pottery was found loose in the backfill. This sporadic material—on the whole better preserved than that on the surface—was collected; like the groups from the surface, it is indicative in a general way of the chronological range of the small necropoleis (see Ch. 29, Selective Catalog of Sporadic Finds: sites 736 and 737).

The few burials that remained in situ appeared immediately at the interface between the disturbed soil and the virgin soil into which the tombs were cut. The backhoe had sheared off some of the tile-covered burials (site 737, T 28) and probably crushed others (site 736, T 28). Of the sole sarcophagus to survive in either necropolis (site 737, T 19), only the floor and one corner of the north end were found in situ. The fine banded-ware amphora at its south end, reassembled from many fragments, provides evidence for what is probably the oldest burial along this linear feature at Pizzica. The unique plaster-based *fossa* burial (site 736, T 19) was miraculously intact. The teeth of the backhoe had shaved off part of the plaster coating—it was well-preserved at the end and along the edge—but missed most of the vessels at the foot, head, and along the side of the skeleton. The grave goods are among the finest red-figure vessels from any burial in the chora and were beautifully preserved, apart from a few missing pieces from the vessels that were standing upright. Among them was an unusual form, a black-gloss *dinos* with ivy leaf and berry decoration on its shoulder.[24]

The only burial to escape damage of any sort from the earth-moving operations of 1999 and previous years was the very deep fossa burial, site 736, T 22, which contained the best-preserved skeleton. This is the only one of the 12 burials surviving at sites 736 and 737 without grave goods. That the skull was located in the northwest end of the grave suggests that it was one of the latest, as such orientation was the one preferred in the period 325–275 BC.[25]

A summary of the major data about the recovered burials from these two small necropoleis is contained in Tables 28.1–2. Bearing in mind that the burials recorded in these tables represent only the lowest level of burials in what would have been larger groups, some general observations can be made about the nature of these isolated rural necropoleis. The variety of tombs in both areas is remarkable, considering the small number of burials recorded. All of the types found between the two sites here occur among the 320 burials of the great rural necropolis at Pantanello. This includes the simple fossa grave in the bare earth,

[23] The necropoleis are first reported in the *Atti Taranto* of 1999 (Nava 2000, 685–86).

[24] All burials and grave goods are fully documented in Chapter 29, following the format introduced in an earlier volume in this series (*The Chora of Metaponto: The Necropoleis*). The vases from T 19 are analyzed and attributed to vase painters by Lucilla Burn.

[25] Carter 1998f, 225.

Tomb no.	Tomb type	Orientation (degrees W of N)	Location of skull	Date of burial	Sex/age of deceased
18	*a cappuccina*	62°	SE	440–370 BC	F 25–35
19	plaster-lined fossa	52°	SE	416–382 BC	F ? 25–30
20	fossa	17°	NE	300–275 BC	F 25–35
21	*a cappuccina*	39°	SE	320–280 BC	F 50–X
22	fossa	56°	NW	—	M 40–50
23	*a cappuccina*? (destroyed)			—	—

Table 28.1 Site 736 burial data.

Tomb no.	Tomb type	Orientation (degrees W of N)	Location of skull	Date of burial (weighted)	Sex/age of deceased
18	*a cappuccina*	45°	SE	–	F ?
19	sarcophagus	47°	SE	500–480 BC	F ? 40–50
20	*a cappuccina*	50°	SE	460–440 BC	M ? 40–X
21	fossa	50°	SE	460–440 BC	M 50–60
22	slab-covered fossa	43°	SE	448–433 BC	? 10–14
23	tile-covered fossa	56°	SE	451–429 BC	F 40–45

Table 28.2 Site 737 burial data.

(site 736 T 22), which could also be lined with plaster (site 736, T 19) or covered with slabs (site 737, T 22) or tiles (site 737, T 23), as well as the sarcophagus and the *a cappuccina* type. The *a cappuccina* burial at site 737, T 20 is an elaborate one, where the joints along the ridge and between the two flat pan tiles (tegulae) are capped by short-radius cover tiles (imbrices);[26] that at site 736, T 18 is the more common later type, though, like the older one, it had a floor comprised of two flat pan tiles. The monolithic sarcophagus (site 737, T 19) measured 0.72 x 1.95 m in plan. This is an average-sized sarcophagus by Pantanello standards. At Pantanello, sarcophagi were not often used before the 5th century BC (though there were notable exceptions, like T 292), and few can be dated much later than the mid-5th century.[27]

The grave goods from the burials at sites 736 and 737 are mostly familiar from other necropoleis in the chora, though there are some that are not so well known, such as the *exaleiptron* from site 737 (T 23-03). Examples of this ritual vessel have been published from a number of sites elsewhere in the Greek world. The askos from site 737 (T 18-12) is a rare design of a shape that is not at all common in funerary contexts. Like the hydria (T 20-01) and the fragmentary Corinthian type A' amphora (T 18-13) from site 737, it was imported. All the other ceramic vessels seem to have been locally made, including the two Attic-looking squat lekythoi from site 736 (T 19-01 and T 19-03).

The study of the black-gloss and other vessels from the Pizzica burials (Ch. 29) is largely the work of Francesca Silvestrelli, who has a remarkable knowledge of the production of the Metapontine kerameikos. Parallels with the ceramics found in the Pantanello necropolis have been helpful in determining the

[26] Compare T 311 at Pantanello, which dates to the same period as the Pizzica example (Carter 1998c, 94, Fig. 3.41).

[27] Carter 1998c, 71–79.

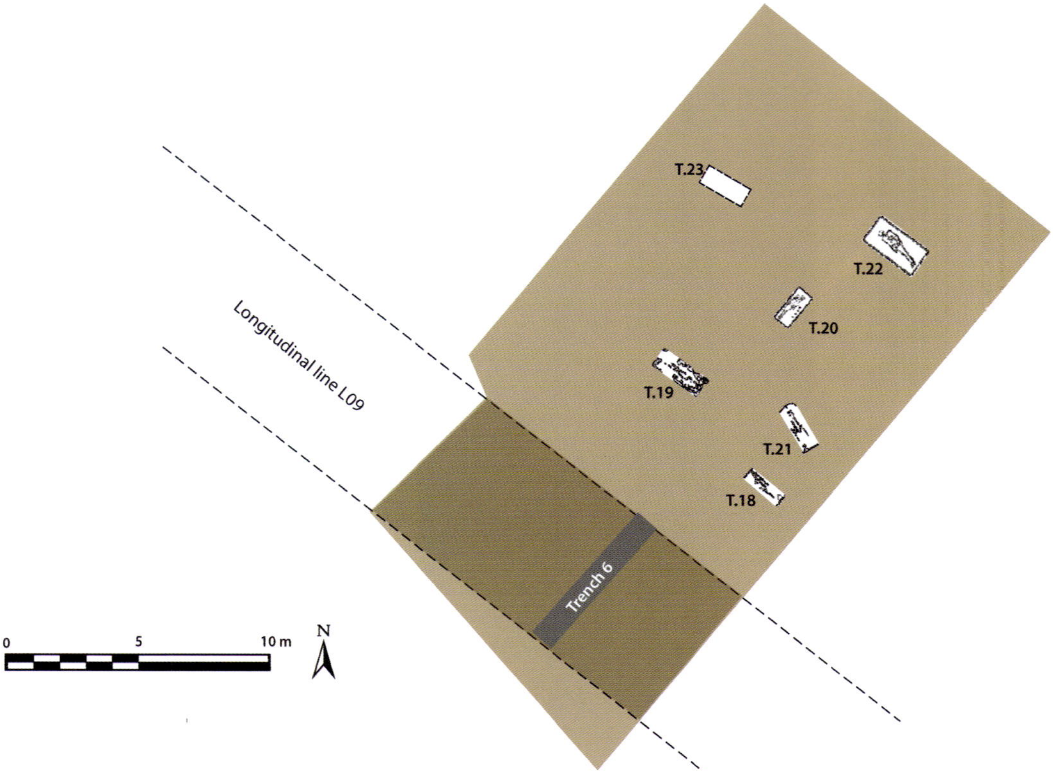

Figure 28.17 Plan, site 736 necropolis with T 18–23, trench 6. (CW)

chronology of the burials. Almost all of the ceramic forms and non-ceramic objects have parallels there, except for the unusual early iron strigils from site 737, T 20 and T 21. These also occur frequently in closed burial contexts with figured pottery, the chronology of which is, in most cases, firmer and more narrowly defined. Apart from a fragment from T 18 at site 737 (a badly preserved burial), figured pottery occurs only in T 19 at site 736.

Thanks to the detailed study of the skeletal remains carried out by Maciej and Renata Henneberg, it is possible to characterize the remaining burials at these two sites in human terms. Though the small size of the sample precludes even the most general considerations about the population of this area of the chora, the anthropological study by the Hennebergs (Ch. 30) nevertheless offers some tantalizing observations. Study of the morphological characteristics of the teeth shows some unexpected differences. Two of the females—one at site 736, T 20 and one at site 737, T 19—do not have the characteristic traits of the rural population that are shared with other Italic populations as defined by a large-scale study of over 600 burials in the chora and asty.[28] The others buried here (such as the individuals in T 19 and T 22 at site 736) are genetically closer to the rural population at Pantanello (the principal rural site studied by Renata Henneberg). This line of research has the potential to reveal much about the social interaction of chora and asty, and the ethnic identity of the population of the colony in general.

The necropolis at site 736

Though similar in many ways, the two necropoleis represent different phases in the history of the chora. Site 736 yielded 6 burials (*Fig. 28.17*), although as mentioned previously, there was possibly a much larger group originally, perhaps 23 or more, insofar as can be known given that only the deepest burials escaped destruction. The surviving examples belong to two different, chronologically separate groups: one dates to the late 5th and early 4th centuries BC, the other to the late 4th and early 3rd centuries BC. There may have been more continuity originally than appears now, though it should be noted that the sporadic material also falls into roughly similar chronological groups (see Selective Catalog of Sporadic Finds in Ch. 29).

[28] Henneberg 1998.

Figure 28.18 Site 736, plaster-lined fossa burial T 19.

Figure 28.19 The grave goods from site 736, T 19 after restoration.

To the early group belong the burials of two young women, the occupants of T 18 and T 19. One of these, T 18, was badly damaged by the backhoe. This burial was extremely modest, though if the red-figure pelike that was outside it can be associated, it had an elegant grave marker.[29] By contrast, the burial in the unusual, covered, plaster-lined fossa (T 19) is distinctive in every way. The collection of red-figure vessels that were arranged mostly around the feet of the deceased is of exceptional quality (*Figs. 28.18–19*).

Even more interesting is the iconography: the scenes on two Attic-style squat lekythoi, as Lucilla Burn points out, represent two aspects of marriage, while the skyphos, pelike, and *dinos* are regularly associated with the world of the symposium, an all-male activity, at least in Athens. The tomb type, along with the prominence of Eros in the iconography and the mirror in which the bride admires herself, was found in a group of burials in the Pantanello necropolis; it has been argued these were burials of women (and a few males) who held a set of beliefs about the afterlife termed "Orphic-Dionysiac."[30] If this also is the case here—and T 19 is a very convincing demonstration—then these beliefs were more widely shared by female inhabitants of the chora than previously thought.[31]

The Hennebergs' study shows that the young woman in T 19 suffered from caries, which seems to have been a common problem in the Pizzica population to judge from these dozen burials. She also had teeth with Italic features, decidedly more characteristic of the rural than the urban population. The occupant of T 20, a young woman buried a century later, seems instead to have belonged to the urban population on the basis of her teeth. She displayed no metopism. Metopism, a failure of the metopic suture of the skull to close, is a trait that is more common in the chora than the asty (and more common in southern Italy than in other parts of the Greek-dominated world).

Most of the five surviving burials and the destroyed sixth burial (T 23) were oriented either parallel to longitudinal line L09 or perpendicular to it (T 20). The burial of the older woman in T 21 at site 736 did not conform. Could it be that she was not a member of this group, or that the linear feature at this time (320–280 BC) was not as universally respected as it had been in earlier periods in the necropolis at site 737 or at Pantanello? The single burial without any grave goods (T 22)—also the best preserved skeleton (*Fig. 28.20*)—is located at the northern extremity of this little group in perfect alignment with L09. This individual was a male in his forties, standing about 1.7 m, who died a violent death, the victim of an antagonist holding a sword in his/her right hand, according to the Hennebergs. He was placed in a simple earthen grave with his head at the northern end. This practice, we found at Pantanello, dates from the end of the 4th to the beginning of the 3rd century BC, and was also typical of indigenous burials in the interior of Lucania. His teeth were sound, with no caries and only a touch of periodontal disease. If he were a member of a family group buried here, why were there no grave goods? Perhaps he was a victim of family violence? This is a mystery that remains tantalizingly unsolved.

[29] On the use of large figured and black-gloss vessels as grave markers, see Carter 1998d, 124–29.

[30] Carter 1998c, 67–69. The correlation of mirrors or representations of mirrors on vases, representations of Eros, and plaster-lined fossa burials is statistically significant.

[31] The male figure on the lebes gamikos was originally misidentified as Orpheus, but see comment by Burn (Ch. 29). See also the tomb, found in the chora between Basento and Cavone, which contained a representation of "Helen and the Egg:" A. Bottini 1988b; 1992, 64–85.

Figure 28.20 Site 736, deep fossa burial T 22. (JCC)

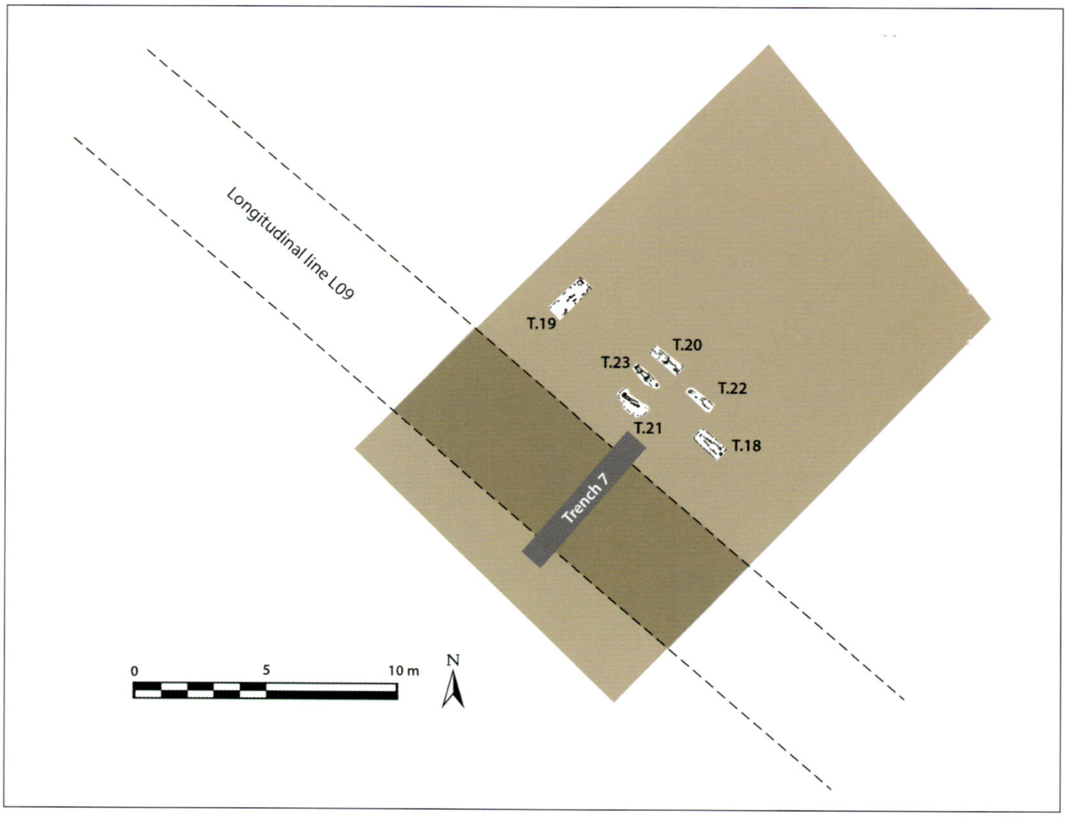

Figure 28.21 Plan, site 737, showing necropolis with T 18–23 and trench 7.

Figure 28.22 Looking NW, site 737 burials. (JCC)

The necropolis at site 737

The remains of this small necropolis, originally about the same size as that at site 736, gives the impression of a close-knit group (*Fig. 28.21*). The tombs were closely spaced (*Fig. 28.22*) and aligned with longitudinal line L09, with the skulls of the deceased regularly in the southeast end of the tomb. All the burials date around the middle of the 5th century BC, with one (T 19) perhaps earlier, and two (T 22 and T 23) somewhat later.

There is also similarity among the tomb types. Half were fossa burials, which had either a local conglomerate slab cover such as T 22 (see discussion above of the crossroads area), a tile cover (T 23), or no cover (T 21). The preserved *a cappuccina* burial (T 20), as noted above, was unusually elaborate, with carefully covered joints (*Fig. 28.23*), a design that seems to have been more common farther south along the Ionian coast at Locri than at Pantanello.[32]

The occupants of these tombs, with the exception of the adolescent child (T 22), lived long lives, well into their forties. At least one of the males reached an advanced age. The women were slightly less long-lived, but easily surpassed the average age of 39 years for the females in the Pantanello necropolis.[33] Considering the dates of birth of the individuals here, calculated with the dates of burial (weighted) and working back using the mid-point of the life span, it is possible to suggest some family relationships, shown below in Table 28.3.[34]

The earliest burial is of a female (T 19) whose teeth indicate that she was biologically closer to the population of the asty than that of the chora. With all the caution urged by the Hennebergs in mind, it still might be suggested that her sarcophagus burial denotes relatively elite status (a matriarch); the tomb type also supports the early 5th-century BC date. She would probably have been born in the middle decades of the last quarter of the 6th century BC and is thus the oldest person at site 737 (and site 736). The sclerotic hyperostosis present on her skull indicates that she, like a number of persons buried in urban necropolis at Crucinia, suffered from malaria, and possibly from trepomatosis. If the males in T 21 and T 20 were her older and younger sons respectively, then the woman in T 23 could have been a daughter-in-law (more likely the wife of the younger son), or her daughter; the child in T 22 could have been the offspring of the younger son (and his hypothetical wife, whose tomb had been destroyed or was elsewhere).

[32] One of the rare examples from Pantanello, T 169, has been dated to the early 4th c. BC. See Carter and Hall 1998, 415.

[33] Henneberg and Henneberg 1998, 509–12.

[34] Carter 1998e.

Tomb no.	Date of birth (mid-point of burial date plus mid-point age at death)	Age (mid-point age at death)	Sex	Hypothetical family relation
18			F	
19	535	45	F	Matriarch ("urban" type)?
20	495	45	M	Younger son?
21	505	55	M	Older son?
22	452	12	?	Child of younger son and wife?
23	483	43	F	Wife of younger son?

Table 28.3 Site 737, hypothetical family relationships.

Figure 28.23 Site 737, *a cappuccina* burial T 20. (JCC)

The individuals in T 21 (older male) and T 23 (female), have no metopism, a genetically transmitted characteristic that is relatively common in the rural population. Malaria may have affected the older male in T 21, as well as the female in T 19. These two had extensive caries, as did the female in T 23. The dental health of the people who were buried at site 737, whether or not they were related, was not good even by Metapontine standards.[35]

[35] Henneberg and Henneberg 1998, 535.

The close relationships of these individuals are supported by the similarity of the grave goods. The middle-aged man in T 20, with iron strigil, aryballos, and palmette lekythos, is characterized as an athlete and habitué of the palaestra. The older male occupant of T 21 also was given an iron strigil, a relatively rare type, especially in the 5th century BC. The amphora between his thighs was a somewhat earlier type than that which accompanied the child in T 22, who was buried perhaps a decade later. The miniature lebes and the lekythos—a reproduction on a tiny scale of the palmette type in T 20—are appropriate to a youth (father and son?), though they are also found in adult graves at Pantanello.

The Division Lines Reconsidered

The division lines have been a topic of debate since their discovery by aerial photography in the 1950s.[36] The anomalies in the photos seemed to some scholars (including this writer) as the manifestation of a rarely observed but significant achievement of the colonial Greeks: organizing and managing the landscape around their urban centers. The prime surviving examples of this undertaking are the chorai of Metaponto and Chersonesos and several other Greek settlements in western Crimea[37] on the Black Sea, two cases that subsequent research has shown to be similar in some aspects, but quite different in others.[38]

Recent discoveries in the chora of Metaponto have demonstrated the intimate connection between the

[36] For a thorough review of the problem, including bibliography, see Prieto 2005 and Ch. 31.

[37] Chtcheglov 1992, 224–78.

[38] For Chersonesos, see Carter et al. 2000.

line features and the management of water.[39] Earlier studies of the lines, before excavations began exploring their physical properties, had considered them as a problem not in topography but in plane geometry, and as a complete and fully functional system (or at most, two systems) imposed on the landscape at one time.[40] The temptation still exists to wrap up the pieces of evidence with a single, all-encompassing solution. This urge should be resisted, as each new discovery will require further modifications. The interaction of water and the land is unpredictable:[41] channels cut by man are not always taken.[42] The regular spacing of the canals is not as regular as it appears. The relationship of these canals to the settlement pattern is complex. The early explanations of the lines were bound to be wide of the mark, and the historical conclusions drawn from them, thus, misleading. For one, they failed to consider, even theoretically, the dynamism of this landscape. This study is a step forward, but surely not the last.

Aerial photos

The principal evidence for the lines is a single series of aerial photographs taken in the 1950s. These display the area before deep plowing over large parts of the chora irrevocably compromised the underlying physical causes of anomalies visible in them (see Ch. 31). We know with certainty now that the underlying physical entities were ditches filled with colluvium—soil mixed with pottery—from the adjacent fields. In spring, after the winter rains, the crops or grasses grew faster and greener over the ditches, creating a dark anomaly against the lighter background of the field. Roads below the surface were also detected in the photographic study, though the mechanism of differential crop growth should have worked differently. The compacted road surfaces (like that of the Basento Road) when unused not for just several hundred years but for many centuries or even millennia, should, in the handbook explanation, have inhibited growth of plants in the soils that covered them and produced a sort of negative effect, the reverse of that made by the ditches.

Ancient stream beds have also left their traces on the air photos of the 1950s. The reasons for the constant shifting and silting up of the stream beds have been discussed at length by Abbott (Ch. 2 and below). The Bradano and Basento were blocked by heavy deposits of rich erosional soils from the interior of Basilicata, and they migrated in both instances from north to south along the coastline. The fill of the old course should have produced an effect similar to that of the fill in the drainage channels.

Prieto's study of the photos has revealed dramatically more of these features than have been previously known (or at least published); even more, his study has made a genuine quantum leap in our knowledge of what the photographs show by connecting them very precisely with the landscape in all of its topographical and geomorphological complexity.[43] The principal tool of his investigation has been the GIS that he helped create using the results of many collaborators in multiple disciplines (including geography, geomorphology, and archaeology). The lines have been referenced to a new generation of images of this landscape taken from space and given real-world coordinates. It is now possible with GPS to move, with a degree of accuracy unimaginable in archaeology a decade ago, from the photographs to the area they document in the landscape. The overall error is on the order of 10 m, and in some instances even less. Inside the GIS, the topography in three dimensions and the physical world are the new contexts for analyzing the division lines.

Excavations 2000–2002

The archaeological excavations that resulted from the construction of the oil, gas, and water pipelines in the years 2000–2002 not only revealed the true nature of many of the linear anomalies, but also their wide distribution throughout the chora. In excavations directed by Antonio De Siena, a series of parallel drainage channels, separated by 200–215 m, was uncovered in the Pizzica area.[44] Their position had previously been accurately predicted using the georeferenced aerial photographs in conjunction with a GPS survey of the chora carried out by ICA in collaboration

[39] De Siena cited in Nava 2001, 946–50.

[40] Prieto 2005; Guy 1995.

[41] "E la terra che comanda," as Dinu Adamesteanu (quoted in De Siena 2005: 433 n.1) observed.

[42] De Siena on San Teodoro, San Salvatore, cited in Nava 2003a, 670–71, 675–76.

[43] Prieto 2005 and Ch. 31.

[44] Nava 2001, 949–50, Pls. 28.2, 29.1; 2002, 735–43, Pl. 48.2; 2003a, 664–76, Pls. 35.1 and 40.1.

Figure 28.24 Plan of the Pantanello necropoleis showing division line L21 and Basento Road.

with specialists from CSR.[45] The areas of the chora that produced the fullest evidence for the system of drainage channels were Pizzica and San Salvatore in the area between the Bradano and Basento, and San Teodoro in the Basento-Cavone watershed. The regular profiles and homogenous fills of the channels (the channel at site Γ, for example) strongly suggest the water that drained off the fields and into the ditches did not move much and may well have been filled with aquatic plants.[46] The ditches were flanked by burials. The numerous sarcophagi by the channel at site β and others suggest that the channels were in use at the time of many of the earliest burials next to them, the first half of the 5th century BC. This area, as noted, is relatively very flat and slopes slightly toward the coast. Additionally, the impermeable clays that underlie the terrace would cause rainwater to collect on or near the surface (see Ch. 1). The longitudinal line L09 visible in the aerial photograph drops 15 m over a distance of 1 km, but in the excavation area the slope is much more gradual: 1 m in the 200 m between the two necropoleis.

In the other areas of the chora with greater topographic relief and more natural drainage, the ditches would have served more to channel the water than simply to collect it from the surface. This is evident particularly at San Salvatore and San Teodoro, where channels were built to direct the water resulting from

45 *Ancient Territories* 1999, 18–20.

46 Nava 2001, 946–47. This has been confirmed by pollen analysis of five samples taken in stratigraphic sequence from the fill of one of the drainage canals. The results will appear in a subsequent volume of this series. A.M. Mercuri, unpublished report, 2009.

torrential rains away from the fields towards the natural drainage provided by the river valleys. Channels were dug and carefully lined with river cobbles to prevent the flow from excavating through it to layers of easily erodible sand beneath, and also perhaps to prevent it from silting up (similar to the type of drain known as a French drain today). At the San Salvatore site, a votive deposit containing Ionic cups and miniatures, dating to the very late 6th or early 5th century BC, commemorated the completion of this work, but the ditch was apparently never functional.

Besides drainage channels, excavations by the Soprintendenza in these years have documented an approximately 5 m wide thoroughfare on the terrace known as San Teodoro. It was oriented parallel to the division lines but for some reason did not appear in the aerial photos. It was flanked on both sides by drainage channels, just as the Basento Road excavated at Pantanello.[47]

The immediate effect of these discoveries is clear, and they have forced a reexamination of the results of earlier excavations of the linear features and their interpretation. These were concentrated in the area of Pantanello in the 1980s. The most thoroughly documented feature is the road excavated at Pantanello that corresponds to longitudinal line L21 (*Fig. 28.24*). There can hardly be any doubt that this piece of land, flanked on both sides by six distinct groups of burials or nuclei, was a road or empty strip. It came into use as a place of burial probably in the early 5th century BC, the date of the earliest burials beside it. There are some indications, however, that the axis of the road may have been established as early as the late 6th century BC. Along the road, far from its intersection with the Basento Road, to one side, was a ceramic deposit (CD 37) containing 6th-century BC pottery, including fragments of an elegant Attic black-figure krater. The E Group vessel dates to ca. 510 BC. It could be either the remains of a burial or a votive deposit. The traces of earlier destroyed burials were located along the northern edge of the strip, the orientation of which was respected by all the intact burials.[48]

Geomorphology

Expert knowledge in the recognition and interpretation of changing land forms is essential to the archaeologist, and nowhere more so than for those investigating the chora. The geomorphological study of the Metapontino carried out between 1990 and 1992 by James Abbott is an invaluable aid in understanding the dynamism that shaped the landscape, particularly in the periods of intense human settlement from the Neolithic (6th–3rd millennia BC) to the present. A synthesis of that study appears as Chapter 2.[49]

The frequently occurring episodes of massive erosion in the interior of Basilicata and alluviation in the Ionian coastal areas had been known and even investigated,[50] but not in such a complete and well-documented way by a geomorphologist with extensive archaeological training. With Abbott's work the various phases and details of these episodes have at last been documented and securely dated by a large number of C14 dates of the sediments.

Flooding of the Metapontine river bottoms and plains began in the Neolithic, with the cutting of the primal forests to clear fields for the earliest agriculture in this area. Abbott has placed the first major episode in the period 4000–2000 BC, and the second in the Classical period, after which a stable regime without flooding permitted the formation of soils in the Roman Imperial and Late Roman periods (late 1st–6th centuries AD). A shift in economic activity from agriculture to pastoralism, it is suggested, could have been partially responsible for this change.

Excavations in the ancient city and the chora illustrate how these long-range developments in river drainages affected human settlement at specific sites along the coast. A relatively sudden increase in the water table in the first half of the 5th century BC resulted in the abandonment of the 6th-century BC structure in the Pantanello sanctuary and its replacement by a new cult building on higher ground, which came into use during the 5th century BC.[51] This was not an isolated instance. Virtually all of the low-lying settlements in the river valley and valley edges were affected. The natural drainage of the rivers, as Abbott shows, was also affected as their mouths began to silt up and create pools of stagnant water. This was always a problem at the urban site of Metaponto on the coast, but it was exacerbated at this time and intervention

[47] Carter 1998b.

[48] Carter 1998b, 47, Fig. 2.18, and 29–30, Fig. 2.2A-B.

[49] Abbott 1997; 1998; Abbott and Valastro 1995. Abbott's work was supported by grants from the National Science Foundation and the National Endowment for the Humanities and by donors to ICA's Metaponto project.

[50] By Neboit, Brückner, and Boenzi.

[51] Carter 1994a.

was required to shore up the great Archaic temples.[52] Not surprisingly, the frequency of malaria, a natural consequence of an expanding mosquito habitat, increased markedly (see Ch. 30).[53]

The situation—a serious threat to livelihood and, it may well have been perceived, to the health of the human population—called for action, and the call was met by the massive construction of drainage channels, and other public works. As argued earlier in Chapter 23, this community undertaking by the Metapontine democracy was perhaps not unique, but it is the one example for which there is now overwhelming evidence. The importance of this response was marked and the event celebrated by the striking of the first double-relief silver stater of the city, with a representation of the river god, Acheloos, the patron deity of such undertakings.[54]

The landscape of the chora today retains some surface evidence of this great project. The correspondence between the positions of the lines on the aerial photograph and linearly aligned swells and dips in the marine terrace were noted by the first students of the photographs, Schmiedt and Chevallier. In several instances these correspondences are quite impressive.

Geologist Robert L. Folk observed in Chapter 1 that the lines have the same orientation as the underlying geologic fracture pattern and argued that they are surface manifestations of that natural phenomenon. The question was first raised by Folk during his study of the geology of the chora and surrounding areas in the early 1980s and is discussed in depth by Abbott in his detailed investigation of the linear feature to the north of the Basento Road in the Pantanello necropolis, as noted above.[55] South of the Basento Road, this feature was either a road or a strip that was, perhaps like the *antomoi* mentioned in the Herakleia Tablets, an empty strip that served as a divider between fields and properties.[56] To the west, above the Basento Road, there is a very evident depression that aligns exactly with the linear feature. Excavation there in 1986 failed to find any trace of an extension of the tomb-lined road. In three trenches dug transverse to the projected line of the lower road, only one burial, T 335, was discovered and no trace of a road surface.

Three more transverse trenches (a total of 7 since 1986) were excavated in 1993 with the express purpose of documenting the depression and describing its geomorphological characteristics. Abbott concluded that it was a natural drainage depression carved through the marine terrace first in the prehistoric period. A second cut intersected the main channel at an angle. Both of these features contained fragments of roof-tile and pottery. Their rough profiles strongly suggested a natural drainage that was filled with cultural material, with black-gloss pottery dateable to the 5th and 4th centuries BC, but the position of perfect alignment with the road to the south left open the possibility that the natural feature may have been elaborated, deepened, or straightened by human agency. This seems reasonable, though the reverse, an eroded man-made feature, is also conceivable. The regular spacing of the channels every 200 m or so meant that while in some cases natural drainages could be exploited, in most instances they had to be excavated ex novo. This was particularly the case in flat areas like Pizzica, with little topographic relief.

Working hypotheses

Further discoveries like those of 1999–2002 will no doubt add to the archaeological map of features related to drainage. They will require reinterpretation of previously known ones and perhaps lead to a whole new formulation of the related questions of drainage and land division. Nonetheless, at this point it may be useful to offer some working hypotheses.

It is safe to assume that many if not all of the linear features appearing in historic aerial photographs are channels, the primary purpose of which was drainage. It is also clear that not all channels are represented in the photographic documentation. Excavations have added completely unsuspected channels (San Teodoro) or extensions of known ones in areas where no anomaly appears in the photographs. A characteristic of the linear features in the photographs is that they are discontinuous. Some of the interruptions are clearly the result of the dissected topography, caused by streams and drainages. Should other gaps, where topography is not a factor, be interpreted as real gaps in the channels, or unevenness in the research results? Can gaps have been caused by the subsequent history of the

[52] Mertens 1985.

[53] Henneberg and Henneberg 1988.

[54] Uggeri 1969, 63–66; Noe and Johnston 1984, 55–56, 70, Pl. 23, Nos. 311, 311½.

[55] Abbott 1998.

[56] Uguzoni and Ghinatti 1968.

particular area of the chora? There are, moreover, whole areas where there are no linear features at all.

Perhaps the search for a uniform system extending over the whole chora, and a comprehensive explanation for that system, is mistaken at the outset. Since the digging of channels of the dimensions encountered represents an enormous cost and labor, presumably these channels would only have been dug where they were needed. If their purpose, or at least primary purpose, was drainage, then there was no need for them to be continuous if they achieved that objective anyway. The channels would have been dug where present conditions or concerns for the future required them. Not all the constructed channels were actually used, as noted in the case of San Salvatore.[57]

In many cases, the areas where there are few lines resemble each other. There are no lines in Giardinetto in the extreme northeast of the chora. The areas of the surveyed chora that are farthest from the coast and the ancient city, at the highest elevations, and relatively more dissected than the lower lying areas, are almost devoid of linear features. The central line (L15) passes through Campagnolo, but it is otherwise unique in this area (Ch. 31). Lower down, on the south side of the chora, Avinella has a few lines, but the Venella valley none. The areas of greatest concentration of the lines are San Salvatore below the Fosso Appio, Pizzica, and Lago del Lupo.

This selectivity may well reflect a greater concern for drainage in areas where the topography is suited for plow-based agriculture. Conversely, with the higher areas and certainly the Venella valley better suited to animal husbandry and grazing, there would have been no pressing need to drain, since animals can generally survive on eroded hillsides and thrive in lush and even marshy valley bottoms, such as that of the Venella.

In light of the geomorphic history of the area, it is not surprising that no lines in the valley bottoms were detected on the aerial photographs. The major valley bottoms were the first areas to be settled in all probability, as these were potentially the richest agricultural soils. It would be reasonable to expect drainage channels there, and in fact one was detected during the excavation in the path of the oil and gas pipeline. It need not follow, though, that the drainage ditches necessarily conformed to a rectangular grid. Certainly less labor would have been expended by channeling the water along the shortest path to the banks of the rivers. This seems to have been so at the Pantanello sanctuary, but perhaps that was a special case.

In earlier investigations the lines on the aerial photos were interpreted as roads or as channels flanked by roads.[58] We now know that most were certainly channels, and that no obvious traces of lanes or paths beside them have been discovered recently does not prove that they did not once exist. They would not have left much physical evidence, but Uggeri saw them and his testimony at an early stage in the deep plowing of the chora must be taken very seriously. Certainly some stretches of roads do appear in the aerial photos (Ch. 31). These were principal thoroughfares and would have left more consistent traces in the landscape. Other excavated examples of roads, such as that in San Teodoro, which was flanked on both sides by large drainage ditches (like the Basento Road excavated at Pantanello), do not appear in the photos, yet they follow the orientation of the lines.

Unpaved lanes or paths (or as suggested above, an *antomos*,[59] an empty unplowed strip) surely provided communication within the chora from one farmhouse to the next, and to the urban center on the coast. They must have existed from the time of the earliest settlements in the early 6th century BC. At first the very early roads along the main river valleys would have sufficed as the settlements clustered along the valleys, but as the interior between the valleys began to fill up, a path at least would have been required to get the products of the chora to the markets. The fortunate discovery of wheel ruts under the drainage channel defined by longitudinal line L09 proves that some sort of road, connecting Pizzica presumably to the coast and traveled by wagon, was in use by the 6th century.[60] In how many other cases, one wonders, did channels replace existing roads, or were these cases where channels were dug parallel to roads?

The most puzzling fact about the system of drainage channels is unquestionably their very regular spacing. The hydrological conditions by themselves could not have dictated such a regular spacing wherever channels were dug. If economy of labor were the major concern—and it was certainly always an

[57] Nava 2003a, 676, Pl. 40.1-2.

[58] By Uggeri 1969, 53; Adamesteanu 1973, 1974a, 80–88.

[59] The term appears in the Herakleia Tablets.

[60] *Ancient Territories* 1999, 24, Fig. 38; Nava 2000, 688.

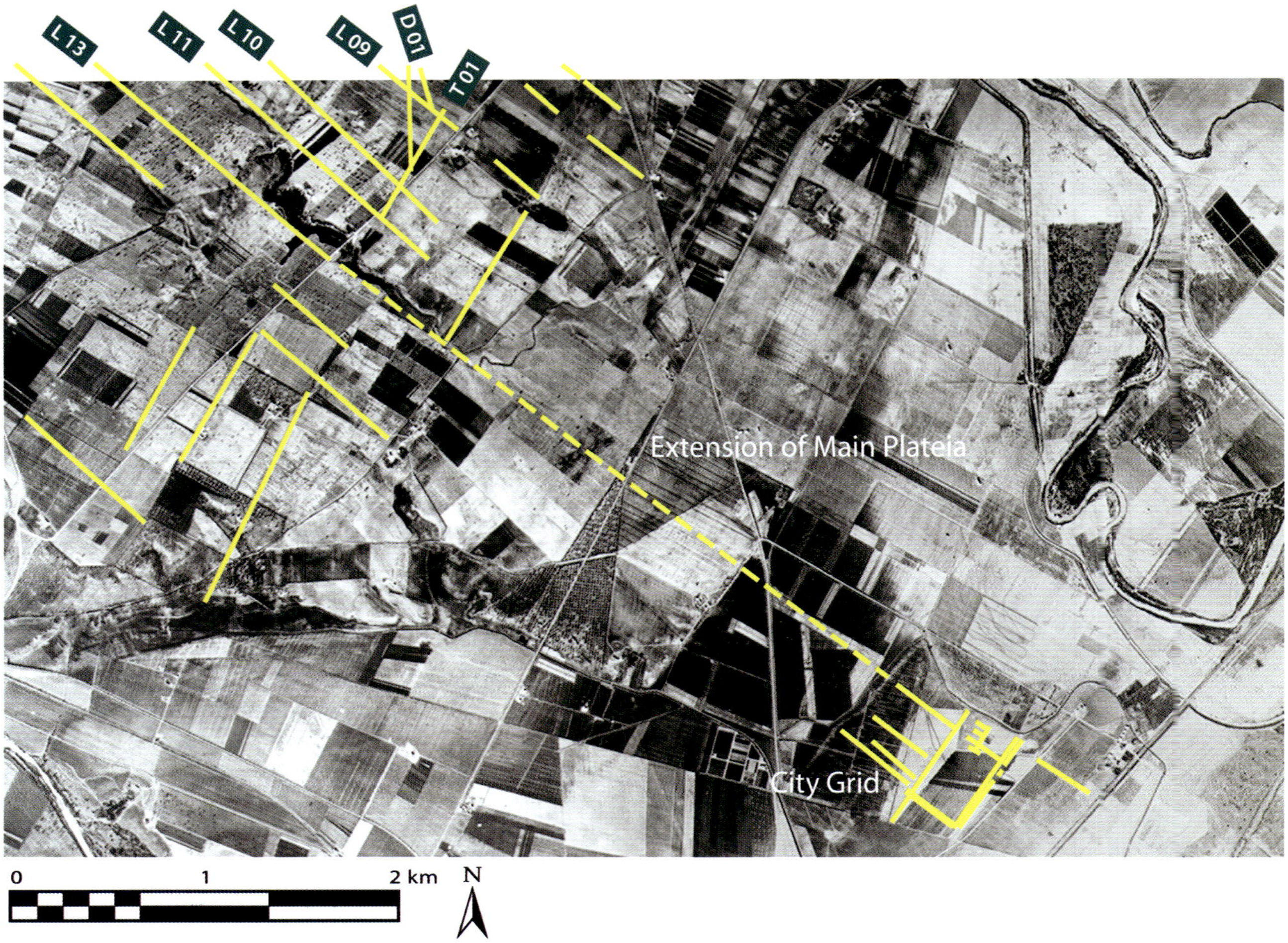

Figure 28.25 Aerial photo showing longitudinal, transverse, and diagonal lines at Pizzica in context of larger area. Note the apparent alignment of the major E-W plateia in the grid plan of the city with the central longitudinal division line L15. (GAI Pcs.5518/Neg.31661, reproduced with permission of the Aerofototeca, Rome, 2004)

important one—then other solutions could certainly have been found.

Coordinating a system of county lanes with the system of drainage channels would have been a necessity. Channels laid out beside roads would not have cut across them, or required bridges. The farmer would have wanted minimum interference in fields to be plowed. With the exception mentioned above, the lanes have left few traces, but surely they were there.

The regular spacing of 200–215 m between canals is not a requirement of drainage. These measurements were early recognized as very close to standard Greek measurements in land division.[61] If the system of channels reflects a division of property in the chora, when was this division made? We cannot automatically assume that it dates to the period of the construction of the channels, though these could have reinforced any division, earlier or contemporaneous. If roads predated the channels, then it is likely that a division did as well.

Analogy with the City Plan

The exploration of the residential area of the ancient city has only just begun.[62] The excavations so far have revealed the continuation of the NS plateia that divides the agora and sanctuary. The intersecting *stenopos* has been dated to the mid-6th century BC. It has been suggested by Dieter Mertens that the oldest part of the grid plan of streets, which appears in the aerial photographs, is to be found in the south side of the city.[63] This is where the earliest residences were located, the capanne of the mid-7th century BC, which

[61] Carter et al. 2000, 732.

[62] Nava 1999, 698–703.

[63] Mertens 1998, 129–33.

were abandoned about the time that the Achaian colonists arrived.[64] This is also where the stenopoi of the mid-6th century BC, part of the widespread new plan of the period, should be found. The definition of the sectors of the city of the Classical period, as De Siena and Mertens have emphasized, goes back to the earliest settlement.[65]

Mertens has further suggested that the Archaic phase of the city plan, which unfortunately has left little actual trace in the form of dateable road surfaces, took its inspiration from the chora.[66] Lending weight to the hypothesis of a precocious development in the chora are the survey results which document fully functioning residences, necropoleis, and sanctuaries, already by the mid-6th century BC (see Ch. 20).

If, in fact, the main divisions of the Archaic residential quarter date to the mid-6th century, and, if Mertens is correct in seeing their inspiration in the chora, then at least the major axes marking the division of the chora ought to be at least as early. Further indication of a close relation between the mid-6th-century division of the city plan and the division of the chora—at least in its initial phase—is the fact of their nearly identical orientation (*Fig. 28.25*). If De Siena is correct in placing the date of some of the stenopoi in the mid-6th century also,[67] that would open up the possibility that a division of the chora into strips was at least as old. Despite some progress, this thorny question is still far from a definitive solution.

[64] De Siena 1990a, 85.

[65] The only surviving features of the new plan of ca. 550 BC in the fully excavated parts of the city, the sanctuary and the agora, were Temples A2 and B2 (Mertens 1998, 126–28).

[66] Mertens 1998, 127.

[67] Cf. the plan in Nava 1999, 701, Fig. 3.

29
Catalog of the Objects from the 1999 Pizzica Excavations

Joseph Coleman Carter, with contributions by Lucilla Burn, Anna Rita Parente, Francesca Silvestrelli, Keith Swift, and Eloisa Vittoria

Excavations conducted at Pizzica following the survey in the path of the SNAM pipeline yielded a considerable number of artifacts. This chapter provides an overview of the finds along with a detailed analysis of the tombs and grave goods. Abbreviations are explained in the general list on page ix. The discussion is divided into three sections.

Section A lists the materials collected from the surface during the initial survey of the deeply disturbed area of Plot 1999.096 on July 20, 1999. This includes a selective catalog of red-figure vessel fragments.

Section B documents the burials with grave goods recovered from sites 736 and 737, and also includes the sporadic finds (SF) discoverd outside the tombs. The grave goods are listed by tomb number along with the accession numbers for the Museo Archeologico Nazionale di Metaponto (MM). Site 736, tomb 19, has an unusual number of grave goods and that section includes a special analysis of the high quality red-figure vases by Lucilla Burn.

Section C covers the materials found in the cuts through the division line features (trenches 6 and 7 through line L09 at sites 736 and 737) and other exploratory trenches. The objects are listed in tabular form, and again there is a selective catalog of red-figure vessel fragments.

JCC

A. Surface Finds: Sites 736 and 737

During the survey of Plot 1999.096 it was clear that the backhoe had severely damaged two small necropoleis. The object clusters on the surface suggested individual burials and so each group was given a tomb number (by coincidence, site 736 and site 737 each had seventeen). The excavated burials at both sites begin with 18. Surface finds of particular interest have been illustrated in a selective catalog. Other finds, consisting of more common vessel types in fragmentary condition, have been simply listed in Table 29.1. Both the selective items and those in the list are part of the database of survey materials and are included in the survey GIS analysis. The catalog and list here, however, help to illustrate the relationship between surface indicators and sub-surface archaeological remains. The fragments of large vessels T2-01 through T2-04 are almost certainly not grave goods, but rather grave markers—*semata*—whose dates closely match those of the objects excavated in tomb 19. This correlation between surface scatter and substantial underlying material is not surprising and would be a common occurrence if more survey sites could be excavated.

Site 736 T 2 (Prob. destroyed tomb, collected 7-19-99)

Site 736

T 2-01 RF bell krater, rim

Dimensions: h 4.7 w 24.6 rim diam. (est.) 38.0

Condition: Surface intact; thick and lustrous black gloss inside and out.

Description: Clay pink (5YR 7/3). Miltos (red wash) on reserved surfaces. Laurel wreath oriented to right with pointed and veined leaves. For profile, see the bell krater by the Amykos Painter Dresden Dr. 353: *CVA* Dresda 1, Munich 2003, Pl. 50, Fig. 18.3 (R. Hurschmann).

Date: Last quarter of 5th–beginning of 4th c. BC.

Site 736
T 2-02 RF pelike, rim
Dimensions: h 2.7 w 9.34 rim diam. (est.) 18.5
Condition: Lustrous black gloss inside and out.
Description: Clay reddish-yellow (5YR 7/6).
Date: For the rim, see T 18-03 (last quarter of 5th c. BC).

1:3

Site 736
T 2-03 RF bell krater, body fragment
Dimensions: h 4.5 w 4.8 th 0.4
Condition: Thick and lustrous black gloss inside and out.
Description: Clay pink (5YR 7/3). Miltos on reserved surfaces. Upper part of a column with capital bearing an egg-and-dart decorative band.
Date: Second half of the 5th c. BC?
Observations: Perhaps part of the same vessel as the rim fragment T2-01.

1:1

Site 736
T 2-04 RF krater (?), body fragment
Dimensions: h 2.8 w 2.5 th 0.3
Condition: Surface intact; lustrous black gloss inside and out; fresh breaks.
Description: Clay pink (5YR 7/3). Thick, lustrous black gloss on interior and exterior. Central part of the leg of a female figure standing to left. She wears a dress decorated with a double vertical stripe. This type of decoration is frequently used by Lucanian painters; see, for instance, the bell krater of S. Petersburg inv. 4744 by the Amykos Painter (*LCS* 43, No. 205 Pl. 16.4). A close comparison is the bell krater by the Creusa Painter in the Loiudice Collection, inv. 7 (Depalo 1997, 22–3, no. 9, figs. 13–14).
Date: Last quarter of the 5th–beginning of the 4th c. BC.
Observations: Perhaps part of the same vessel as the rim fragment T2-01. (FS)

1:1

Tomb #	Objects	Date
736 T 1	1 PW closed body fragment (unguentarium)?	
736 T 2	See selective catalog below	
736 T 3	2 BG closed body fragment 1 CW open body ring base	Classical
736 T 6	1 BG krater rim (Kr 5) 1 BG open body fragment 1 BG closed body fragment	310–250 BC Classical Hellenistic?
	2 amphora body fragments	
736 T 10	1 BG closed body fragment	Classical
736 T 13	1 BG closed body fragment	Classical
736 T 15	1 BW one-handler (SB/02)	Arch.–Hell.
736 T 16	1 BG open body fragment 1 CW open body flat base 4 CW closed body fragments	Hellenistic
736 T 17	1 CW body fragment (jug?) 1 CW open body base	
737 T 2	3 amphora body fragments	
737 T 16	2 amphora body fragments	

Table 29.1 Surface finds from sites 736 and 737.

B-1. Burials and Grave Goods: Site 736

For a general discussion of the necropolis at site 736 and plans, see Ch. 28, Figs. 28.6 and 28.17.

Site 736 T 18
Type: *A cappuccina*
Orientation: 62
Skull: SE
Date range: 440–370 BC
Basis for date: BG pottery
Sex: Female
Age: 25–35
Dimensions: l 1.64 w 0.47
Ceramic Summary:

T 18-01	BG skyphos (femur)
T 18-02	RF krater base (outside tomb)
T 18-03	BG pelike rim (outside tomb)
T 18-04	RF pelike frag. (outside tomb)
T 18-05	BW or PW ring base (outside tomb)

Comments: The sloping pan tiles of the north side of the cover collapsed over those of the south side of the tomb. Both ends were closed by vertical pan tiles. The body lay directly on the earthen floor of the tomb. The skull, long bones of the left side (including the hand), and lower body (including the feet) were preserved. The skull was placed in the center of the upper end of the tomb, but the feet were in the NE corner as if to make room for grave goods. In fact, a black-gloss vessel, a skyphos, was found in fragments beside the left knee. (JCC)

Site 736
T 18-01 BG skyphos (MM 323.062)
Dimensions: h 6.5 base diam. 4.15 rim diam. 8.3
Condition: Restored from many fragments.
Description: Horseshoe handles, ring base.
Date: Type B2 skyphos. Cf. S49, Elliott 1998, 679, 681. 440–370 BC.

1:3

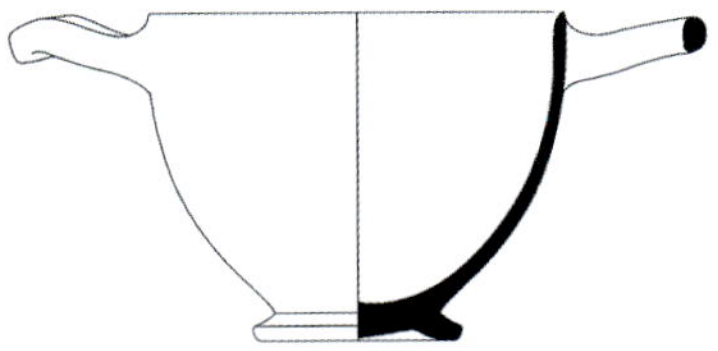

1:3

Site 736
T 18-02 RF krater base (MM 323.063)
Dimensions: h 5.5 base diam. 18.3
Condition: Lustrous black gloss; fresh breaks.
Description: Reserved wide band around top of base, which is offset, and thin reserved band where vertical and horizontal parts of base meet.
Date: Feet of similar shape are usually associated with kraters of the second half of the 5th and beginning of the 4th c. BC; see, e.g., the volute krater from Ruvo (*LCS*, 47–48, No. 243, Pl. 19), the bell krater by the Dolon Painter Louvre CA 2193 (*CVA* Louvre No. 38, Pl. 25, 3–4 Fig. 6), and the bell krater by the Creusa Painter Taranto 140198 (*LCS Suppl.* III, 3/C6). End of the 5th c. BC. (FS)

Site 736
T 18-03 RF pelike, rim (MM 323.063)
Dimensions: h 4.2 rim diam. (est.) 18.1
Condition: Surface intact; lustrous black gloss inside and out; fresh breaks.
Description: Slight trace of red-figure wreath of laurel, upper edge (see T 18-04 below).
Date: The rim can be attributed to a type of pelike derived from Attic models of the third quarter of the 5th c. BC (see, e.g., the pelikai by the Achilles Painter: Oakley 1997, 123, No. 63, Fig. 5a, 445–440 BC) that were used by Apulian painters (Hoffmann 2002, type 311/1, 400–380 BC) and Lucanian painters (Silvestrelli 2008, 282, Fig. 2.8) at least until the first quarter of the 4th c. BC. The similarity with the above mentioned Attic examples and comparison with the example from T 259 at Pantanello (Elliott 1998, 670, P27, 440–400 BC) suggest a date for the fragment in the final decades of the 5th c. BC. (FS)

1:3

Site 736
T 18-04 RF pelike, fragment (MM 323.063)
Dimensions: h 7.0 w 7.5
Condition: Surface intact; lustrous black gloss inside and out; fresh breaks.
Description: Fragment of pelike with wreath of laurel; strap handle attached at level of leaf band encircling, which is continued on fragment T 18-03 on the opposite side of the vessel. There is an unglazed area here where the other handle was attached.
Date: Unknown.

1:3

Site 736
T 18-05 PWBW ring base (MM 323.064)
Dimensions: h 2.2 base diam. 3.9
Condition: Very powdery surface, interior decoration has disappeared, leaving raised surface.
Description: Two bands on the interior.
Date: Late 5th c. BC, from context.

1:3

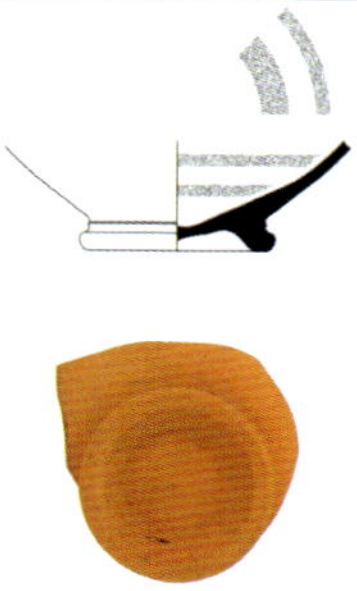

Site 736 Tomb 19
Type: Plaster-lined fossa
Orientation: 52
Skull: SE
Terminus date range: 430–360 BC (see Morter and Hall 1998, 449–51, for an explanation of the dating system)
Weighted date range: 416–382 BC
Basis for date: Figured vases. A separate calculation of dates using only the black-gloss vessels (less accurately dateable) gives a terminus date of 420–360 BC and a weighted date range of 431–374 BC.
Sex: Probably female
Age: 25–30
Dimensions: l 2.10 w 0.80

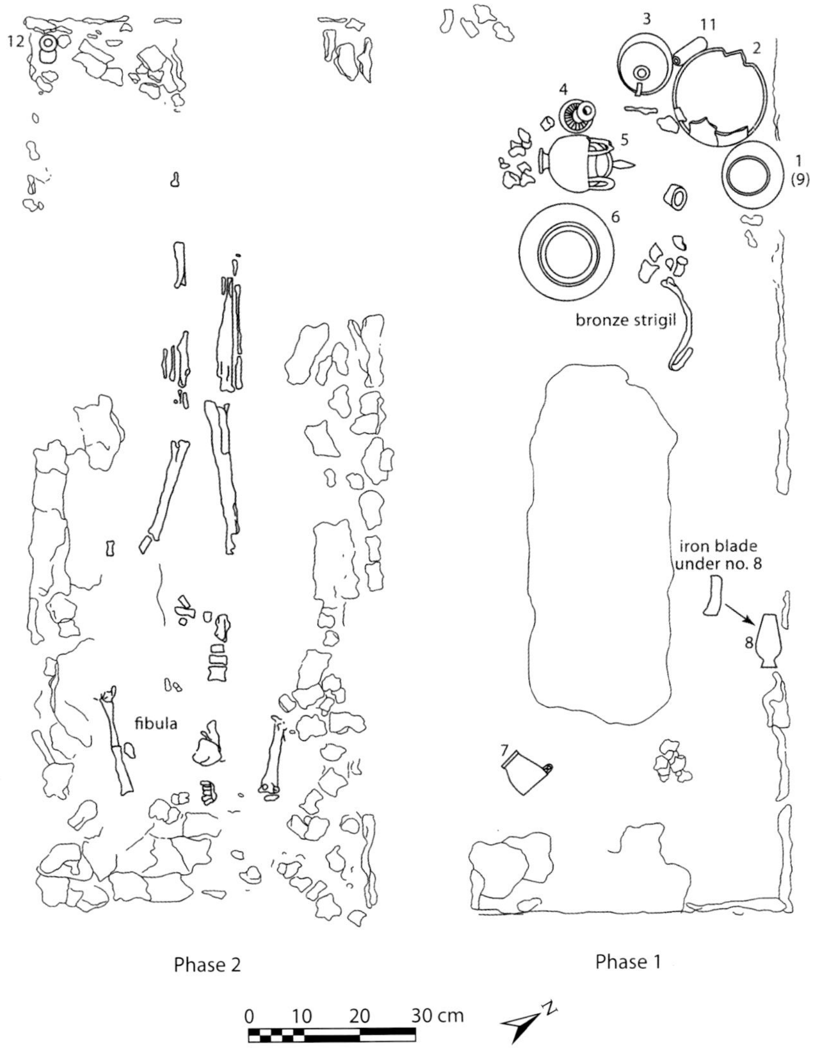

Ceramic summary:

T19-01	RF squat lekythos (foot)
T19-02	RF pelike (foot)
T19-03	RF squat lekythos (foot)
T19-04	BG pyxis (foot)
T19-05	RF lebes (foot)
T19-06	RF *dinos* (tibia)
T19-07	BG skyphos (head)
T19-08	PWBW olpe (humerus)
T19-09	RF skyphos (tibia)
T19-10	BG skyphos (foot) by head?
T19-20	BG skyphos (foot)
T19-21	BG skyphos (foot)

Non-ceramic objects:

T19-11	Alabastron (foot)
T19-12	Alabastron (foot)
T19-13	Strigil, Bronze (tibia)
T19-19	Iron blades (humerus)
T19-14	Fibula (humerus)
T19-15	Fibula (around upper torso)
T19-16	Fibula (around upper torso)
T19-17	Fibula (around upper torso)
T19-18	Fibula (around upper torso)

Comments: The burial was almost certainly completely covered with a blanket of plaster (*intonaco*); considerable expanses of it were preserved above the skull, over the lower torso and legs, and over the feet and grave goods at the NW end. It forms a very straight, almost continuous edge along the north side of the fossa and clearly defines the adjacent sharp angles. The SW corner of the burial was also preserved and was covered with a large expanse of plaster. Over the skeleton lay a stone in the region of the pelvis. The skeleton was supine with arms extended along sides. The long bones below the waist, some vertebrae of the lower lumbar region, the upper arms (humeri) on both sides, and some fragments of the skull survived.

Most of the grave goods were located in the NW end of the burial around the feet and in the NW corner. The strigil was placed on the right side of the figure, below the level of the right hand and along the right tibia. The *dinos* was placed directly over the tibiae; the fibulae were distributed around the upper torso. (JCC)

1:3

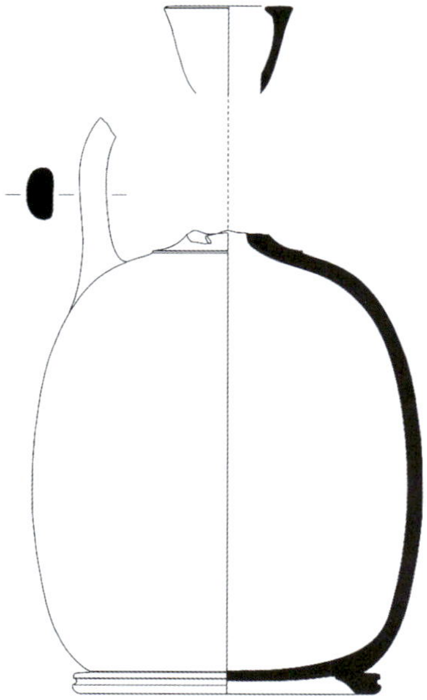

Site 736

T 19-01 RF squat lekythos (MM 322.344)

Dimensions: h 19.1 base diam. 10.0 d 13.0

Condition: Restored from numerous fragments; the neck, mouth, upper part of handle, and small areas of the body, including from the figured scene, are missing.

Pattern decoration: Tongues at junction of neck and shoulder, ovolo below figured scene.

Figured decoration, attribution, and date: Woman and Eros, attributable to Cyclops Painter, ca. 430–400 BC. See more photos and discussion below.

1:4

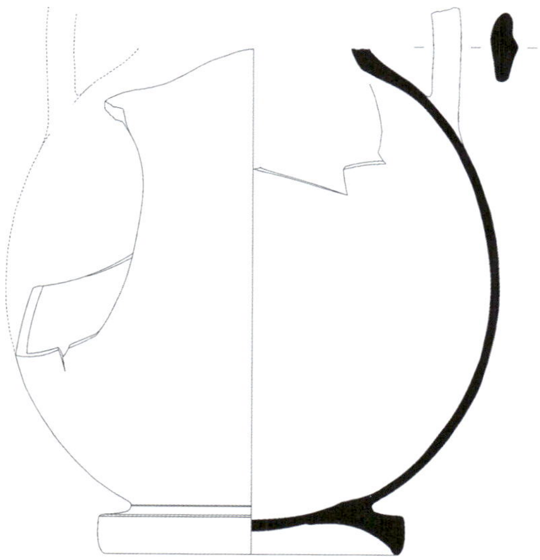

A

Site 736

T 19-02 RF pelike (MM 322.345)

Dimensions: h 28 base diam. 15.5 d 25.1

Condition: Much of the upper part of the vase is missing, so that in total about half of the figured scene on side A and one-third of that on side B is missing, along with all but the start of one handle and all of the neck and mouth.

Decoration: Meander and saltire-cross border below each figured scene.

Figured decoration, attribution, and date: side A, seated woman, youth, and woman; side B, three draped youths. Workshop of the Creusa and Dolon Painter ca. 400 and 360 BC. See more photos and discussion below.

Site 736, T 19-02 cont.

1:4

B

C

1:3

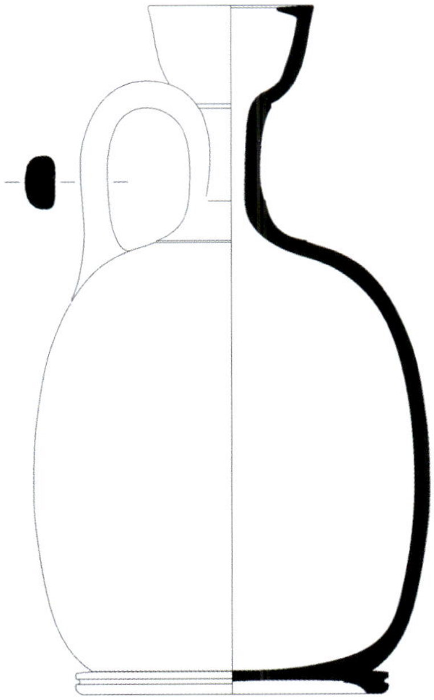

Site 736
T 19-03 RF squat lekythos (MM 322.346)
Dimensions: h 23.8 base diam. 10.9 d 13.7
Condition: Complete profile preserved; restored from numerous fragments; several small areas from the surface are missing, as well as many whole fragments from the body of the vase, including from the figured scene.
Decoration: Tongues at junction of neck and shoulder, ovolo below figured scene.
Figured decoration, attribution, and date: Mistress and maid. Attributable to Cyclops Painter, ca. 430–400 BC. See more photos and discussion below.

1:3

Site 736

T 19-04 BG pyxis with lid (MM 322.347)

Dimensions: h 7.8 base diam. 8.0 d 9.0
lid h 2.8 lid diam. 4.8

Condition: Intact with a few nicks.

Description: Ribbed decoration on shoulder except around handles. Body reserved below projecting flange and upper surface of the broadened base, which is covered with black gloss. Bottom reserved, with red slip. Some miltos (red slip) is still visible on the reserved part of the body.

Date: The shape and decoration, with minor variations, is closely paralleled by a group of pyxides from tombs dating between about 430 and 370 BC. Particularly close is one of a pair PY8 (T 212-4) from T 212 in the Pantanello necropolis (*Necropoleis* 353; Elliott 1998, 676). See also PY10 from T 222. These comparisons suggest a date for the pyxis in the last decades of the 5th c. or first decade of the 4th c. BC. This type of pyxis (for which see also Morel 1981a, 319 Serie 4471, Pl. 110) is frequently recovered in funerary contexts at Metaponto and appears to have been introduced into the Metapontine morphological repertory after the middle of the 5th c. BC. Besides the abovementioned T 212, it appears also in T 358 from Pisticci (Destra Basento, SS 106) in association with a Panathenaic amphora attributable to the Amykos Painter. The form is encountered in the black-gloss production of the workshop of the Creusa and Dolon Painters, attested by an unpublished example from scarico 4, and in funerary assemblages containing other vases by these painters (e.g., T 45 in proprietà Sellitti at Metapontion: Nava 2002, 742, Pl. LXIX). The examples from this phase seem to prefer the impressed palmette decorative motif. The pyxides of the first quarter of the 4th c. (e.g., those from T 306) are taller in relation to their bases, a process that seems to become more pronounced in the course of the 4th c. BC, as in the case of the vase from T 1 in the Basento valley, part of the ENEL excavations (Nava 2003a, 668, Pl. XXXII), and the example from T 12 at Montescaglioso (Lo Porto 1973, 193, No. 17, Pl. XLIV,2). The form is attested, albeit more rarely, in Tarantine funerary assemblages of the A0 phase, both ancient (425–400 BC) and recent (400–380 BC): Hoffmann 2002, 4[illegible], Pl. 78. (FS)

Site 736 1:3

T 19-05 RF lebes gamikos (MM 322.348)

Dimensions: h (with lid) 22.2
base diam. 6.9 d 13.2

Condition: Excellent, intact apart from a few chips and scratches.

Decoration: On the lid, lines radiating from the base of the knob, then ovolo; on shoulder, ovolo with tongues beyond; below figured scenes, ovolo; palmette complex under each handle.

Figured decoration, attribution, and date: Side A, youth with a lyre accompanied by a girl with pipes and a torch. Side B, woman seated on a block, mounted on a base, perhaps representing a sepulchral stele, addressed by a youth leaning on a thin and knotted staff. See more photos and discussion below.

Site 736, T 19-05 cont.

1:3

1:3

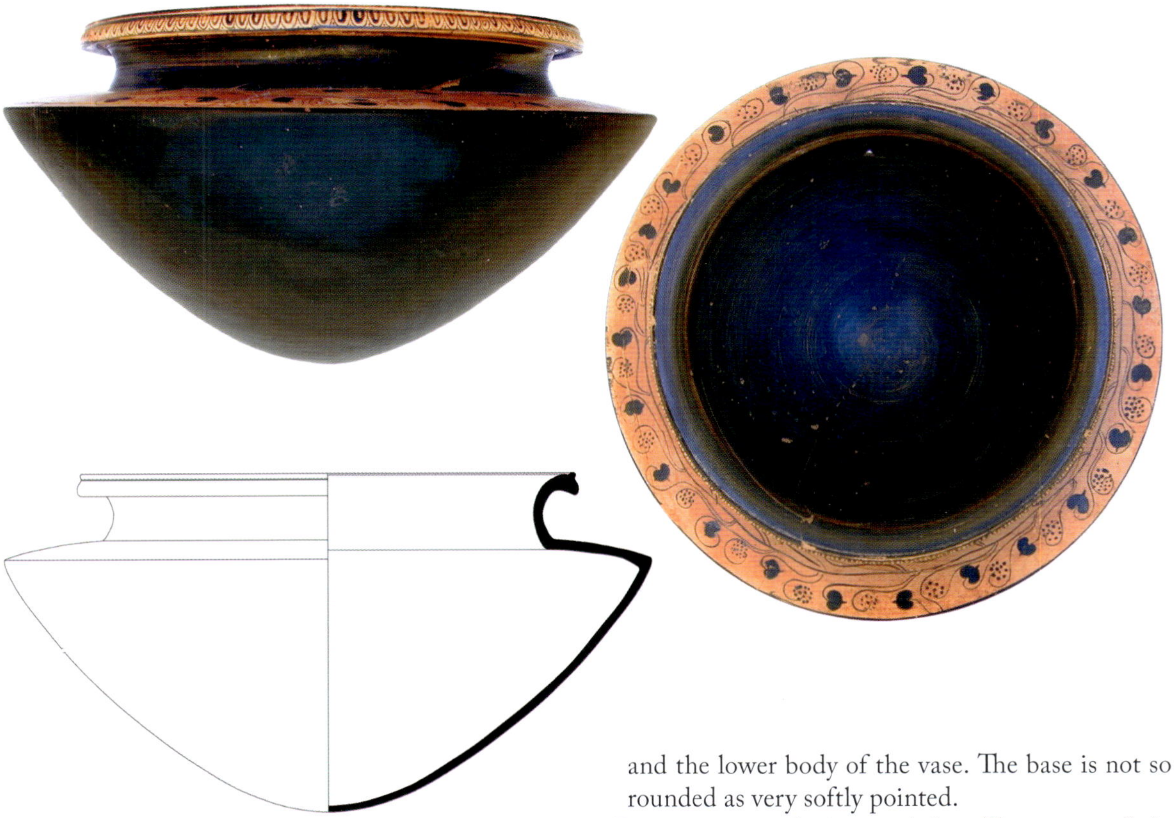

Site 736

T 19-06 BG *dinos* (MM 322.349)

Dimensions: h 13.3 rim diam. 20.1 d 25.8

Condition: Complete, excellent, with a few chips

Shape: The *dinos* has a broad mouth, a concave neck, and a slightly convex shoulder sharply offset to both the neck and the lower body of the vase. The base is not so much rounded as very softly pointed.

Decoration, attribution, and date: Two areas of the *dinos* are reserved for pattern decoration: the outer edge of the overhanging rim, which has a tongue pattern, and the shoulder, where an elegantly designed, calligraphic chain of ivy leaves and dotted berries encircles the vase, a painted image of a garland that such a vase might wear at a feast. See more photos and discussion below.

Site 736
T 19-07 BG skyphos (MM 322.350)
Dimensions: h 7.8 base diam. 4.7 rim diam. 9.0
Condition: Reconstructed from many fragments.
Description: Base flaring with reserved narrow band on resting surface. Miltos on reserved center. Horseshoe handles.

1:3

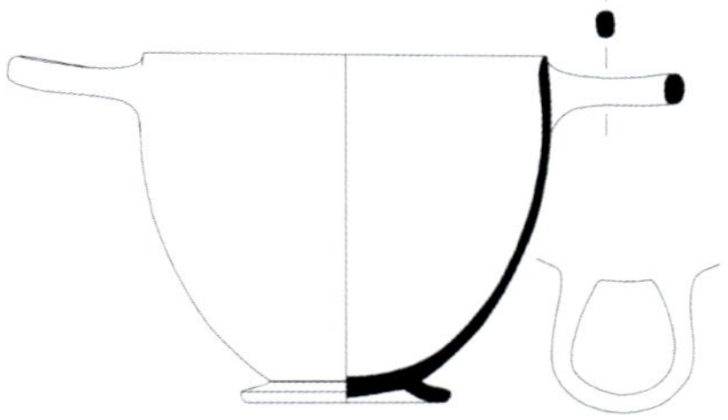

Date: Type B2 skyphos, Elliott 1998, 678. See site 736, T 18-01 above, 440–370 BC. Very similar in shape and dimensions to the example from scarico 1 at Metaponto, inv. 125957: Silvestrelli forthcoming a, Fig. 10.9. (FS)

Site 736
T 19-08 BW olpe (MM 322.351)
Dimensions: h 14.3 base diam. 3.7 rim diam. 4.2 d 6.7
Description: Band around the flat base and the shoulder. The interior of the rim is also banded, and the outside of the handles.
Condition: Complete except for small section of shoulder.
Date: The shape is comparable to two black-gloss olpai from Pantanello (T 336-2) dated by the context 470–440 BC, but in this one the shoulder and center of gravity are higher (*Necropoleis*). The olpe was popular in Athens around 480 BC. Its presence here adds to the impression that some vessels (e.g., the squat lekythoi) belong stylistically to an earlier group that reflects a backward-looking, Athenian-influenced taste that is apparent even in the stylistically more advanced pieces (e.g., the lebes) from this burial group. There can be no question that these grave goods were placed around the body and covered with plaster at the same time.

1:3

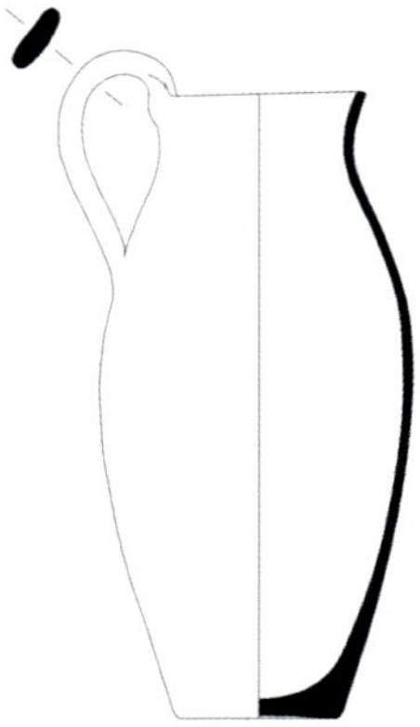

1:3

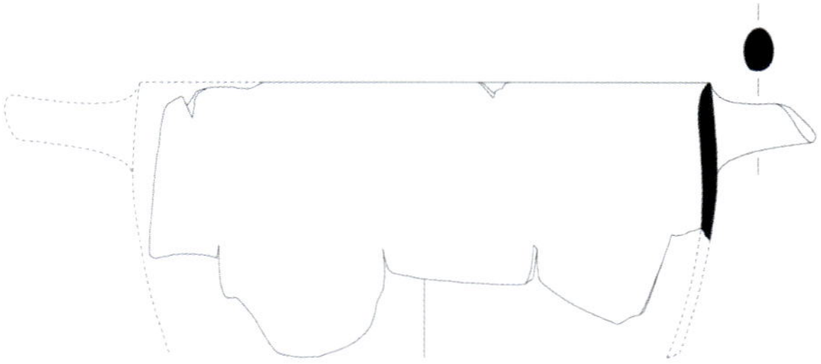

A

B

Site 736
T 19-09 RF skyphos of Corinthian type (MM 323.352)
Dimensions: h (partial) 7.6 rim diam. 16.1
Condition: About three-quarters of the rim is preserved, along with the upper one-third or so of the walls; the full w of Side A remains, rather less of Side B, and parts of the palmette complex beneath each handle.
Figured decoration, attribution, and date: Side A, woman with oinochoe and phiale beside a column; Side B, woman who stands fully in profile to the right: both hands outstretched; in her right, a fluted phiale. Workshop of the Creusa and Dolon Painters. See more photos and discussion below.

Site 736
T 19-10 BG skyphos (MM 322.353)
Dimensions: h (partial) 6.1 base diam. 4.9
rim diam. (est.) 9.4
Condition: Restored from nine fragments, including base and two body sherds which join and two horseshoe handles and rim sections (two joining fragments each).
Description: Base flaring with reserved narrow band on resting surface. Miltos on reserved center.
Date: A matching pair with T 19-07.

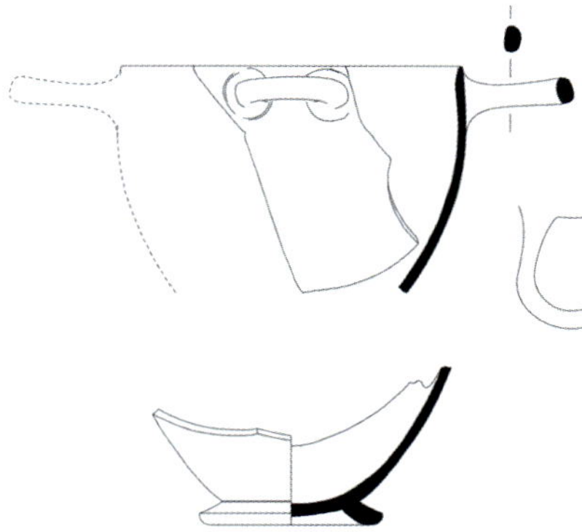
1:3

Site 736
T 19-20 BG cup-skyphos, fragmentary
Dimensions: rim diam. 12.0 handles project from rim 3.0 handle h 3.0 handle w 3.0
Condition: Three fragments.
Description: Two horseshoe handles attached to rim.
Date: Late 5th c. BC.

Site 736
T 19-21 BG skyphos, fragmentary
Dimensions: h (partial) 7.0 d (partial) 9.1
Condition: Three large joining fragments, two smaller ones.
Description: Good-quality black gloss. Very similar to T 19-07 and T 19-10, but perhaps slightly larger.
Date: Late 5th c. BC.

Site 736
T 19-11 AL alabastron (MM 322.354), two pieces.
Dimensions: h (est.) 16.5 rim diam. 4.7 body d 4.1
Condition: Restored from eight fragments; lower part complete, no trace of the long handles
Description: This single piece type A alabastron has a max diam. to h ratio (w/h) of 0.27, which at Pantanello would correspond to a mid-point base of ca. 450 BC (Carter and Toxey 1998b, 759). Its shape is very close to the alabastron from T 350 but compares also to that from T 208 (Carter and Toxey 1998b, 767, 765).
Date: Both of these contexts indicate a date for the alabastron in the period 440–400 BC.

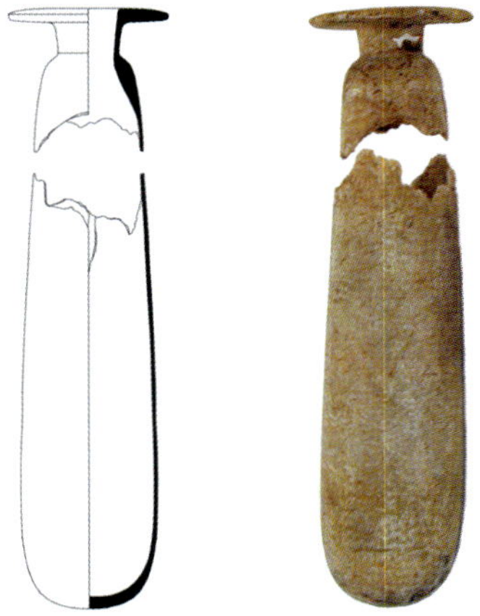
1:3

Site 736
T 19-12 AL alabastron (MM 322.355), one piece (complete)
Dimensions: h (est.) 16.5 rim diam. 4.5 body d 4.6
Condition: Intact.
Description: Practically identical to T 19-11.
Date: This alabastron is practically a twin of T 19-11.

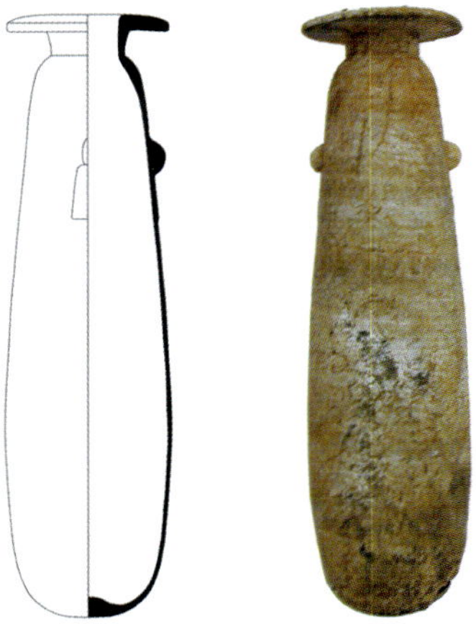
1:3

1:3

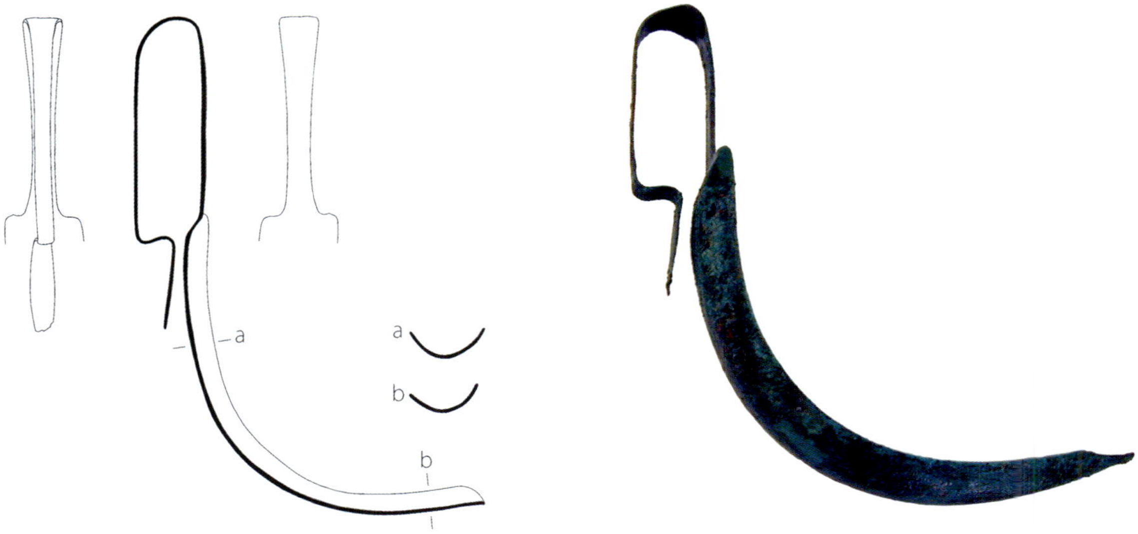

Site 736
T 19-13 AE bronze strigil (MM 322.356)
Dimensions: l (tip of blade to tip of handle) 22.5 w (max) 3.1
Condition: In two pieces, otherwise excellent condition.
Date: Compare the B4 type strigil from Pantanello T 106, *Necropoleis*, which is dated 420–360 BC.

1:1

Site 736
T 19-14 Fibula, iron with coral spool (MM 322.357)
Dimensions: l 4.0 w 2.2 l spool 1.5 d spool 1.0
Condition: Iron badly corroded; coral powdery.
Description: Iron bow fibula with a coral spool attached to the catchplate.
Date: This is a type 5; compare the examples from Pantanello T 95 dated 440–400 BC (Prohászka 1998, 814).

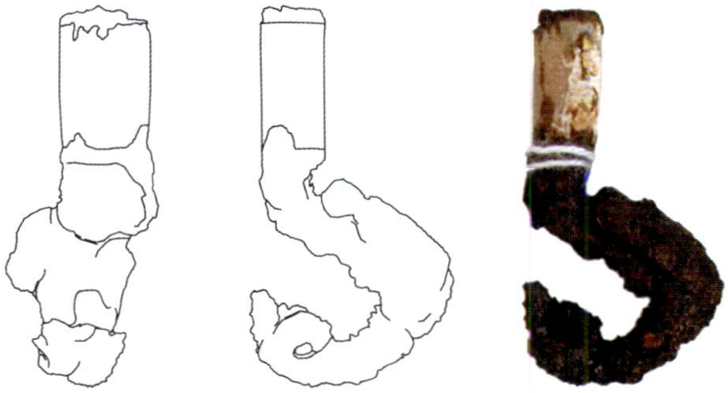

Site 736 1:1
T 19-15-18 Fibulae, like T 19-04 but less well preserved.

1:3

Site 736
T 19-19 Iron blade (MM 322.362)
Dimensions: l 15.7 w 5.1
Condition: Incomplete, badly corroded, with bent edge.
Description: Possible knife or scraper blade.
Date: Late 5th–early 4th c. BC, from context.

The Figured Vases from Tomb 19, Site 736

Lucilla Burn

The five figured vases from tomb 19 comprise two squat lekythoi, one lebes gamikos, one pelike, and one skyphos; with these may be classed, although strictly patterned, rather than truly figured, the pointed *dinos* with its black-figure ivy-leaf and berry chain. All are of high quality in terms of potting, decoration, and firing. Although the occupant of the grave is known to have been female, as regards both shapes and scenes, the selection of vases would seem suitable for either a male or a female burial (an impression reinforced by consideration of the rest of the grave goods—the strigil, the fibulae, the iron blade, and the alabastra). While the smaller figured vases (the squat lekythoi and the lebes) are shapes used primarily by women, whose perfumed oils and ointments they contained, the skyphos, the pelike, and above all the ivy-wreathed *dinos*, are shapes associated rather with the exclusively male symposium.

Much of the iconography of the vases is dominated by female figures, and several of the objects or activities shown reflect traditional female activities and preoccupations. The kalathos on the first lekythos (T 19-01) for example, symbolizes the woman's primary function of spinning and weaving wool to provide her household with all its textile needs; the caskets offered to two of the women (on T 19-01 and T 19-05), the mirrors held by the seated woman on the second lekythos (T 19-03) and hanging on the wall beside the woman on the first lekythos (T 19-01), the alabastron held by the standing woman on T 19-03, and the plemochoe balanced on the casket on the first lekythos (T 19-01) all allude to women's interest in their appearance, and to their fondness for jewelry and perfume. Elements in the scenes on the two lekythoi (T 19-01 and T 19-03) suggest that they may both relate to wedding rituals and customs.

The muffled woman approached by the Eros on the first squat lekythos (T 19-01) recalls brides about to set out on the journey to the bridegroom's house, while the casket and plemochoe offered by the Eros often appear in gift-giving scenes of the epaulia. Similarly the "Aphrodite in the Gardens" pose of the woman on the second lekythos (T 19-03) is that often assumed by brides. These allusions might even suggest that the vases were given as gifts to women who died before marriage. The women on the skyphos (T 19-09), on the other hand, may be shown fulfilling another important female role, that of tending the family graves. The function or identity of the women on Side B of the lebes (T 19-05) and Side A of the pelike (T 19-02) is less clear. There are obvious affinities between the two in terms of their block-like seats and the wreath each holds, but it is uncertain whether their seats are tomb or other markers (perhaps boundary), and whether they are giving the wreath to, or receiving it from, the youths who approach them. The scene on the lebes (T 19-05) stands out from the rest in its very clear representation of a typically male-centered activity, the komos or procession that could follow a symposium. The female figure here must certainly represent either a slave or a paid musician—in any case, someone of lower social status than her aristocratic male companion. The composition, however, gives equal weight to the male and female figures, and the elegantly-dressed woman's lively step and cheerful expression suggest she shares the excitement of her male companion as they hurry together through the night.

While the lebes, the pelike, and the skyphos are clearly Lucanian work, almost certainly produced in the early years of the 4th c. BC in the Metapontine workshop of the Creusa and Dolon Painters, and the magnificent *dinos*, while highly unusual in shape and decoration, can also be assigned to a local workshop, the two squat lekythoi have proved much harder to place. Had they suddenly appeared on the international art market without a findspot or other context, the immediate reaction would have been to identify them as Athenian products of the last decades of the 5th c. BC—related, perhaps, to the work of such painters as the Washing, Eretria, or Shuvalov Painters. However, as the discussions of the individual pieces should make clear, close observation suggests that, like the rest of the tomb group, they too are of local manufacture, the work of the early Lucanian painters in imitation of late 5th c. Athenian products and so slightly earlier in date than the other vases in the group.

Site 736, T 19-01 RF squat lekythos. h 19.1, base diam. 10.0, d 13.0.

Site 736

T 19-01 RF squat lekythos (MM 322.3–4)

Description: Figured scene on one side showing woman and Eros. The woman stands facing right. Her right arm is bent across her chest while her left is extended, but both are fully enveloped in her long himation. Under the himation, she wears a finely pleated chiton, visible just above the ankles; her feet are bare. Although there is some damage to the upper front part of her head, her hairstyle appears fairly simple, a thick mass of curls, with, perhaps, the back hair swept up into a bun, the arrangement secured with a simple bandeau. Her face is shown in profile, the eye drawn with a single line for the eyelid and another for the brow above, a long, straight nose, and (probably, as this area is damaged) a full, heavy chin. In front of her on the ground stands a wool basket (kalathos), its wicker material indicated by the interior markings that divide it into two sections horizontally and several vertically. Flying towards the woman from the right-hand side is the winged figure of Eros. His wings are raised, his legs bent at the knee, and in his outstretched arms he carries a chest or casket, probably supposed to be wooden, with a striped, perhaps beaded upper edge. On top of the chest stands a finely drawn perfume vessel (plemochoe) with tall stem, a decorative band around its central part, and an elegant lid-knob. Just above and to the left of the plemochoe, a mirror hangs in the field. The Eros, whose proportions are fairly chubby, is naked, and details of his anatomy have been carefully drawn in a mixture of thick and dilute gloss lines. His wings are shown in some detail, with the different types of feathers indicated by stippling, shorter and longer lines of gloss. The treatment of his eye differs from that of the woman in that there is a double line for the eyelid, with a single brow line above. His hair is finely drawn, with long, curling ringlets falling over his neck.

Close analysis of the subject matter suggests perhaps a conflation or even a confusion of two kinds of wedding scene. The kalathos on the floor and the mirror on the wall combine to fix the setting of the scene indoors, in the women's quarters. The casket-bearing Eros is the sort of figure found on so-called epaulia scenes, representations of the ceremony that took place on the day after the actual wedding, when the bride's friends brought her gifts of vases, caskets, and other items. But in these scenes the bride is normally seated, often in the pose adopted by the woman on the second squat lekythos (T 19-03). The pose of the woman on T 19-01, combined with the tightly muffled mantle that she wears, recalls rather an image of a bride waiting to be led from her father's house to that of her husband, in the first and major stage of the wedding ritual.

Attribution and date: The slight confusion in subject may provide a clue as to the identification of the painter or workshop to which the vase may be attributed. It is undeniable that some details of the painting, especially as regards the figure of the Eros, recall the finest work of the Washing Painter (for the subject, for example, cf. Aurigemma, Spina, I, Pl. 205a; for the casket-bearing Eros, cf. Peredolskaya, Hermitage, Pl. 152.1; for the hairstyle and pose of the Eros, cf. Louvre G 549, Boardman 1989, Fig. 208). The head of the Eros with his tousled curls and ringlets recalls more strongly the work of the Eretria Painter (compare, for example, Leningrad 1867/68.960, Lezzi-Hafter 1988, Pl. 181d); but the treatment of the wings is different from that usually found in his work (see for example the numerous Erotes on a pyxis in Worcester, Mass., Lezzi-Hafter 1988, Pl. 165). Then again, there are also clear echoes of the work of the Shuvalov Painter: for the head of the Eros cf. Louvre G 437, Lezzi-Hafter 1976, Pl. 86. The figure of the woman, on the other hand, is harder to parallel in Athenian work. Occasional echoes are found in the work of the Washing Painter: cf., for example, Peredolskaya, Hermitage, Pl. 154.2. But in general terms her face and figure are too angular and elongated for Athenian work, and her closest affinities lie instead with the male mantle figures who populate the backs of Lucanian (or Apulian) vases; she is basically one of these (Trendall's type IBl, see *LCS* 11), modified to the extent of letting a chiton emerge at the lower edge of the mantle and placing a female, rather than a male, head on top. Compare, for example, for the arrangement of the mantle with its stiff folds at the upper edge and cascade hanging from the left arm, the central figures on the reverse of two Pisticci Painter bell kraters, Denver, AN 108, *LCS* Pl. 3.2 (No. 14), or (Once Rome Market) Pl. 52 (No. 49), or one of the figures on the reverse of the Cyclops Painter's name-vase, London, British Museum, GR 1947.7-14.18, *LCS* Pl. 8.2 (No. 85), or, later, in the work of the Palermo Painter, for example on his name vase, Palermo 955, *LCS*, Pl. 22.2, No. 266.

Taken altogether, and combined with the slight oddness of the subject matter already noted, these factors suggest the possibility that the painting on this vase is not the work of an Athenian painter, but instead one influenced by or deliberately copying Athenian styles and motifs. If we hypothesize that the painting of the vase should in fact be attributed to an early Lucanian painter, the two most likely candidates are two of the very earliest, the Pisticci and Cyclops Painters. Martine Denoyelle has demonstrated that these two were such close associates that their work can be difficult to distinguish and that there are places where they can be detected to have collaborated on a single vase (1992). She has also shown that both had extremely close contacts with Athenian workshops, and she is currently researching an example of their work in which they seem to "cut and paste" motifs from Athenian vases to form their own design. Denoyelle herself suggests that this vase (and also T 19-03) should be attributed to the Cyclops Painter, an artist whom she has vividly characterized as *protéiforme* on account of the elusive and changeable nature of his style (1992, 27). While much of his early work is fairly pedestrian, the Cyclops Painter's name-vase, the large British Museum calyx krater with its scene of the blinding of Polyphemus (*LCS*, Pl. 8.1–2), shows that in his mature years he could be a very daring and highly innovative artist. A.D. Trendall, in his original characterization of the Cyclops Painter, observed that one detail of female dress is consistently characteristic of his work: the "vertical folds of the chiton are always clearly shown below the himation, but sometimes there is a gap between them and the hem" (Trendall 1960). This detail is clearly visible here at the lower edge of the woman's inner garment. The likely date of this vase will be discussed in connection with that of the next.

Site 736

T 19-03 RF squat lekythos

Description: Figured scene on one side showing mistress and maid. A woman, a slave or other companion, stands facing right, her right hand outstretched in a gesture that may be conversational, while from her lowered left hand an alabastron, similar in shape to the two calcite examples from the actual tomb (T 19-11 and T 19-12), hangs by a string or cord. She holds the cord through a cloth, either to protect herself from the oil or in readiness for its application. She wears a peplos, belted over the overfall, and with thick black borders decorating both the edge of the overfall and the lower edge of the garment above her ankles. Facing her, a second woman, the mistress, sits on a high-backed chair (klismos). Her left arm is propped over the chair-back, in the pose believed to have been adopted by Alkamenes' statue of "Aphrodite in the Gardens" and very popular for women on late 5th century Athenian vases (cf. Ferrara T 512, Lezzi-Hafter 1976, Pl. 115e; for the motif see Burn 1987, 27, with refs). In her right hand she holds a large mirror upright on her knee, and probably she is to be thought of as gazing at her reflection in this. She wears a finely-pleated, sleeved chiton with a himation draped over her lower body and legs. Her hair is more elaborately arranged than that of the maid, with a fringe of short corkscrew ringlets framing her face. Above the mirror, and apparently unseen by either woman, a small Eros figure hovers in front of the mistress's head; stretched between his hands is a fillet or necklace that he is about to place on her head or neck. The presence of the Eros, the perfume, the mirror, and the elaborate arrangement of the seated woman's hair combine to suggest that she may be a bride; rather than an epaulia scene, the accoutrements suggest the seated woman is being prepared for marriage by her attendant.

Attribution and date: Like that of the last vase, the painting of this one suggests at first glance the possibility of Athenian workmanship. The pose of the seated woman—the patterned borders of the standing woman's garment, the hovering Eros—all of these echo elements of scenes attributed to, once again, such late 5th century Athenian artists as the Washing Painter. (For the pose of the seated woman, cf. Copenhagen, National Museum, 153; Boardman 1989, Fig. 209; for the thick black borders on the standing woman's garment, cf. those on three garments worn by women on New York, Metropolitan Museum of Art 16.73; Boardman 1989, Fig. 207; for the little hovering Eros, cf. a similar figure on the same vase.) However, once again there are also anomalous details that are hard to reconcile with an attribution to any Athenian painter. The gesture of the standing woman and the large, clumsy hands of both, for example, are better paralleled in a Lucanian context; nor do the women's heads with their heavy chins and, in the case of the seated figure, the fringe of stylish ringlets, look truly Athenian; the thick-handled mirror held by the seated woman is clumsier than contemporary Athenian examples. In short, although the figures are very competently drawn, there is a heaviness and awkwardness about them that once again suggest the painter is copying Athenian models rather than being himself Athenian.

Again, the closest parallels lie with the work of the Cyclops Painter, to whom, as already mentioned, Martine Denoyelle has attributed this vase. The gesture and drawing of the outstretched hand of the standing woman is very characteristic of the painter, finding parallels on bell kraters in the Louvre (G 500, *LCS* 28, No. 99, *CVA* Louvre 25, Pl. 7), in an American private collection (Denoyelle 1992, 23, Fig. 16) and in Christchurch, New Zealand (Trendall 1989, Fig. 1 O). The relaxed hand and wrist of the seated woman, lying over the back of her chair, although copied from Athenian models, is very close in its drawing to that of the somnolent Cyclops on the painter's name-vase, *LCS*, Pl. 8.1, No. 85. The fringe of sausage-like ringlets, as yet unparalleled, is perhaps an experimental development of the dotted edge of hair sometimes favored for both male and female characters by many early Lucanian painters, including the Pisticci, Cyclops, and Amykos Painters (cf., from many possible examples, Burn 1998, 598, fig. 13.10, T 212-2).

In terms of shape, technique, and style the two squat lekythoi from tomb 19 are exceptional, and yet they have much in common with each other. This suggests the strong likelihood that they were made and painted in the same workshop and probably by the same painter, who was in all likelihood the Cyclops Painter, pupil, collaborator, and follower of the Pisticci Painter. The discovery at Metaponto of kilns with fragments and almost complete examples of painted pottery attributable to the second generation of Lucanian red-figure painters, notably the Creusa

Site 736, T 19-03 RF squat lekythos. H 23.8, base diam. 10.9, d 13.7.

Site 736, T 19-03 Detail.

and Dolon Painters and artists associated with them, has not only made it very clear that these painters worked at Metaponto but also thrown light on possible workshop connections between them (see *Metaponto* I, esp. 355–442). The three dominant painters of the first generation were the Pisticci Painter, generally believed to be the founder of the school, the Cyclops Painter, and the Amykos Painter. That the last-named also worked at Metaponto has been demonstrated by the excavation of fragments of at least three of his vases in kiln dumps in the northwest zone of the city (see *Metaponto* I, 447–52). But although numerous vases, including early-looking pieces attributable to the Pisticci and Cyclops Painters, have been found in tombs in and near Metaponto, supporting the hypothesis that they too are likely to have worked here (see Trendall, *LCS* Supp.3, 4, and the lists of vases attributable to these painters both in *LCS* and in Supp.3; also Burn 1998, 615, T 69-3, 616, T 88-1, T 95-7, 617, T-136-6, 619-20, T 209-7, 620, T 212-2, 622-3, T 354-2—all by or near the Pisticci Painter), unless more relevant kiln material turns up, this will remain unproven. Further questions as to the dates, origins, and inspirations of the earliest Lucanian red-figure painters remain similarly subjects for debate, but perhaps it is a discourse to which the two squat lekythoi from T 19 can at least contribute.

Until relatively recently, it was assumed that the foundation of the Lucanian red-figure school was associated with that of the Athenian colony of Thourioi in 443 BC (for the origins and development of this theory see Trendall 1989, 17). Although no evidence has, in fact, turned up to suggest that Thourioi had anything to do with red-figure vase painting, the idea that the Lucanian school began at around this time appears to have become entrenched. The likely starting date for the Pisticci Painter's career must of course depend in part on that of the Athenian painters or workshops with which he seems most closely connected. But although the Lucanian painter's debt to Athenian models has often been observed, the question of which group of Athenian painters he drew his inspiration from has been debated. Trendall believed that the Pisticci Painter's work shows the influence of painters associated with the Group of Polygnotos or the Achilles Painter, generally reckoned to have been operating in Athens in the 430s BC (see Trendall 1989, 18). But more recently N.R. Jircik and M. Denoyelle have suggested that it was rather from the Niobid Painter and his associates, operating in the later 450s and early 440s, that the Pisticci Painter learned his craft (Jircik 1991; Denoyelle 1997). (I have not been able to consult Jircik's thesis.) It might thus be suggested that if the Pisticci Painter was at work for about 30 years, his main period of activity in Metaponto could lie between about 445 and 415 BC. Then, if the evidence of the two squat lekythoi from T 19 can be taken to suggest that the Washing and Shuvalov Painters, generally believed to have been at work between about 440 and 420 BC, were influential on the Cyclops Painter, and if it is further agreed that the British Museum Cyclops krater is likely to postdate the production of Euripides' Cyclops in ca. 406 BC (see Trendall 1989, 19–20), it might be suggested that the Cyclops Painter was active in Metaponto between about 430 and 400 BC.

How, though, was influence transmitted from Athens to Metaponto? The hypothesis that the Pisticci Painter was an Athenian emigrant (an assumption often made—see, e.g., MacDonald 1981, 159–60) has, I think, been very firmly stamped out by Denoyelle, who has argued very strongly that his style is too different in character from that of his Athenian counterparts, essentially too colonial, for him to have been an Athenian (see Denoyelle 1997). We must still ask, though, if he actually trained in Athens, or if it was sufficient for him simply to observe and imitate Athenian wares as they arrived in Italy. Either suggestion is a hypothesis impossible to prove in light of current knowledge. His skill points in the direction of a proper Athenian apprenticeship, as was claimed by Trendall: "he must almost certainly have been a Greek himself, trained in Athens" (*ESIV*, 22–23). But the continuing absence of any pot attributable to him found in mainland Greece might point in the opposite direction, as might the fact that his work shows, perhaps, the influence of more than one Athenian master. The evidence of the Cyclops Painter's style further complicates the issue, for no one has ever suggested that he was anything other than a native Italiote Greek, and yet he too shows clear evidence of learning from Athenian wares. So how did he come into contact with Athenian styles and subjects? These issues of influence and contact are intriguing: did the Cyclops Painter visit Athens? Did enough Athenian imports arrive in Metaponto, or southern Italy in general, for him to absorb the styles and subjects so completely? Although museums such as Metaponto and Taranto do contain considerable quantities of mid- to late

Site 736 T19-02 RF pelike, side A. H 28, base diam. 15.5, d 25.1.

5th c. Athenian pots, the numbers seem to decline after about 430 BC, presumably as the result of a combination of social and economic factors in Athens and the growth of local production (see MacDonald 1981, 160 n. 8). In terms of imports into Italy, it seems as though it was chiefly at the north Etruscan city of Spina in the delta of the River Po that large quantities of the types of vases admired and imitated by the Pisticci and Cyclops Painters continued to be imported in the last decades of the 5th century.

Although Trendall lists just one Lucanian vase in Ferrara from a tomb in the Valle Trebba (a Rustic Group krater, *LCS* 78, No. 393), is it possible that there could be others there as yet undetected? Might there have been more contact between the northern Etruscan cities and the southern Greek colonies than has previously been suspected? (For a different sort of connection, see Denoyelle's account of a painter who seems to have learned his craft in the workshops of southern Italy before heading north to found his own business in Etruria: 2008, 281–93). We might also want to ask if more examples of copying from Athenian models on the part of the early Lucanians might be detected. In the Pantanello necropolis publication, two red-figure squat lekythoi were classed as Lucanian although there were good grounds for supposing they might, rather, be Athenian (see Burn 1998, 615, T 69-3, and 619, T 209-1); in the light of the two squat lekythoi from Tomb 19, the status of these two should perhaps be reconsidered. And finally it must surely be the case that in museum collections throughout the world there are many pots whose nationality is not securely or finally ascertained.

Site 736

T 19-02 RF pelike (MM 322.345)

Description: Side A shows a seated woman, youth, and woman. Seated at the left on a block-like seat, and facing right, is a woman who holds out a wreath in both hands. She wears a himation over a finely-pleated chiton; both the upper and lower border of the himation and the lower edge of the chiton are decorated with a single thick line of black gloss, while two vertical lines run down the right side of the chiton from below the woman's arm. Her hair is worn in a bun at the back, supported by a wide sakkos; she also wears a simple beaded necklace. Little of her face is preserved apart from her down-turned mouth. Standing in front of her, apparently shown in three-quarter view to the left, is the figure of a youth leaning on a staff, preserved to a height just above the genitals. What survives of his figure suggests the likelihood that he is naked apart from a small himation that he most probably wears draped across his lower back, with the loose ends hanging over each forearm. One free end appears between his right thigh and the staff; the other, beyond his left leg, suggests his left arm was considerably outstretched. It is unclear from the surviving portion of the scene whether the woman is offering him the wreath or receiving it from him. Traces of a third figure, another woman, appear behind the youth, closing the scene on its right side. She faces left and was probably holding the spotted fillet that appears in the field behind the further free end of the youth's himation. Side B shows three draped youths.

Attribution and date: This piece may be attributed to the workshop of the Creusa and Dolon Painters, and perhaps the Dolon Painter himself. For the youths on side B, cf. Bari 6264 and Paris, Louvre CA 2193 (Trendall 1989, Figs. 76, 78). For the spotted fillet held by the woman on the right, cf. London, British Museum F 157 (*LCS* Pl. 52.3–4); the woman's down-turned mouth and the swirl of her chiton at its lower edge are also very characteristic of this painter. The Dolon-Creusa workshop is dated between ca. 400 and 360 BC (for chronology, see Trendall 1989, 270).

Site 736 T 19-02 RF pelike, side B.

Site 736

T 19-05 RF lebes gamikos (MM 322.348)

Description: Side A shows a youth with a lyre, accompanied by a girl with pipes and a torch. The youth is shown moving to the right, his rapid progress suggested by the way a loose end of his himation flies out in a sweeping diagonal between his legs. He is naked apart from a wreath on his head and the himation draped across his back, the ends hanging over each forearm. He carries his lyre tucked under his left arm, while in his right hand he holds a plectrum. He looks back to the figure of the girl hurrying behind him. The position of her right foot suggests she is running to keep up with him: in her right hand she holds a torch up against her shoulder, while in her lowered left is a pair of pipes. She wears a peplos belted over the deeply frilled double overfall and a dot necklace. Her hair is arranged in a bun with a narrow bandeau at the front. These two figures are beautifully fitted into the small field at the painter's disposal; the girl's torch overlaps the topmost element of the palmette decoration, while her right heel springs away from the lowest spiral; at the far right the youth's lyre is neatly fitted right up against the spiraling tendrils.

On side B a woman seated on a block, mounted on a base, perhaps representing a sepulchral stele, is addressed by a youth leaning on a thin and knotted staff. The woman wears a chiton with a single gloss stripe down the side and a necklace with a single pendant bead. Her hair is arranged in a bun above a striped band that encircles her head. Her left hand rests on the block, while in her right she holds a wreath that she may be offering to the youth. He stands with his weight on his left leg and his right leg slightly bent at the knee. He is naked apart from the small himation that is pulled across his back with the loose ends hanging down over each arm. His body is shown in three-quarter view to the right, his face in profile. His right hand holds the top of his staff, while in his left he holds a decorated box or basket, which he is apparently offering to the woman. J.C. Carter originally identified the figures on side A as Orpheus and Eurydice, Orpheus leading his wife back from the Underworld, turning to catch a glimpse of her and so losing her a second time; Eurydice hurrying behind, her torch lighting the journey up from the dark subterranean kingdom of the Underworld. This is an attractive interpretation and one which fits in well with the theory that those ancient inhabitants of Metaponto who chose to be buried in plaster-lined tombs like Tomb 19 were initiates or devotees of the Orphic-Pythagorean-Dionysiac complex of beliefs (see *Metaponto* I, 67–69, with further references). It has to be said, however, that there are no real grounds for supposing that the two figures represent anything other than a reveler (komast) with a flute girl, hurrying to or from a party. The iconography is popular on mid 5th century Attic vases (cf., e.g., a krater by the Kleophon Painter, Munich 2414; Matheson 1995, Pl. 126), and apart from any other considerations the pipes the woman holds make her an unlikely Eurydice. The language and meaning of iconography, however, can be both subtle and capable of many layers or levels of interpretation, and it is undeniable that there is a sense in which every image of a youth with a lyre, followed by a young woman, partakes in the iconography of Orpheus and Eurydice and might suggest this tale to a contemporary viewer.

Attribution and date: The shape of the lebes is very similar to various of the black-glossed examples from Pantanello,

Site 736 T 19-05 RF lebes gamikos, side A. H (with lid) 22.2, base diam. 6.9, d 13.2.

including LG6 (T 51-2) and LG41 (T306-2), both from tombs that included red-figure vases attributed to the Dolon Painter or his circle (for the black-glossed examples see Elliott 1998, 659, 663; for the red-figure vases, T 51-1 and T 306-12, see Burn 1998, 614, 621). There are also affinities both in the shape of the vase and in details of the figure style with red-figure vases of the Intermediate Group dated to the years around 400 BC: cf., e.g., for the shape of the vase including its elegant lid-knob and for the frilled peplos of the woman on Side A the shape and female figure on Taranto 52558 (illustrated *LCS*, Pl. 30.4). The knock-kneed pose of the youth on Side B is also found in this group; cf., for example, in addition to Taranto 52558, Reggio C. 7004 and Schwerin 720 (*LCS*, Pls. 30.2, 4, 31.6). However, the piece also recalls aspects of the slightly later Dolon and Creusa workshop, characteristic of whose style (especially that of the Creusa Painter) are both single pendant and multiple-dotted necklaces (for both single pendant and multiple dots cf. Prohászka 1998, 818, T 165-1, with further references), as well as the pose of the youth on Side B (cf. Vienna 653; Trendall 1989, Fig. 73; *Metaponto* I, 399, Fig. 47a; Naples 2894; Once Swiss Market, Trieste S420; and Auckland, New Zealand 29700, *LCS*, Pls. 40.2,5, 42.5, 45.4, etc.); also the profile and hairstyle of both women (cf. Burn 1998, 618, T 165-1(a); Rome, American Academy 1840; Trendall 1989, Fig. 70; Toledo 81.110, *LCS* Supp.3, Pl. 71).

Site 736

T 19-09 Skyphos of Corinthian type (MM 323.352)

Description: Side A depicts a woman with oinochoe and phiale beside a column. She stands with her head and breasts shown in profile to the left, but her body is basically frontal. Both her arms are outstretched; in her right hand she holds a fluted phiale, while hanging from her left is an oinochoe. These vessels and the column suggest she may be preparing to pour a libation at a grave. Her hair is worn in a bun at the back, supported by a simple bandeau; a corkscrew curl, shown in dilute gloss, escapes in front of her ear. She wears a sleeved peplos belted beneath the overfall, the lower edge of which is decorated with a band of dilute gloss. Around her neck she wears a necklace shown as a series of short, straight strokes, and she has bracelets on each arm. Her right hand stretches in front of the column, which is thin and fluted, topped with Ionic-style volutes.

Side B has no column, just a woman who stands fully in profile to the right: both her hands are outstretched. In her right hand, she holds up a fluted phiale, very like that held by the woman on side A, but her left appears to be empty unless a vertical gloss line running down over her thumb is part of an otherwise now missing object. She wears a sleeved garment similar to that of the first woman, poking forward slightly at the neck, and her hair is similarly drawn back into a bun; her bandeau is slightly wider and twisted over at the back, and also has a marked-off upper and lower border. In front of her ear, too, is a corkscrew curl painted in dilute gloss. *Attribution and date*: This piece may be a product of the workshop of the Creusa and Dolon Painters, perhaps by the Creusa Painter, typical of whose style are the prominent breasts of the women, the corkscrew curl before the ear (cf. Taranto 22492, *LCS*, Pl. 43.1,2), the way the lower edge of the peplos overfall on side A billows up (cf. that of the woman on the right on Vienna 1159, *LCS*, Pl. 43.1, and of the woman in the same position on Vienna 653, *LCS*, Pl.

Site 736 T 19-05 RF lebes gamikos, side B.

46.2), the way the neckline of the peplos shown in profile on side B pokes forward (cf. Taranto 22492, *LCS*, Pl. 43.1, 2), the slender and rather spindly column (cf. Naples 2181, *LCS*, Pl. 42.3), and the pose and attributes of the woman on side A (cf. Vienna 1112, *LCS*, Pl. 44.3); cf. also the women on two fragments by the Dolon Painter, *Metaponto* I, 390, Fig. 40, and 393, Fig. 95. For the form, see the skyphos from scarico 1 at Metaponto inv. 29063: D'Andria 1980a, 402n165, *LCS* Suppl. III, 3/D48. For the design cf. Silvestrelli 2008, 272, Fig. 1.12.

Side A

Side B

Site 736 T 19-09 RF skyphos of Corinthian type. H (partial) 7.6, rim diam. 16.1.

Site 736 T 19-06 BG *dinos*. H 13.3, rim diam. 20.1, d 25.8.

Site 736
T 19-06 BG *dinos* (MM 322.349)
Description: Two areas of the *dinos* are reserved for pattern decoration: the outer edge of the overhanging rim, which bears a tongue pattern, and the shoulder, where an elegantly designed, calligraphic chain of ivy leaves and dotted berries encircles the vase, a painted image of the actual garland that such a vase could well have worn at a feast.
Date and attribution: In the absence of figured scenes date and attribution are difficult to determine. The ivy-leaf and berry chain is most commonly used for the decoration of the neck of column kraters, not a common shape in early Lucanian workshops. The early Lucanian painter in whose work most examples of the pattern are found is the prolific Amykos Painter. However, it is really the shape and the potting of this elegant and perhaps unique vessel that is most remarkable; for the shape and quality the closest parallels are found in Athenian pottery of the Late Archaic period. It is possible that the shape owes its inspiration to native Italic or Lucanian potting.

For the ivy-leaf and berry chain on the neck and upper surface of a column krater formerly on the Swiss Market attributed to the Amykos Painter, see Trendall 1989, Fig. 18; cf. also the decoration on fragments of another column krater and a nestoris attributed to the same painter in the Louvre, *CVA* Louvre 25. Pls. 14.1–3 and 10.1. For Archaic Athenian examples see Miller 1997, 139, Fig. 40. (LB)

Site 736
Tomb 20
Type: Fossa
Orientation: 17
Skull: NE
Terminus date range: 300–275 BC
Weighted date range: 300–275 BC
Basis for date: BG pottery (the date of the burial as in Prohászka 1998, 830–33, is useful in establishing the chronology of the coins)
Sex: Female
Age: 25–35
Dimensions: l 1.60 w 0.80
Ceramic summary:

T20-01	PWBW one-handler (tibia)
T20-02	BG stemmed cup (femur)
T20-03	BG ribbed globular unguentarium (ulna)
T20-04	PW shallow plate from fill

Non-ceramic objects:

T20-04	AE Bronze coin (hand)
T20-05	AE Bronze coin (hand)

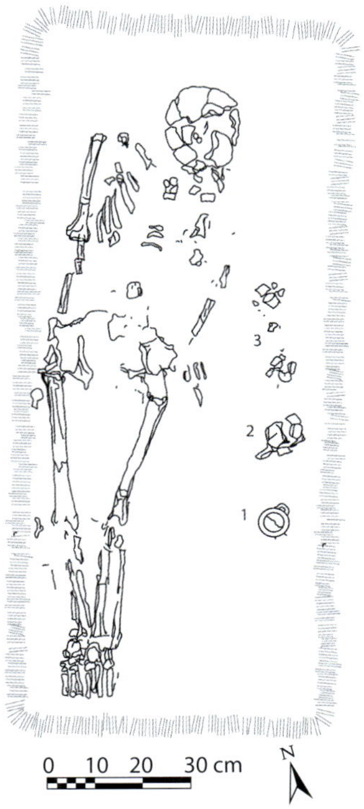

Comments: The body was laid in the simple earthen grave with the skull centered along the central axis but the pelvis and lower legs occupying the left side. The skull, to judge by the position of the teeth, was turned to the left, facing east. The skull, rib cage, upper right arm, and rib cage, pelvis, lower legs, and bones of the feet were reasonably well preserved. The left arm and left side of the torso were not. Except for two coins placed by the right hand, the other grave goods, consisting of three small vessels, were located on the left in the nearly empty east side of the grave.

Site 736
T 20-01 PWBW one-handler (MM 323.296)
Dimensions: h 2.3 base diam. 2.4 rim diam. 6.1
Condition: Chipped rim.
Description: Red-brown gloss on rim, narrow interior band of same color, and a touch at the tip of the handle.
Date: The shape is close to an early 3rd c. BC BW one-handler from Pantanello, such as T 199-4, but the interior narrow stripe is not typical of Group II one-handlers, which are generally undecorated (Carter and Toxey 1998a, 702).

1:3

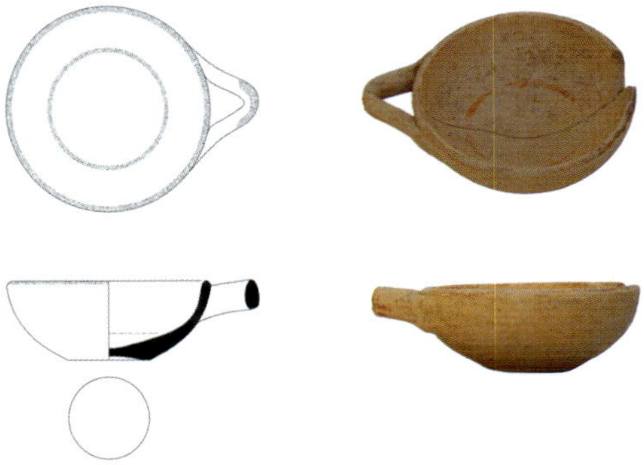

1:3

Site 736
T 20-02 BG small stemmed cup (MM 323.297)
Dimensions: h 5.5 base diam. 3.35 rim diam. 6.8
Condition: The base, one handle, and most of the body survive; restored from small fragments.
Description: Gloss inside and out; the stem and the resting surface only are reserved.
Date: A stemmed cup of identical shape and decoration was found in the Pantanello necropolis. It is not a common shape there. See the example from T 94, dated 300–275 BC. *Necropoleis* 302, Elliott 1998, 651. Also Pianu 1990, type 1 (330–280 BC).

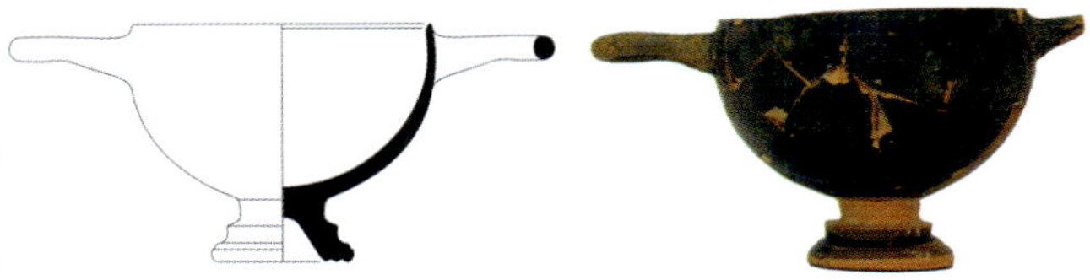

Site 736

1:2

T 20-03 BG globular unguentarium (MM 323.298)

Dimensions: h 7.5 base diam. 3.4 rim diam. 3.4

Condition: Restored from small fragments.

Description: Ribbed; lower body below ribbing reserved except for band around base.

Date: This type of unguentarium or "feeder" (for which see Morel 1981a, 389, *Serie* 5814, Pl. 191; and Labellarte 1992c, 132) is principally attested in funerary contexts of the Ionian poleis. The example from T 20 closely resembles examples of the late 4th or early 3rd c. BC from the Pantanello necropolis, Group III. Compare those from T 194, especially T 194-3. The burial is dated 300–275. *Necropoleis,* 422; Elliott 1998, 654. Very close, albeit with molded foot, are the vases from T 18 in proprietà D'Onofrio at Metaponto (A. Bottini 1993b, 185, No. 13, last quarter of the 4th c. BC), T 7 from contrada Ricotta at Metaponto (Lo Porto 1966, 194, Fig. 48, 325–300 BC), and Pantanello (Morard 2002, Pl. 1.1, dated 325 ± 5 BC). For the attestations at Taranto see Hoffmann 2002, type 116/5 (325–275 BC). (FS)

Site 736

1:3

T 20-04 PW shallow unglazed dish from fill (MM 323.299)

Dimensions: h 2.6 d 14.1

Condition: Missing base.

Description: Deliberate perforation in center of dish.

Date: Late 4th–early 3rd c. BC?

Site 736

2:1

T 20-05 AE bronze coin (MM 323.300)

Dimensions: d 1.6 th 0.3 weight 3.3 g

Orientation: 270°

Condition: Badly corroded

Description: Obv.: Male head rim diam.. with pointed beard. Rev.: M (to l.); E (to rim diam.). Barley ear.

Date: Last quarter of 5th to middle of 4th c. BC, from Johnston 1989, Cat. no. 14, and Rutter 2001, 1650. (ARP)

(See Ch. 15 for further discussion of coins.)

Site 736

2:1

T 20-06 AE bronze coin (MM 323.301)

Dimensions: d 1.6 th 0.3 weight 3.9 g

Orientation: 180°

Condition: Badly corroded

Description: Obv.: Head of Demeter rim diam., hair gathered under barley wreath. Rev.: Barley ear with leaf to rim diam.; above, thunderbolt.

Date: Second half of 4th to first quarter of 3rd c. BC, from Johnston 1989, Cat. no. 25, and Rutter 2001, 1660. (ARP)

(See Ch. 15 for further discussion of coins.)

Site 736
Tomb 21
Type: *A cappuccina*
Orientation: 39
Skull: SE
Date range: 320–280 BC
Basis for date: Pottery
Sex: Female
Age: 50–x
Dimensions: l 1.60 w 0.57

Ceramic summary:
T21-01 Hatched lekythos (head)

From fill:
2 BG fragments of a large vessel
1 BG base, bowl (diam. 4.5)

Comments: The tomb, constructed with three vertically placed pan tiles on a side (0.56 m wide), was almost completely destroyed by the backhoe, and it was possible to recover only the lower parts of the tiles (preserved to a h of 0.20 m). Fragments of the skull and a few teeth were all that remained of the upper part of the skeleton. The upper part of the arm (humerus) on both sides was destroyed by the backhoe, but much of the lower part of the skeleton including the forearms and finger bones, but not the left lower leg and toes, was intact. A single vessel, a hatched lekythos, in the extreme SE end of the tomb was preserved and recovered in situ.

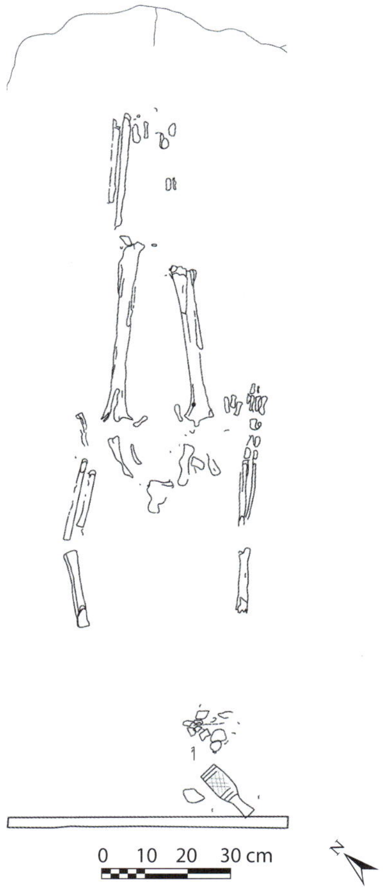

Site 736
T 21-01 BG hatched lekythos (MM 323.303)
Dimensions: h 13.1 base diam. 3.4 rim diam. 3.5 max diam. 4.0
Condition: Handle missing, otherwise intact. Chipped lip/mouth.
Description: Hastily and crudely applied black gloss. Blob of black gloss around the point where the handle attaches.
Date: Hatched or net-decorated lekythoi are rarely found in tomb groups with other dateable objects. What evidence there is suggests that they belong to the third quarter of the 4th c. See Elliott 1998, 668. Compare the example from T 74 at Pantanello, which is virtually identical and dated 320–280 BC. *Necropoleis* 270. The example from T 20 at Metaponto is dated to the last quarter of the 4th c. BC: Lo Porto 1966, 206. Hurschmann (1995, 677–79) has classed very similar lekythoi in his type III, attested in Apulia and western Lucania in funerary assemblages of 330–320 BC. For the form and dimensions cf. also the example from the Banca Intesa Collection, inv. 680: *Collezione Banca Intesa* 740, No. 355 (350–325 BC). (FS)

Site 736
Tomb 22
Type: Fossa
Orientation: 56
Skull: NW
Date range: Undated, 4th century BC?
Sex: Male
Age: 40–50
Dimensions: l 2.20 w 1.05 d 0.45-50

Comments: This simple earthen grave was considerably more spacious than it had to be to contain the body of this larger-than-average occupant. He measured 1.8 m in the grave. The lower part of the body occupied the west half, while the head and torso were mostly in the east. The skull faces towards the SW. No trace of grave goods. The skeleton, which was interred at a level about 0.50 m below that of those in the other burials in this area, was exceptionally well-preserved, far better than any of the others.

Grave goods: None.

Fragments from fill:
8 BG
1 fine handle (a 4th c. type)
13 BW or PW
2 CW
1 tile

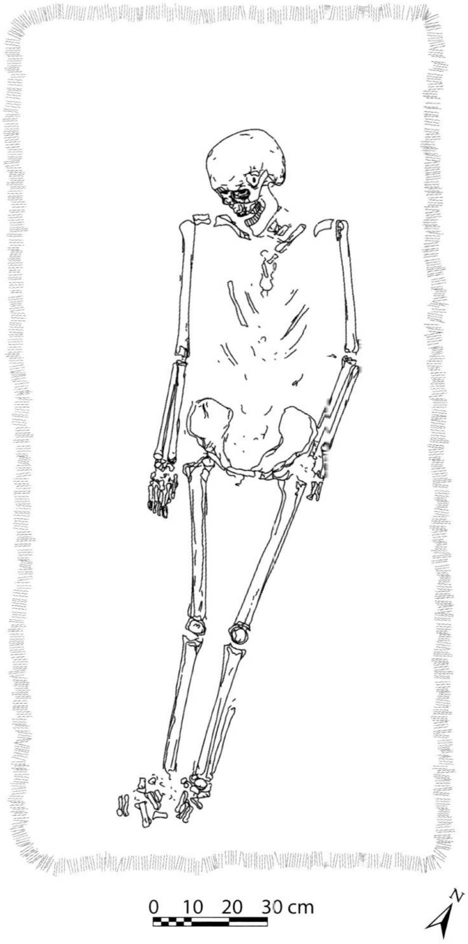

Site 736
Tomb 23

Comments:
Destroyed. No grave goods can be associated with certainty.

Fragments from fill:
6 BW or PW, including rim of a plate like that from T 20
2 CW
2 amphorae
5 tile
9 human bone

Selective Catalog of Sporadic Finds: Site 736

Site 736
SF-07 BG unguentarium (MM 322.365)
Dimensions: h 9.8 base diam. 2.2 rim diam. 1.9
Condition: Complete but for the missing vertical handles on either side.
Description: Dull black gloss; lower body reserved, but foot dipped.
Date: 325–250 BC. Cf. Ampullae 58, Pl. 4, type A 13.7.

1:3

Site 736
SF-08 BG unguentarium (MM 322.364)
Dimensions: h 12.5 base diam. 2.6 rim diam. 2.5
Condition: Missing most of base.
Description: Poor-quality black gloss, turning red in places; parallel grooves on top of shoulder; foot unglazed.
Date: Cf. Elliott 1998, 686–87, esp. U44. Group IV is dated to the late 4th–early 3rd c. BC. Cf. *Eraclea* 100, 229, Pl. 42, type 7. Also *Ampullae* 62, Pl. 8, type 23.2 (example from Toloe).

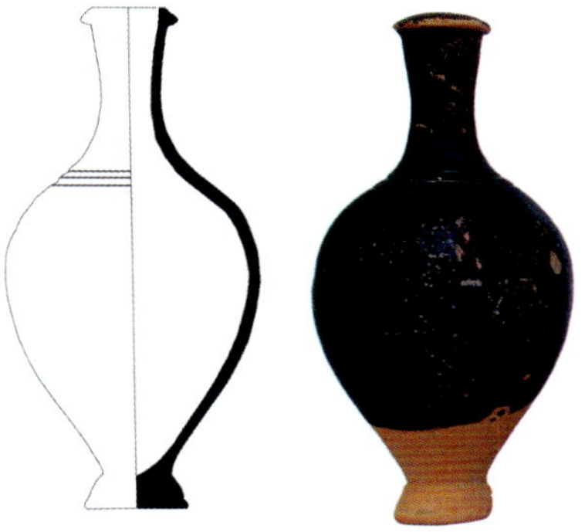

1:3

Site 736
SF-14 GN pelike, neck and shoulder (MM 322.367)
Dimensions: h 13.2 rim diam. 12.5
Condition: Powdery surface with the dull, poor-quality black gloss missing in some places, while the white decorative overpainting has vanished entirely. Three fragments have been found, but not the other handle.
Description: Preserved decoration around the neck consists of a single meander between two narrow bands, with a row of dots below. For these decorative motifs, attested in Middle and Late Gnathia, see Curti 1998, 88–89; clay color 5YR 6/6 reddish-yellow.
Date: The fragment belongs to a type 2b pelike present in the morphological repertory of red-figure and overpainted polychrome ceramics of the second half of the 4th c. BC (cf. Morel 1981a, Serie 3671, Pl. 110; Rheo Bumbalova 1979, 106–08, Fig. 5; and Lanza 2005, 31–32). The type, attested at Taranto (Hoffmann 2001, 38, type 311/14, Pl. 66: phase A2, third quarter of the 4th c. BC), was adopted also by the Metapontine workshops, as indicated by an example from scarico 3 (inv. 147121+146698) that can be compared with the overpainted pelike from T 14 at Pomarico (Carando 1997, 1.294, No. 62, Pl. 109, third quarter of the 4th c. BC) The fragment, which has a flared rim and a fairly narrow neck, is comparable to pelikai of the last quarter of the 4th c. BC, such as a vase from T 16 in Via Agrigento at Herakleia (Pianu 1990, 85, No. 1, Pl. IV, 340–320 BC), one from the Museum of Antiquities at Turin (published in Lanza 2005, 31, No. 1, Pl. VII.1, 330–310 BC, perhaps of Lucanian production), and one in Madrid (Pérez Ballester 2002, 63, No. 7, 330–300 BC). Finally, cf. the rim of the black-gloss pelike type P38 from the field survey: No. 254, BG 471-13. (FS)

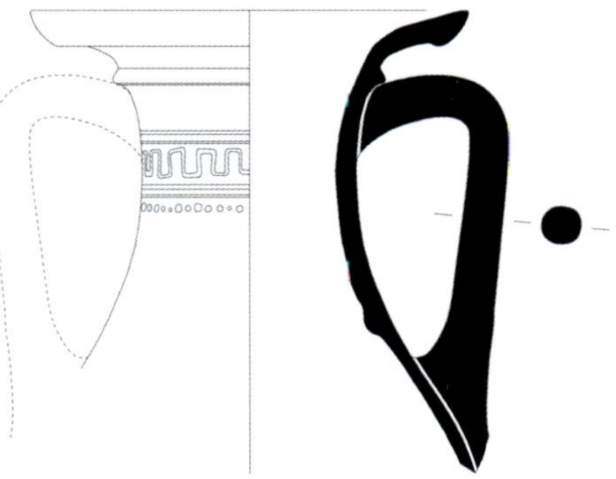

1:3

Site 736

SF-15 BG dish (MM 332.363)

1:3

Dimensions: h 3.4 base diam. 6.0 rim diam. 14.6

Condition: Intact, with only a few nicks.

Description: Dull black gloss; foot and interior of base partly glazed.

Date: Late 4th–early 3rd c. BC. Elliott 1998, 652–53. Cf. especially D2 from Group VI.

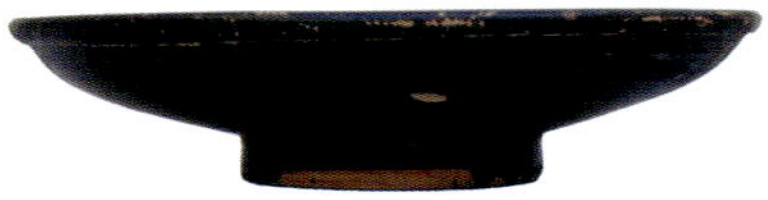

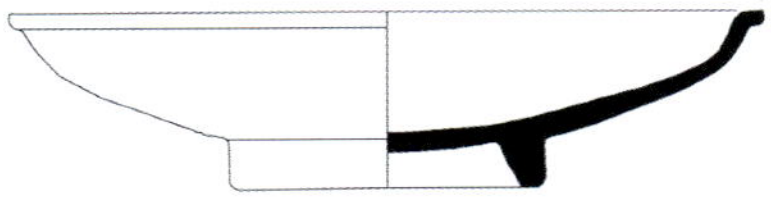

Site 736

SF-18 RF bell krater fragment (no inventory No.)

1:2

Dimensions: h 4.8 w 6.3 th 0.4

Condition: Lustrous black gloss inside.

Description: The fragment, which belongs to the back side of a bell krater, depicts the central portion of the body of a draped youth of type H (*LCS* 85), in front of whom stood a second figure (not preserved). Clay color 5YR 6/8 reddish-yellow.

Date: This type of draped youth was frequently employed by the workshop of the Creusa and Dolon Painters; very similar is the drapery of the draped youth on the Naples bell krater (*LCS* 90 No. 445, Pls. 42.3–4), attributable to the Creusa Painter. Early 4th c. BC. (FS)

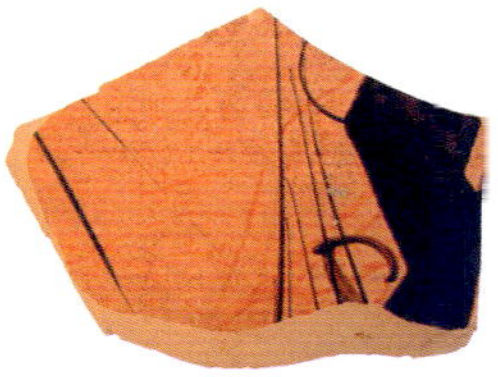

No.	Description	No.	Description
SF-01	BG cup base	SF-11	BG large vessel fragment
SF-02	BG rim frag. with flaking surface, large vessel or plate	SF-12	3 Ck and PW fragments
SF-03	BG fragment, same fabric as 2 above	SF-13	Human bone fragment
SF-04	BG fragment, same fabric as 2 above	SF-14	BG pelike, with neck, lip (see selective catalog)
SF-05	Amphora neck fragment	SF-15	BG dish, intact
SF-06	CW fragment	SF-16	BG very large vessel shoulder fragment, exterior gloss only
SF-07	BG unguentarium (see selective catalog)	SF-17	3 BG large vessel fragments (2 rim fragments of a pelike; red-brown gloss)
SF-08	BG unguentarium (see selective catalog)	SF-18	RF bell krater fragment (see selective catalog)
SF-09	BG fragmentary base, lebes?	SF-19	Amphora fragments from one large coarse vessel
SF-10	Unglazed spout		

Table 29.2 Site 736 tombs, list of sporadic finds.

B-2. Burials and Grave Goods: Site 737

Site 737
Tomb 18
Type: *A cappuccina*
Orientation: 45
Skull: SE
Date range: See below
Basis for date: Pottery
Sex: Unknown
Age: Adult
Dimensions: length 1.48 w 0.60 (approx.)

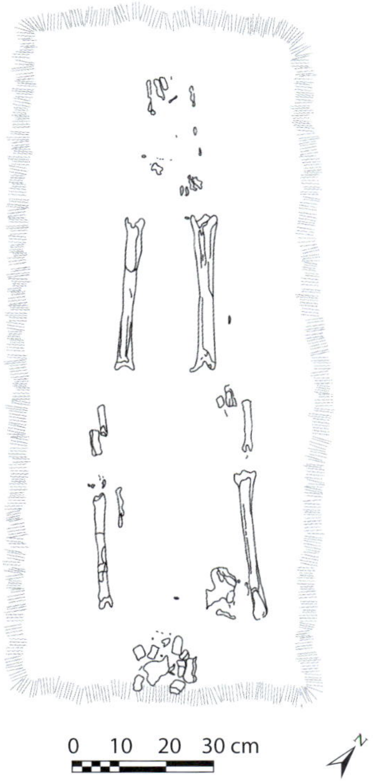

Ceramic summary: The association of the pottery collected around the site of the tomb and the skeleton is not secure. Conclusions about the date of the burial must be tentative at best, and the placement of the vessels is completely unknown. The pottery will, nonetheless, be discussed as a group around T 18 in the section on the grave goods. There appear to be at least two chronological groups: to the first half of the 5th c. BC belong the skyphos (T 18-01), lebes base (T 18-07), amphora base (T 18-03), and skyphos with inward-curving rim and reserved base (T 18-02); the mug (T 18-05), skyphos bases (T 18-04 and T 18-06), and two-handled cup (T 18-08) belong to the end of the 5th c. BC.

T18-01	BG skyphos
T18-02	BG skyphos
T18-03	BG amphora base
T18-04	BG skyphos, base
T18-05	BG mug
T18-06	BG skyphos base
T18-07	BG lebes, base
T18-08	BG two-handled cup
T18-09	BG dish
T18-10	RF fragment
T18-11	BG miniature lekythos base
T18-12	unglazed askos

Comments:
This tile tomb was damaged by the backhoe, which reached down to the level of deposition of skeleton and grave goods. Parts of the skull and most of the long bones were preserved.

Site 737
T 18-01 BG skyphos (MM 323.304)
Dimensions: h 11.7 base diam. 9.8 rim diam. 15.8
Condition: Missing both handles.
Description: Reserved underside of the bottom and resting surface, small hollow circle on the center of the bottom, a very large skyphos.
Date: Type C, Group C (1) skyphos. Elliott 1998, 679. See the two skyphoi S61 and S62 from T 350 (middle of the 5th c. BC). (FS)

1:3

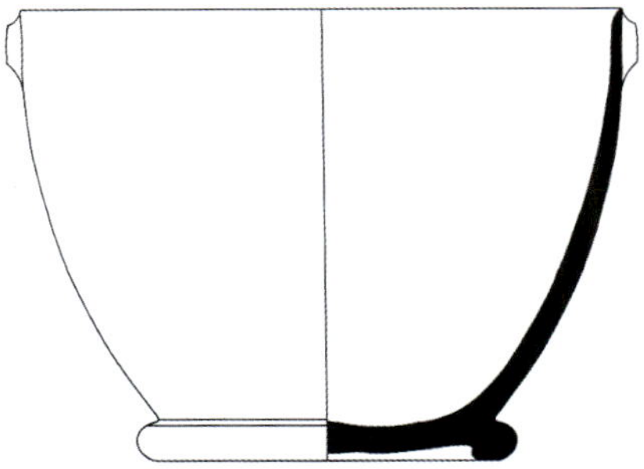

Site 737

1:3

T 18-02 BG skyphos (MM 323.305)

Dimensions: h 8.6 base diam. 6.8 rim diam. 10.6

Condition: Approx. half preserved, including one handle.

Description: 2 cm high reserved band around base, just above the foot. Underside of the bottom reserved, but foot and resting surface covered by gloss. Slightly inward-curving rim.

Date: Type B1(i) cf. S40. Elliott 1998, 678, dated 460–440 BC. See also S56, T 336-9 (weighted date range 468–442 BC). Cf. the Attic skyphos in *Agora* XII, 257, No. 316, Pl. 14 (450 BC) and the skyphos of colonial manufacture from Kaulonia: Gagliardi 2001, 305, No. 52, Pl. 280f, variant B2 (460–430 BC). (FS)

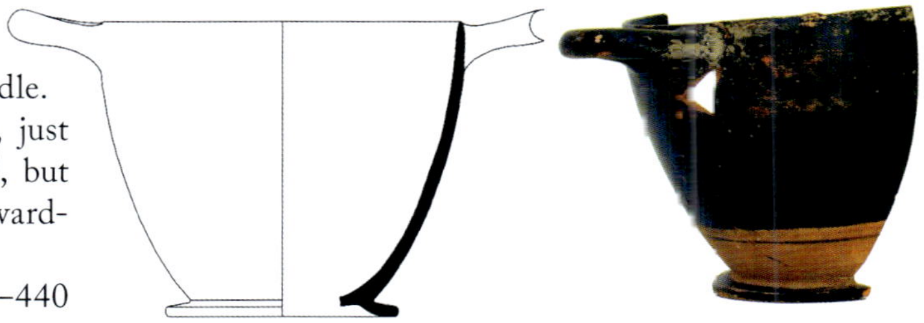

Site 737

1:3

T 18-03 BG amphora (MM 323.312)

Dimensions: h (preserved) 9.6 base diam. 9.2

Condition: Only lower third and base preserved.

Description: Molded foot, the upper side covered with gloss. There is a wide band from foot up that is reserved, as is the whole underside of the foot and the bottom. Gloss on exterior only.

Date: End of the 6th c. BC? Unlike any of the amphorae from the Pantanello necropolis. It resembles more closely than any other A18 from T 339 (dated 480–460 BC), but the underside of the base is quite different. See also the amphora from T 29 at Metaponto: Lo Porto 1966, 209–10, Fig. 64 (dated to the middle of the 6th c., but associated with a completely glossed skyphos with off-set rim). (FS)

Site 737

1:3

T 18-04 BG skyphos, base (MM 323.306)

Dimensions: h 7.4 base diam. 6.25

Condition: Lower part only.

Description: Underside of bottom and resting surface are reserved.

Date: Type B2, like S3, dated 440–400 BC, Elliott 1998, 677–78. Compare also the profile with that of S49, from Pantanello T 303, dated 400–360 BC, *Necropoleis* 322; Elliott 1998, 681. See also the Attic example from the chamber tomb in Via Dante at Taranto: Patera 1986, 46, No. 18, Pl. XVIII.18 (last quarter of the 5th c. BC). (FS)

Site 737

1:3

T 18-05 BG mug (MM 323.307)

Dimensions: h 7.4 base diam. 5.1 rim diam. 5.7

Condition: Chipped mouth.

Description: Reserved are the underside of the bottom and resting surface; small hollow circle on bottom.

Date: For the form, cf. T 21-02. The morphological characteristics of the example from T 18, particularly the lightly accentuated "S" profile, are comparable to those of examples dateable to the last quarter of the 5th c. BC. See, for example, the olpe from T LX = 92 at Ordona: *Ordona* VII, 2.47–48, Fig. 196, dated to 375 BC but probably belonging to the last quarter of the 5th c. (*Forentum* I, 236–237n2). See also the olpe from T 15 at Fratte: *Fratte*, 273, No. 1, Fig. 465.1, end of the 5th c. BC. (FS)

Site 737
T 18-06 BG skyphos, base (MM 323.308)
Dimensions: h (preserved) 2.25 base diam. 5.9
Condition: Entire base and very small part of the body are preserved.
Description: Good-quality gloss on the exterior, underside of the bottom, and the resting surface. Miltos (red wash) over the resting areas.
Date: A B2 skyphos. Elliott 1998, 677–82. Cf. S33, from T 208, dated 460–440 BC.

1:3

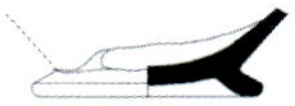

Site 737
T 18-07 BG lebes, base (MM 323.309)
Dimensions: h (preserved) 6.6 base diam. 8.1
Condition: Part of base section with narrow section of wall attached.
Description: Gloss on the exterior only. Reserved are the entire underside of the base and bottom and a wide band from the foot.
Date: Compare type A, group iv plain lebes gamikos LG 44, T 326, dated 460–440 BC; Elliott 1998, 659. (FS)

1:3

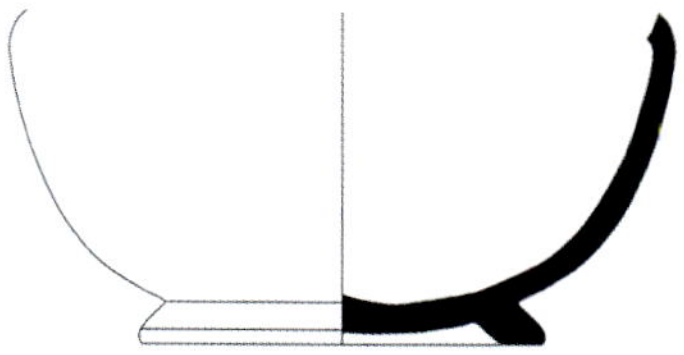

Site 737
T 18-08 BG two-handled cup, rim (MM 323.310)
Dimensions: h 3.5 rim diam. ca. 9.5
Condition: Fragmentary.
Description: Rim with part of one handle, very thin. Projecting element at the base of the body, which has a continuous profile.
Date: The two-handled cup is a Metapontine imitation of the Attic bolsal, which was produced from the end of the third quarter of the 5th c. BC and was quickly imitated in numerous western production centers. At Metaponto these cups appear in contexts of the kerameikos that can be dated to the last quarter of the 5th c. BC, such as scarico 16: Silvestrelli forthcoming a, Fig. 8.3. For the attestations at Pantanello, see B1, group i, T 106: Elliott 1998, 650. End of the 5th c. BC. (FS)

1:2

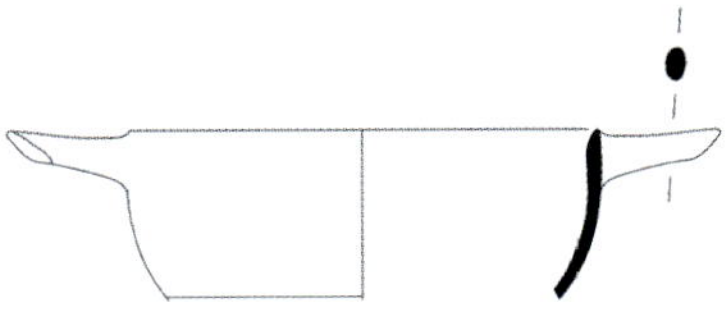

Site 737
T 18-09 BG dish
Dimensions: h (preserved) 3.2 base diam. 5.3 rim diam. 20.0
Condition: Complete profile and base preserved.
Description: Small proportions, gloss all over; orange-red to brown.
Date: End of the 5th c. to early 4th c. BC.

1:3

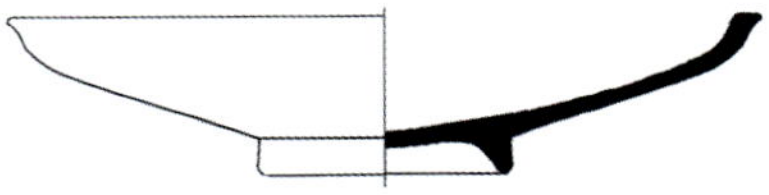

Site 737
T 18-10 RF fragment (MM 323.311)
Dimensions: h 8.3 w 10.5
Condition: The lower border, a tongue pattern, is incomplete.
Description: Wall of a small pelike. Lower part of a female in chiton, tied around the waist by a belt, facing a stele on a square base. The scene is bordered at the bottom by an Ionic cyma.
Date: First half of the 4th c. BC. (FS)

1:2

Site 737
T 18-11 BG miniature lekythos, base (MM 323.313)
Dimensions: h 1.2 base diam. 3.0
Condition: Lower part of vessel only.
Description: Reserved edge of base.
Date: Cf. Elliott 1998, 664. Group 1, ca. 460 BC.

1:2

Site 737
T 18-12 PW unglazed askos (MM 323.314)
Dimensions: h 11.9 base diam. 5.8 d 11.2
Condition: Nearly complete but missing handle and rim.
Description: The fabric is gritty, powdery, and of a greenish white color (2.5Y 8/3 pale yellow). Spout complete, connected to the body by a three-holed strainer for the contents. Very lightweight, slight malformation of the body. An import, like the hydria T 20-01 below. It is unlike askoi of local production, having a large opening in the center and a relatively small spout with strainer on the side. Cf. Toxey and Carter 1998, 721.
Date: First half of the 5th c. BC.

1:3

Site 737
T 18-13 AMP storage amphora, rim and handle fragment (MM 323.315)
Dimensions: h 8.5 w 12.0 rim diam. 12.0
Condition: Part of a rim and handle.
Description: Corinthian type A'. Fabric is Corinthian. For color see Carter 1998f, 194–96. Corinthian rim profile is less angular than any Pantanello necropolis examples. Droopy rim, lines on top of the handle.
Date: First half of the 5th c. BC.

1:3

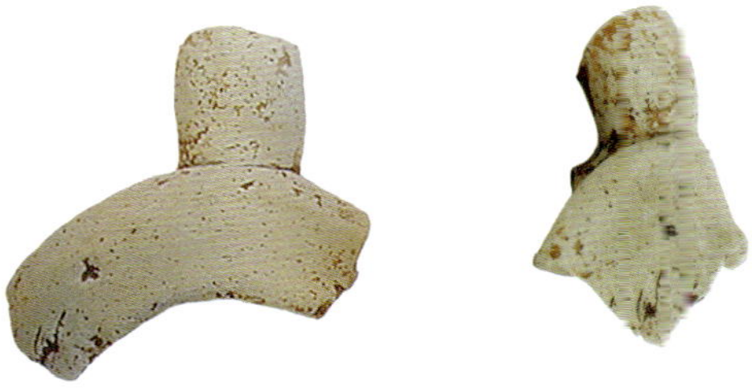

Site 737
T 18-16 PW plain basin, rim.
Dimensions: h 2.2 rim diam. 49.0
Condition: Part of the rim.
Description: Clay pink (7.5YR 7/4), with rare shiny, minute inclusions.
Date: For the profile, cf. the example from Scarico 19 of the Metapontine kerameikos, dated third quarter of the 5th c. BC: Silvestrelli forthcoming a, Fig. 7.23. (FS)

Site 737
Tomb 19
Type: Sarcophagus
Orientation: 47
Skull: SE
Date range: 500–480 BC
Basis for date: BW pottery, tomb type
Sex: Female?
Age: 40–50
Dimensions: l 1.95 w 0.73

Ceramic summary:
T19-01 BW amphora (foot)

Non-ceramic objects:
T19-02 Iron awl or slender knife

Comments: Carved from a single block of calcarenita, the tomb was all but destroyed by the backhoe. Only the bottom remained intact. Very little of the skeleton—some fragments of the skull and the femurs—was found in situ, confirming that the skull was located in the SE end of the tomb. A BW amphora in many fragments was recovered in the NW corner. This type of tomb, it appears, was employed in the late 6th and first half of the 5th c. BC.

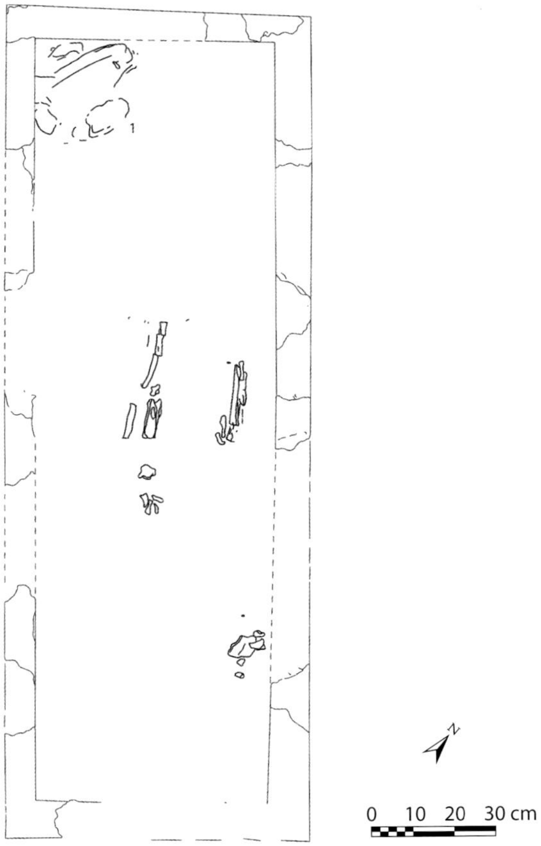

1:6

Site 737
T 19-01 BW amphora (MM 323.317)
Dimensions: h 36.4 w 30.0 rim diam. 14 base diam. 13.7
Condition: Very fragmentary, but profile is complete.
Description: Thick black band around shoulder below vertical strap handle. Above it two thin red bands, below one thin red band halfway down the belly, a thin red-brown band, thin raised band around neck. The buff-colored clay is probably of local origin.
Date: Early first half of the 5th c. BC. Cf. Attic table amphora, *Agora* XII, 351, Pl. 60, Fig. 12, No. 1445, dated 500–480 BC. The Metapontine vessel has a disk base like most Metapontine banded-ware vessels, while the Athenian has a ring base.

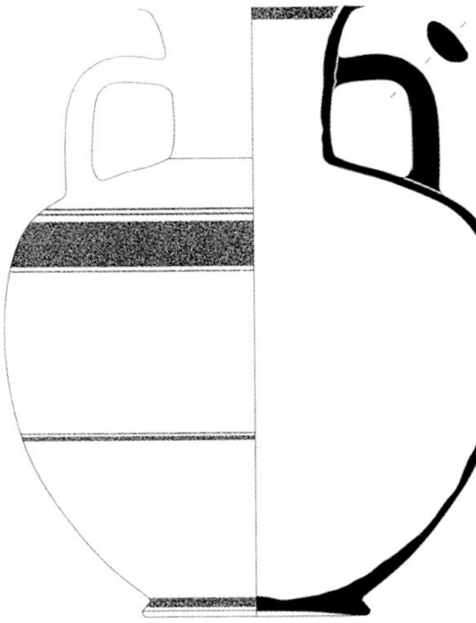

Site 737
T 19-02 Iron awl or slender knife
Dimensions: l (extended) 11.0 w (max) 0.8
Condition: Badly corroded.
Description: The blade with flat rhomboid section narrows and grows flatter towards the point. Bent at the center. The broad end is bent slightly with respect to the central axis, may have had an inserted or attached handle.
Date: First half of the 5th c. BC, from context.

Site 737
Tomb 20
Type: *A cappuccina*
Orientation: 50
Skull: SE
Terminus date range: 460–440 BC
Weighted date range: 460–440 BC
Basis for date: Pottery
Sex: Male?
Age: 40–x
Dimensions: l 1.78 w 0.75 (ext.), 0.55 (int.)

Ceramic summary:
T20-01 Unglazed hydria (outside tomb)
T20-02 Palmette lekythos (shoulder)

Non-ceramic objects:
T20-03 Strigil, bronze (pelvis)
T20-04 Strigil, iron (pelvis)
T20-05 Aryballos mouth, bronze (ulna)

Comments: The tomb was formed of four flat pan tiles laid on their long sides to form its sides. The joints between the pan tiles in the center were covered by sections of gently curving Laconian cover tiles and closed along the ridge by sharply curving semicircular cover tiles. Upright flat pan tiles formed the ends. The tomb contained a very poorly preserved skeleton. Only the left leg, the tibia of the right, and a few scraps of the skull survived to indicate the position of the body and its relation to the grave goods. Two strigils (one in iron, one in bronze) were laid across the waist, a palmette lekythos across the chest at the level of the humerus. An aryballos made of leather, of which the bronze mouth is preserved, was placed beside the left ulna. Outside the tomb at the NW corner a small fragmentary amphora (hydria?) served, perhaps, as a grave marker.

Site 737
T 20-01 BW hydria (MM 323.320)
Dimensions: h (est.) 30.0 base diam. 10.7
lip diam. 12.3 d 25.7
Condition: Many fragments, top half and base restored.
Description: Distinctive fabric (2.5Y 8/2 white) is similar to that of the askos T 18-12. Rounded side handles, flastraps, vertical handle. Thin red band at shoulder just above handles. An import of unknown provenience.
Date: First half of the 5th c. BC.

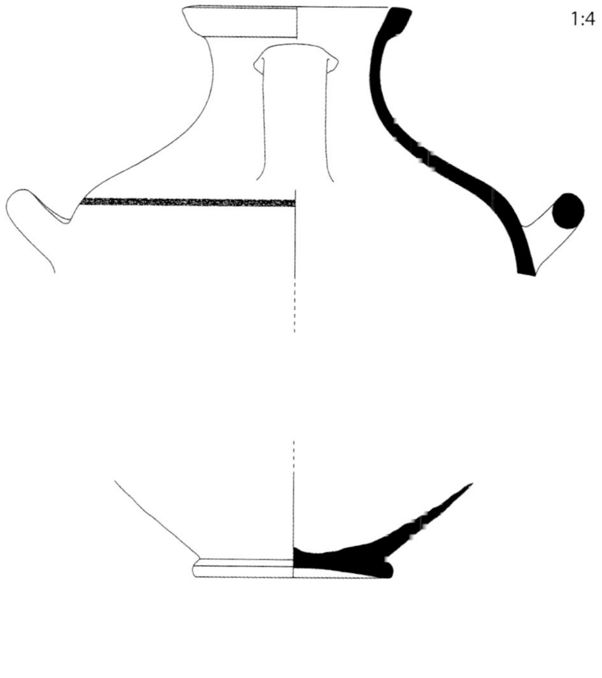

Site 737, T 20-01 cont.

1:4

1:2

Site 737

T 20-02 RF palmette lekythos (MM 323.318)

Dimensions: h 12.85 rim diam. 3.0 base diam. 4.0 d 5.7

Condition: In fragments, but complete and restored.

Description: Red slip on upper surface of lip traces three careless palmettes on the shoulder. A ring of dots at the base of the neck. Neck, shoulder, and upper surface of the lip are reserved. Narrow bands on the base of the foot, whole of the vertical surface and underside of the foot. Traces of red wash near handle.

Date: This lekythos belongs with Group IV from Pantanello, which suggests a date of 460–440 BC. Compare L24 (T 230-7), *Necropoleis* 374; Elliott 1998, 665.

1:4

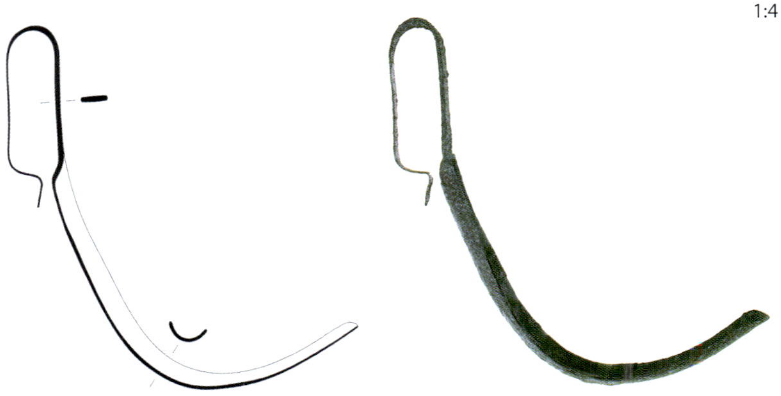

Site 737

T 20-03 Bronze strigil (MM 323.319)

Dimensions: l (contrasted) 23.6 w (max) 2.5

Condition: Restored and almost perfectly complete; handle broken at termination.

Description: A B1 type strigil. See Prohászka 1998, 797.

Date: Compare a strigil from Sant'Angelo Vecchio: Prohászka 1998, 803, No. S31. This type of strigil can be rather precisely dated to 460–440 BC.

1:4

Site 737

T 20-04 Iron strigil (MM 323.320)

Dimensions: l 30.1 w (max) 3.0

Condition: Badly corroded, complete.

Description: The handle has a round section and is formed of the same sheet of iron as the blade. It has been rolled over a cylinder 1.5 cm in diameter and 10 cm long. The two edges do not quite meet. The hollow end has been filled with a circular iron plug.

Date: This is a very distinctive type. The handle is unlike that of any strigil, in iron or bronze, from the Pantanello necropolis, where there is a great variety of types. At Pantanello the iron strigils date mostly to the 4th c. BC, Prohászka 1998, 799. This one should be dated by the context to the mid-5th c. BC.

1:1

Site 737

T 20-05 Bronze aryballos mouth (MM 323.321)

Dimensions: h 1.1 d 3.65

Condition: Missing one half of the overturned edge that joined the leather aryballos body. The edge of the opening at the bottom of the inverted cone extended very slightly into the interior of the container.

Description: The exterior of the rim, a strip 0.6 cm wide, was added separately and was meant to hold the leather(?) container in place, next to the mouth. The bronze sheet used for all is ca. 0.1 cm thick.

Date: Two examples of an aryballos mouth in bronze were found in the Pantanello necropolis: Prohászka 1998, 819, H6 and H7 from T 193 and T 329. The first dates to the early 3rd c. BC, the second to the mid-5th c. BC.

Site 737
Tomb 21
Type: Fossa
Orientation: 50
Skull: SE
Terminus date range: 460–440 BC
Weighted date range: 460–440 BC
Basis for date: Pottery
Sex: Male
Age: 50–60
Dimensions: l 1.80 (approx.) w 0.50 (approx.)

Ceramic summary:
T21-01 BG amphora (femur)
T21-02 BG mug (foot)
T21-03 BG skyphos (foot)
T21-04 unglazed one-handler (foot)

Non-ceramic objects:
T21-05 Strigil, iron (foot)

Comments:
The skeleton was better preserved than most. At the feet were placed an iron strigil and all but one of the ceramic vessels, a small amphora set between the legs.

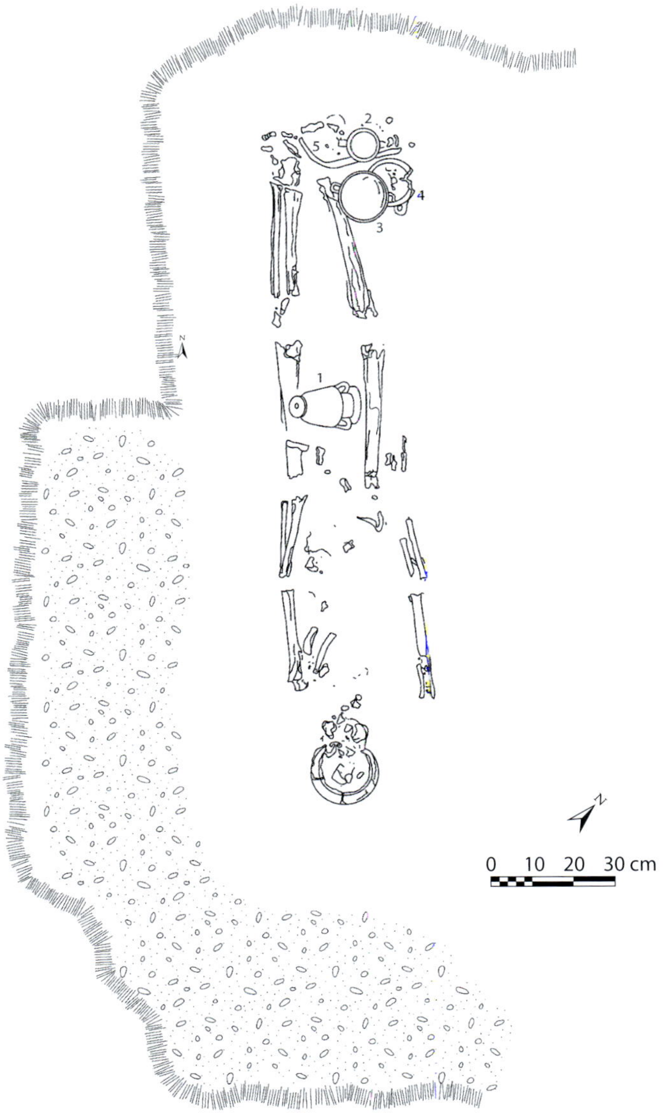

Site 737
T 21-01 BG amphora (MM 323.324)
Dimensions: h 16.4 base diam. 6.6 rim diam. 9.2 d 10.0
Condition: Complete.
Description: Gloss inside the neck and over the body and top side of the foot. A wide band from foot tip and the entire underside of the base is reserved. Lip has a molded profile. Gloss orange to black-brown.
Date: This complete amphora belongs clearly to Group III, which dates to the period 460–440 BC: Elliott 1998, 648–49. Very close is the example from T 8 at Saldone: *Necropoleis*, 432 (second quarter of the 5th c. BC). (FS)

1:3

Site 737

T 21-02 BG mug (MM 323.325) [cf. T 18-05]

Dimensions: h 10.2 base diam. 8.0

Condition: Restored from many fragments, missing parts of rim.

Description: Gloss inside and out. Only the exterior bottom and resting surface are reserved; traces of a hollow red circle in the center of the bottom.

Date: This type of mug derives from the Attic VIIIC oinochoai (Green 1972, 8, and 15n67, Pl. II e) and was imitated at numerous Italiote production centers (for which see *Forentum* I 191 and Labellarte 1995, 108). Its adoption by Metapontine workshops, at least beginning in the third quarter of the 5th c. BC, has been confirmed by the recovery of wasters in scarico 22 (Silvestrelli forthcoming a, Fig. 7.9) and scarico 1 (D'Andria 1980a, 408, No. 205, Fig. 55; and Silvestrelli forthcoming, Fig. 10.11) of the kerameikos. The morphological characteristics of T 21-02 are comparable to Attic examples of the second quarter and middle of the 5th c. BC; see, for example, the red-figure vase from T 619 of the kerameikos (*Kerameikos* VII.2, 151, 470–460 BC) and the one from T 50 at Manicalunga, Selinous (Kustermann Graf 2002, 126, 50/O 496, Pls. XXVI and CXXIX, 450 BC). (FS)

1:3

Site 737

T 21-03 BG skyphos (MM 323.329)

Dimensions: h 9.8 base diam. 6.0 rim diam. 10.3

Condition: Restored from many fragments; small section of the rim missing.

Description: Outward-curving rim, upward-tilting handles; gloss over whole, only the outside of the bottom and resting surface reserved. Traces of red wash on base.

Date: Mid-5th c. BC. This is a type A2 skyphos, 460–440 BC. Cf. *Necropoleis* 523; Elliott 1998, 678, from T 151. The Metapontine production of this type of skyphos has been verified by the recovery of wasted examples in scarichi 12 (second quarter of the 5th c. BC), 22, and 19 (third quarter of the 5th c. BC: Silvestrelli forthcoming a, Fig. 7.6). Rarely attested in Peucetia (Gravina, S. Felice: Andreassi 1979, 438, Pl. XXVII.2, from a funerary assemblage dated to the second half of the 6th c. BC; Valenzano: Ciancio 1985, 77, 101–02, Pl. XXXI, 145, first half of the 5th c. BC, associated with a Late Archaic cup-skyphos of the Haimon Group), these skyphoi are particularly common in funerary assemblages of the first half of the 5th c. BC at Metaponto and in the Metapontino, around Matera, and at Banzi (Matera, Ponte S. Giuliano, T 4: Lo Porto 1973, 221, Pl. LXV 2, late 6th c. BC; Pisticci, tombs excavated in 1910 near the school: Lo Porto 1973, 157–58, end of the 6th c. BC; the type C cup and the Late Archaic black-figure pottery suggest a chronology for the funerary assemblage in the first quarter of the 5th c. BC; Banzi: A. Bottini 1987a, 4, Fig. 3.1, first quarter of the 5th c. BC). It is nevertheless probable that their production and diffusion continued into the third quarter of the 5th c. BC: two examples of probable Metapontine production belong to the unpublished assemblage of T 337/1999 from Via Montegrappa at Montescaglioso which also contained a bell krater by the Pisticci Painter (the author thanks A.M. Patrone for the information about this context). The example from T 21 is close to that from T 1 at Montescaglioso (Lo Porto 1988–1989, 332, No. 16, Fig. 32.9, dated to the end of the 6th–beginning of the 5th c. BC but ascribed to the second deposition that contained a small column krater attributed to the Pittore del Frutteto dated 460 BC). (FS)

1:3

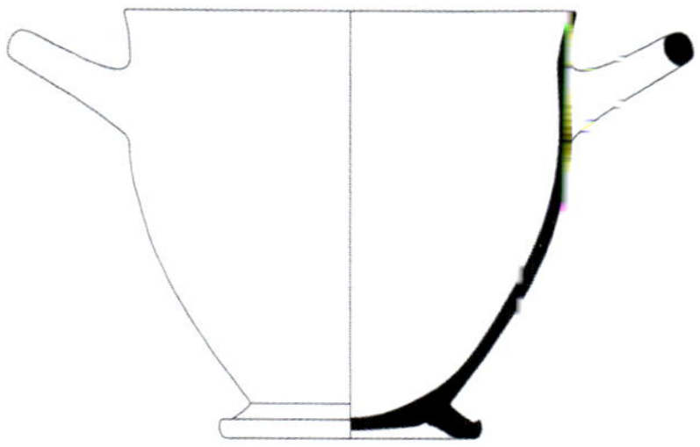

1:2

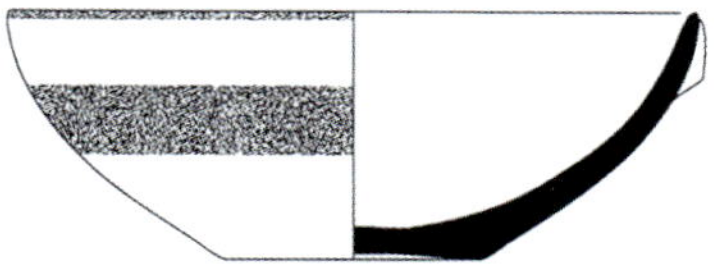

Site 737

T 21-04 BW one-handler (MM 323.323)

Dimensions: h 3.2 base diam. 3.8 rim diam. 9.2

Condition: Restored from many fragments.

Description: Pink-brown clay with traces of red gloss, narrow band inside.

Date: The date range for this type must begin in the middle of the 5th c. BC at the latest.

1:2

Site 737

T 21-05 Iron strigil (MM 323.322)

Dimensions: l (extended) 30.0 w 2.5

Condition: In a dozen fragments, badly corroded.

Description: Unusual cylindrical handle, terminating in a hook. The handle is cylindrical like that of the iron strigil in T 20, but has an additional hook on the end for appending, and it is possible the plug at the end of the other example from this necropolis may have been part of a device for storing the instrument when it was not in use.

Date: Given the similarity of the date of the grave goods and the strigil design, not to mention the close proximity of the graves, it is likely that this is an example of a mid-5th c. BC iron strigil.

Site 737
Tomb 22
Type: Slab-covered fossa
Orientation: 43
Skull: SE
Terminus date range: 440–430 BC
Weighted date range: 448–433 BC
Basis for date: Pottery
Sex: Sub-adult
Age: 10–14
Dimensions: l 1.55 w 0.88

Ceramic summary:

T22-01	BG amphora (foot)
T22-02	BG min. lebes (foot)
T22-03	BG min. lekythos (femur)

Comments: The shallow earthen grave was completely covered with slabs (0.75 m wide) of local conglomerate stone (*puddinga*), a friable material whose disintegration was clearly accelerated by modern agricultural activity. Of the skeleton, only some teeth and fragmentary long bones survived to indicate the position of the body and clarify the position of the grave goods in the burial. At one foot were placed an amphora and miniature lebes, by the right hand a miniature lekythos.

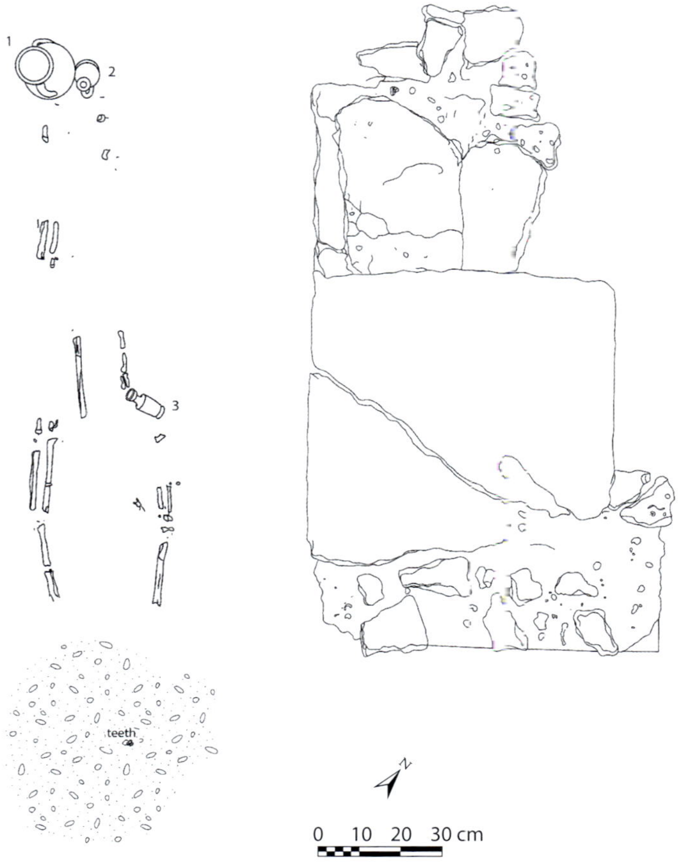

Site 737
T 22-01 BG amphora (MM 323.326)
Dimensions: h 23.1 base diam. 7.95 rim diam. 9.8
Condition: Restored from many fragments; base, neck, lip, and handles complete.
Description: The base has been perforated.
Date: The complete amphora has a pronounced, horizontal shoulder and a relatively long, slender neck. Should belong to Group III and date to the period 460–440 BC. Compare A17, Elliott 1998, 648–49. The general profile of this amphora recalls the Attic nolan amphorae of the second quarter of the 5th c. BC (for which see Euwe 1988, 146, Fig. 1b). (FS)

1:3

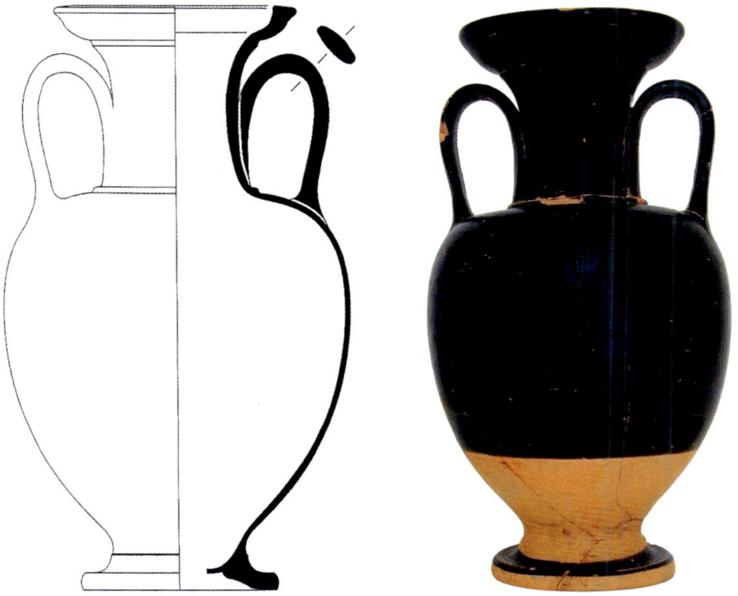

Site 737

T 22-02 BG miniature lebes (MM 323.327)

Dimensions: h (max.) 7.3 (with lid 9.0) base diam. 4.1 rim diam. 2.9 d 6.6

Condition: Complete and intact.

Description: Gloss covers the exterior entirely, but nowhere inside. The inside surface of the lid has clear traces of red wash.

Date: Though there are no perfect matches with the lebetes from Pantanello, this example is close to the plain lebes LG30 and it should date to roughly the same period, 440–430 BC, as LG30 (Elliott 1998, 660).

1:2

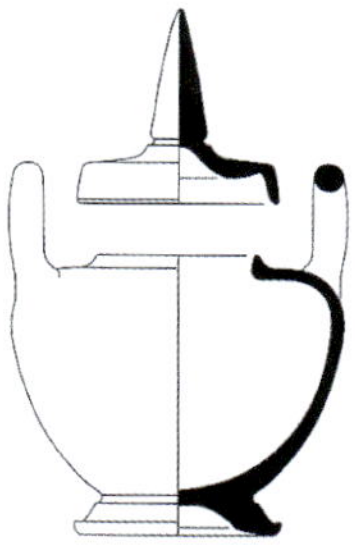

Site 737

T 22-03 BG miniature lekythos (MM 323.328)

Dimensions: h 8.3 base diam. 2.3 rim diam. 2.5

Condition: Complete and intact; gloss has flaked off sides and lip.

Description: Neck and base are reserved. There is a reserved band around the shoulder, middle of the body. Ray pattern on shoulder and dots at the base of the neck.

Date: Compare L34 (T 349-4) from the Pantanello necropolis both for shape and decoration of the shoulder. T 349 has been dated 450–430 BC. Also very similar is the example from scarico 3 in the Metapontine kerameikos (D'Andria 1980a, 434, No. 341, Fig. 68). (FS)

1:2

Site 737
Tomb 23
Type: Tile-covered fossa
Orientation: 56
Skull: SE
Terminus date range: Second quarter of the 5th c. BC
Weighted date range: 451-429 BC
Basis for date:
Sex: Female
Age: 40–45
Dimensions: l 1.98 w 0.60

Ceramic summary:

T23-01	BG pelike (foot)
T23-02	BG squat lekythos (foot)
T23-03	BG exaleiptron (tibia)
T23-04	BG skyphos with off-set rim (tibia)

Comments: Two large pan tiles, measuring approximately 0.45 x 0.80 m., in fragmentary condition, covered the shallow grave containing the skeleton, whose skull and long bones were in reasonably good condition. The grave goods were, as in the case of T 21 beside it, mostly around the feet. An exaleiptron was precisely centered between the knees, in the same fashion as the small amphora in T 21.

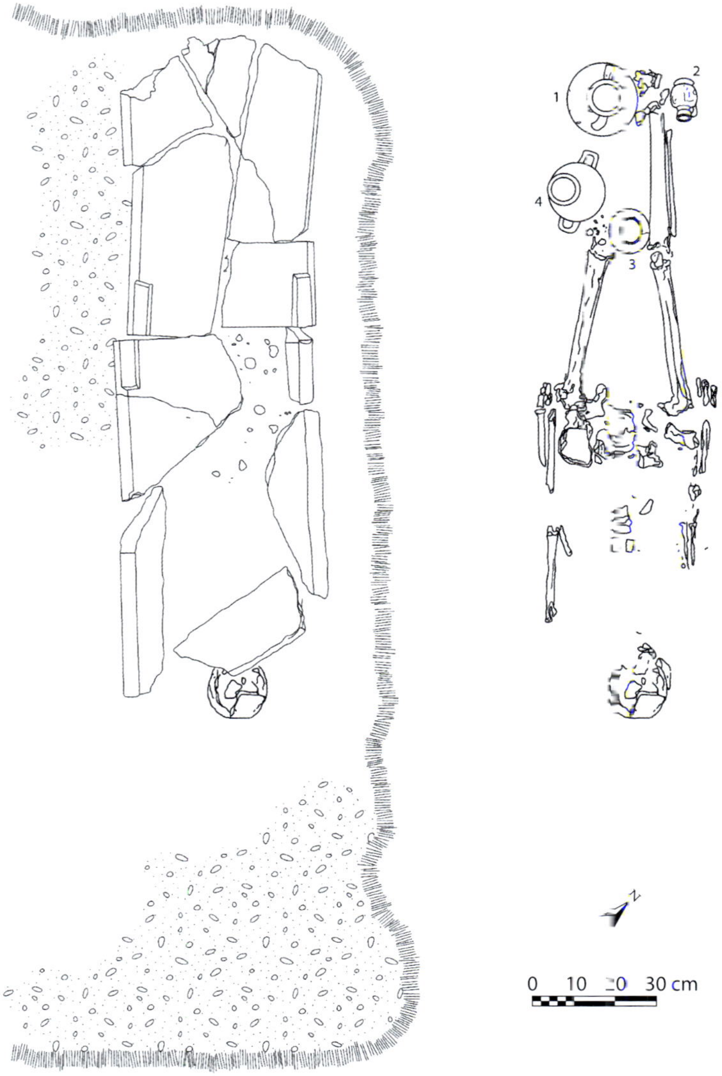

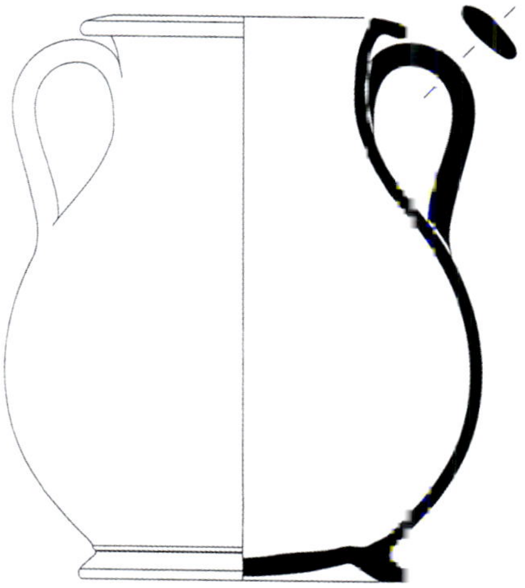

Site 737
T 23-01 BG pelike (MM 323.330)
Dimensions: h 18.1 base diam. 10.8 lip diam. 10.8

Condition: Restored from fragments, virtually complete.
Description: Gloss over the entire body except for a band

Site 737, T 23-01 cont.

above the foot. Gloss on the upper surface of the foot, but the whole of the bottom and the underside of the base are reserved. Gloss inside the neck. Grayish appearance indicates misfiring.

Date: This type of pelike, morphologically close to group vi at Pantanello (Elliott 1998, 671, dated 440–400 BC), is known principally from finds at Metaponto and surrounding areas and probably originated in the production centers of that Achaian colony and its chora. The known funerary assemblages seem to indicate a chronology between the second quarter and end of the 5th c. BC; Metaponto, contrada S. Salvatore, T 4: Lo Porto 1966, 227, Fig. 86.1 (dated to the first half of the 5th c. BC; in the same assemblage there are two lebetes gamikoi that compare with group iv at Pantanello, dated 460–430 BC). Lo Porto 1973, 161, No. 24, Pl. VIII 3.2 (first half of the 5th c. BC. The example is without context). Very similar is the vase from T 19 at Saldone: *Necropoleis* 435, which contained a type B skyphos; the tomb is dated to the last quarter of the 5th c. BC. A pelike similar to that from the tomb was found at Taranto, in a tomb in Via Dante: Patera 1986, 55–56, No. 43, Pl. XXIV 43 (dated generically to the 5th c. BC on the basis of the Metapontine comparanda). (FS)

Site 737

T 23-02 BG squat lekythos (MM 323.331)

1:3

Dimensions: h 8.7 base diam. 5.2 lip diam. 3.2 max diam. 5.0

Condition: Complete and intact.

Description: Gloss everywhere but on the base and a band of the body above the base.

Date: Precise parallels for this type of lekythos could not be found. Two reasonably close comparisons from the Pantanello necropolis are T 266-4, which was found outside the tomb dated 480–440 BC, *Necropoleis* 328, and T 149-3, whose shape and decoration is very close, but with a different mouth. It comes from a well-dated burial of 450–430 BC. Similar lekythoi are attested in funerary assemblages of the central decades of the 5th c. BC: see, for example, the vessel from T 146 at Fratte (*Fratte*, 263, No. 4, Fig. 450.4, 450 BC). (FS)

Site 737

T 23-03 BG exaleiptron (MM 323.332)

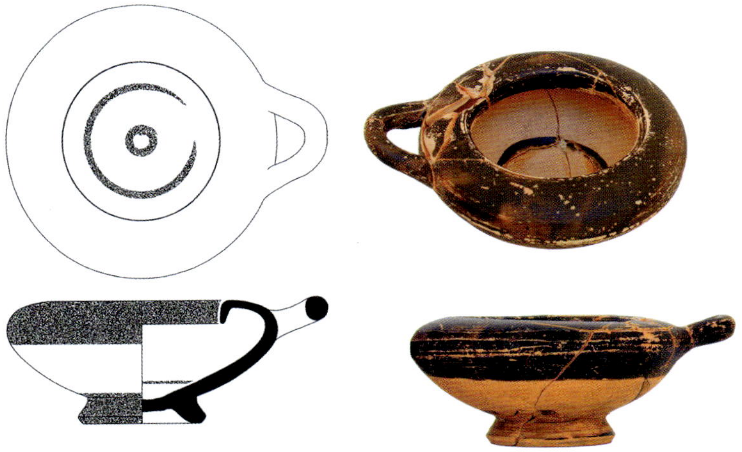

1:3

Dimensions: h 4.85 d 10.8

Condition: Complete, restored from several fragments.

Description: Gloss on the exterior and a circle on the exterior floor beneath the large circular opening. There is a wide reserved band above the foot, the lower surface of the base is reserved. The handle juts and is rounded in section.

Date: This rare shape, linked to the feminine world and employed as a container for scented water (Scheibler 1964, 73–80), is without parallels in the Pantanello necropolis. Comparisons with Athenian examples suggest a date in the first half of the 5th c. BC; very similar, in fact, is the exaleiptron from T 9 in the kerameikos: *Kerameikos* VII.2, 12, No. 7, Fig. 6 (490 BC, dated by the context). The form was produced at Metaponto at least from the end of the 6th c. BC; the same chronology is suggested by an example from Oria (Semeraro 1997, 236, No. 837, Fig. 199) and a fragment from Pomarico (*Pomarico Vecchio* I, 1.55–56, Pl. 28.2). Both seem to evoke Corinthian models of the last quarter of the 6th c. BC: Scheibler 1964, 93 and 90, T5. The absence of documentation obscures the evolution of the form, which is attested, albeit with morphological characteristics different from those present on the Metapontine examples, also in other areas of southern Italy between the 6th and 4th c. BC. (Chiaromonte,

Site 737, T 23-03 cont.

middle of the 6th c. BC: *Greci, Enotri e Lucani* 2.10.28 and 2.16.16; T 9 from Pisticci, dated to the second half of the 6th c. BC: Lo Porto 1973, 175, No. 4, Pl. XXV.2; Satriano, beginning of the 5th c. BC: Holloway 1970, 54, No. 46, Fig. 99. For the attestations in Peucetia, where the form was locally produced, cf. *Collezione Banca Intesa* 880, No. 492. For an example of Locrian production from the end of the 6th c. BC, see Olivero Ferrero 1989, 83). (FS)

1:3

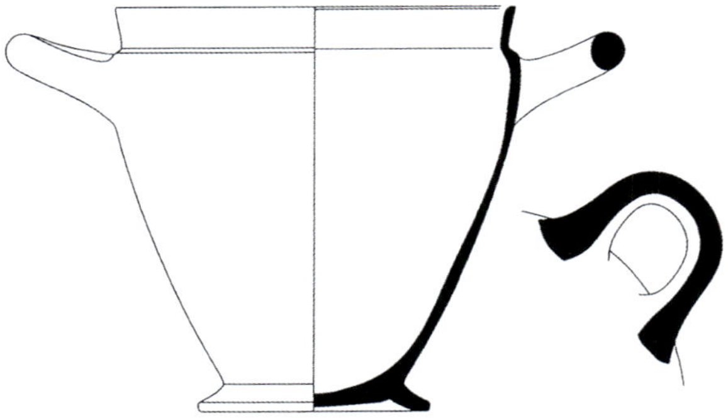

Site 737
T 23-04 BG skyphos with off-set rim (MM 323.333)
Dimensions: h 13.0 base diam. 7.5 lip diam. 13.0
Condition: Restored from many fragments, missing parts of lip and shoulder.
Description: Offset lip. Gloss everywhere over the interior and exterior, except for two narrow reserved bands around the foot; vertical surface of the foot, resting surface, and bottom reserved, with a hollow circle and a dot in its center in the center of the bottom. Red wash appears in reserved areas around the handles and on inner surface.
Date: This is a Group A1(ii) skyphos, a typical local product of Metaponto. This group has been dated 460–440 BC: Elliott 1998, 678. For the evolution of the form, see Semeraro 1997: 228, and above, pp. 152–153. The example from T 23, characterized by complete glossing of the exterior surface, vertical rim, horseshoe-shaped handles, tapered body wall, and disk foot borrowed from the type C skyphoi, is comparable to skyphoi of the second quarter of the 5th c. BC; very similar are the examples from T 11 at Pisticci, Matina Soprano, found associated with Attic vases of the second quarter of the 5th c. BC: A. Bottini 1993b, 145–46, Nos. 22–25. (FS)

Selective Catalog of Sporadic Finds: Site 737

Site 737 SF-01 RF pelike fragment
Dimensions: h 4.5 w 10.0 th 0.8
Condition: Surface intact; lustrous black gloss inside and out; fresh breaks.
Description: Clay pink (7.5YR 7/4) with white stone inclusions. On the neck, laurel leaves oriented to left. On the body, head of a youth facing left. The preserved portion is too small for a safe attribution, although the fragment can be likened to the workshop of the Creusa and Dolon Painters. For the treatment of the hair and eye, see the pelike in the Bersanetti Collection by the Creusa Painter (*LCS Suppl.*3, *C* 106).
Date: Beginning of the 4th c. BC.

1:2

C. Materials from Trenches 6 and 7, Across the Division Line Feature L09

The composition of the assemblage from the fill suggests that some or all of it originated from nearby tombs, rather than a farmhouse. The absence of cooking ware, the presence of a Corinthian amphora, and frequent tomb markers support this. At site 736, the lowest level, level 3, contained a red-figure body sherd (Tr. 6-17) dated to the late 5th c. BC. The feature began to fill up no later than this date. In the uppermost preserved level, level 1 of the feature, the latest date furnished by the pottery is broadly the 4th c. BC. Remains of burials made as early as the late or early 5th c. (e.g., Tr. 6-09) made their way into the fill. Pollen samples were taken but have not as yet been analyzed, though those from another canal have (see Ch. 19).

The pottery from the shallow fill deposited in the division line feature at site 737, especially the BG carinated dish with plain rim (Tr. 7-03), would indicate that the fill was formed no earlier than the late 4th or early 3rd c. BC.

Site 736, Trench 6: Cleaning Surface—Level 1 (Uppermost)

1:2

Tr6-01 Bell krater rim
Description: Four joining sherds and two associated sherds.
Date: Late 4th to mid-3rd c. BC.

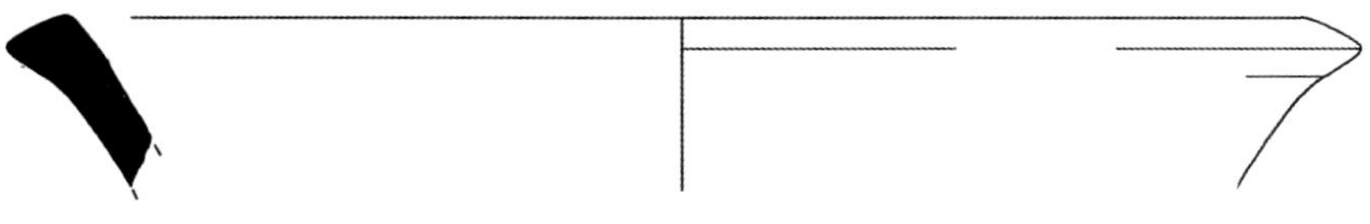

Tr6-02 BG small bowl (SB)
Description: Plain thickened rim. The type is difficult to date more precisely on rim morphology alone, but it is close to type SB8.
Date: General dates for this type are the late 5th–early 4th to mid-3rd c. BC or SB9, with a proposed date range of the last quarter of the 5th to the mid-4th c. BC.

1:2

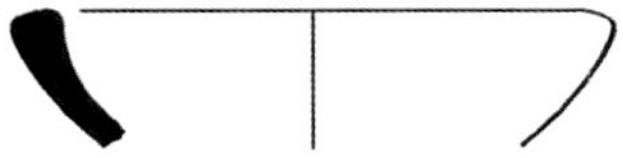

Tr6-03 BG skyphos base
Description: Slipped lower body with reserved foot ring and underside. The morphology of the foot-ring torus, is more gracile than the earliest A-type skyphoi and more robust than the later base types (see Ch. 6, **Nos. 31–39**).
Date: Shape consistent with a date in or around the first half of the 4th c. BC.

1:2

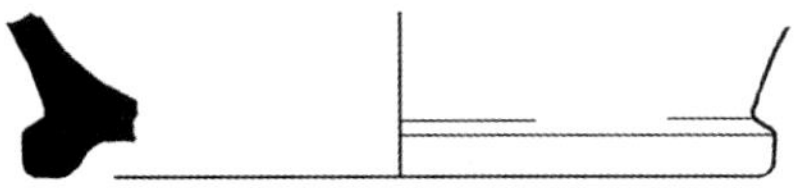

Tr6-04 BW lekythos rim
Description: For the general type, but without close parallel, cf. type J24 (with parallel from Gravina).
Date: 4th–3rd c. BC.

1:2

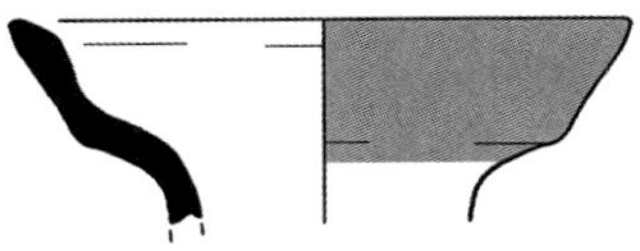

1:2

Tr6-05 PW or BW bowl rim
Description: Pendant rim, diam. ca. 18 cm, unfinished surfaces. The form is generally close to the Plate/Lekane (Pl/Lne), though not closely paralleled in the survey typology.
Date: The general type can be dated only extremely broadly to the Classical and Hellenistic periods.

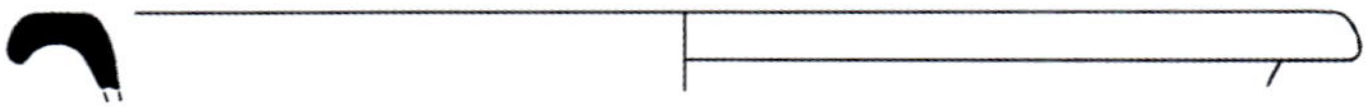

1:2

Tr6-06 PW jug (J17d) with high-swung handle
Description: Triangular rim of medium-large liquid container in a coarse fabric with terminus of high-swung handle arching above the line of the rim.
Date: A broad date range from the late Archaic to late Hellenistic periods, see Ch. 9 J17d.

Tr6-07 Incurved coarse-ware rim

1:2

Description: This is an unusual shape without close parallels. It is probably an irregularly formed coarse-ware bowl rim, the irregular preserved section showing a more steeply incurved angle than may have been the case with the rest of the vessel.

Other:
PW or BW
Skyphos or bowl handle fragment of uncertain form class, fired grey (reducing kiln atmosphere). Fifteen body sherds.
Ck
3 body sherds, relatively thick-walled, from medium-large vessels.
Amp
7 body sherds (previously classified as medium storage), perhaps from regional amphorae or medium-large closed vessel forms.
Tile
2 fragments.

Site 736, Trench 6: Level 1

1:2

Tr6-08 BG Panathenaic amphora neck/handle (736.14)
Description: Relatively thick-walled type, probably an early variant.
Date: Late 5th c. BC, providing a terminus post quem for this uppermost deposit.

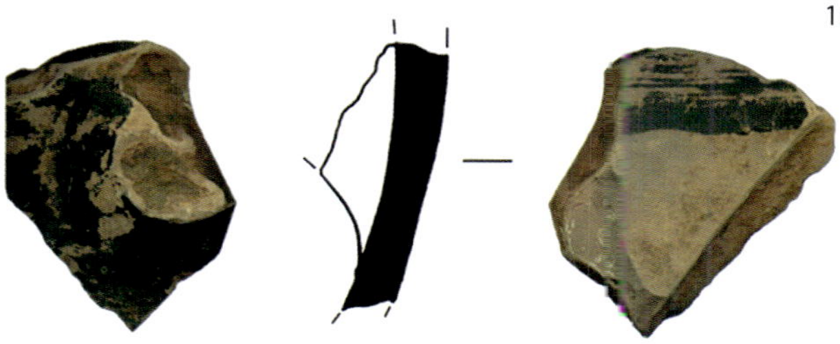

Tr6-09 AFW Ionic cup base
Description: Thin slip, partially preserved, typical of Ionic cups from this region.
Date: Typical base type for Ionic cups of the 6th and early 5th c. BC. This piece is clearly residual within the division-line fill.

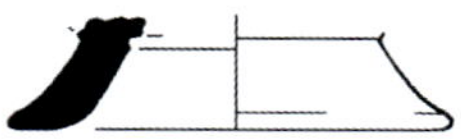

1:2

1:2

Tr6-10 BG skyphos rim
Description: S, Pl. 7.6
Date: It is not possible to date this form very precisely based on extant features, but it is not typical of the later phases of development of the Attic skyphos and can be broadly dated to the 5th and 4th c. BC.

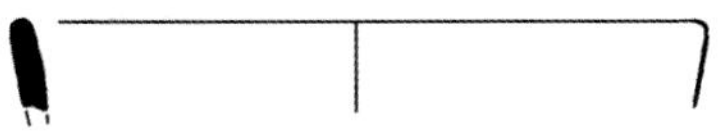

1:2

Tr6-11 BW or PW jug/pelike/table amphora rim
Description: Type J14, Pl. 9.12
Date: Parallels for this survey form type date from the late 4th c. to the Late Hellenistic period, though these are slightly more angular in section. This general rim type is attested by banded-ware pelikai from the Pantanello necropolis dating to the mid-5th c. BC (Carter and Toxey 1998a, 711 T 324-2), and it therefore seems safer to regard the date of this piece as broadly consistent with the Panathenaic amphora from this deposit (see above), rather than providing a later terminus post quem for the fill of the division line.

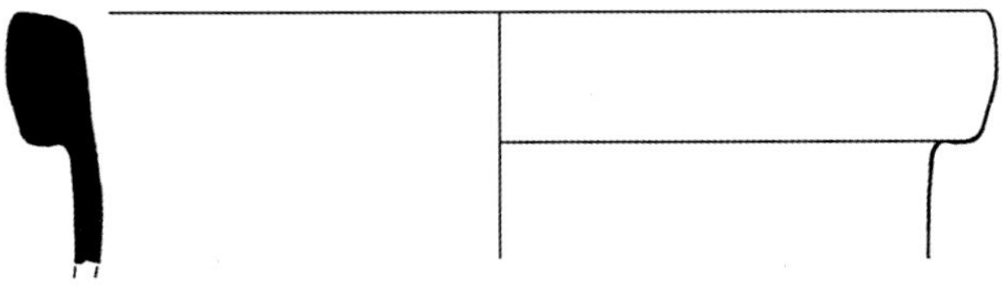

1:2

Tr6-12 BW or PW closed form base (Cl)
Description: Heavily weathered, uncertain form type.

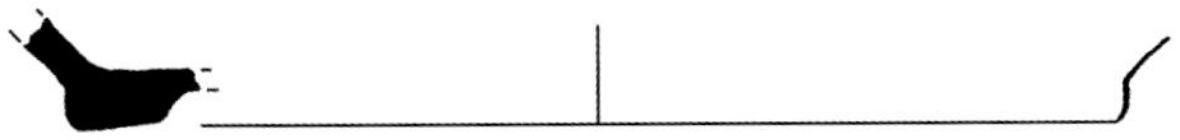

1:3

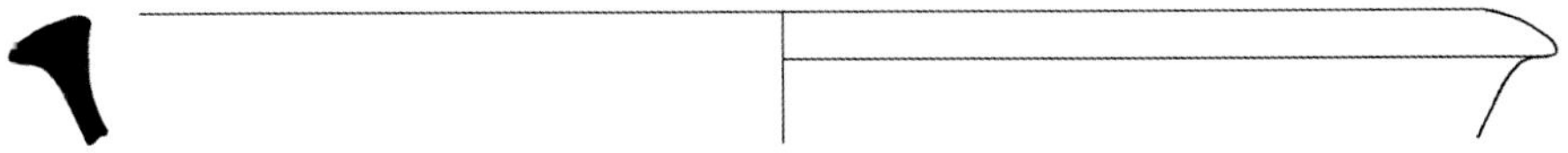

Tr6-13 PW lekane rim
Description: Cf. type Lne/Bas3 (Ch. 9, **No. 19**, for narrower vessel with similar rim form).

1:2

Tr6-14 PW disk base of open form (lekane/mortar?
Description: Concave underside, smoothed interior indicating an open vessel, perhaps from a mortar (e.g. Mor12c, Ch. 9, **No. 188**) or a lekane.

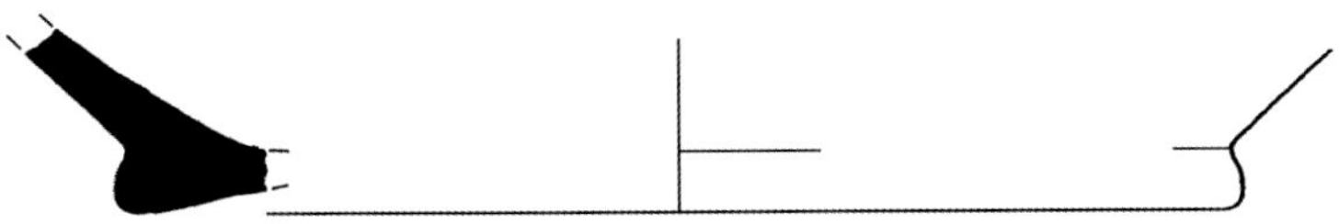

Other:
2 PW or BW body sherds.

Site 736, Trench 6: Level 2

1:2

Tr6-15 BG base of Pheidias-type mug
Description: Flat base with slightly concave underside. This form is not paralleled in the corpus of fine wares from the survey or from the necropoleis.
Date: Pheidias-type mug following the Attic type produced from the second quarter of the 5th c. BC (*Agora* XII, 201–18, dated ca. 460–400 BC) with a much closer parallel from the kerameikos of Metaponto probably dating to the second quarter of the 5th c. BC, the equivalent form in red figure classified as oinochoe type VIIIB (Silvestrelli, pers. comm.).

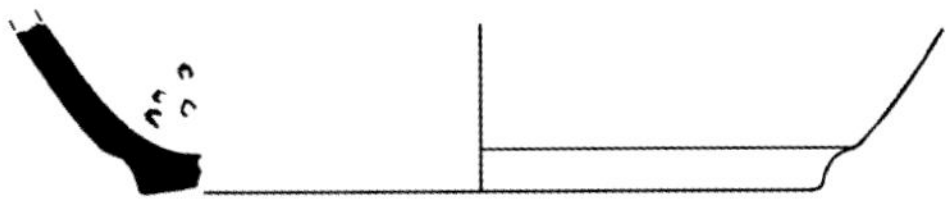

Tr6-16 Lekane rim
Description: Series of loosely formed concentric grooves on the upper plane of the rim, paralleled by type Lne5 from the survey (Ch. 9, particularly **No. 49**), though on this piece the flange is articulated to incline.
Date: Parallels for this form range from the 5th to 1st c. BC.

1:2

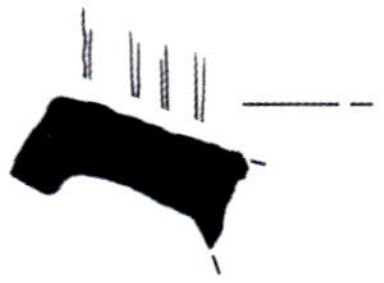

Site 736, Trench 6: Level 3

Tr6-17 RF body sherd
Description: Closed vessel form with floral/palmette decoration. Weathered/over-washed surfaces on areas of reserved decoration. Semi-lustrous, consistent gloss misfired to weak red (10R 4/4).
Date: Late 5th or early 4th c. BC (Silvestrelli pers. comm.). This piece provides the most secure terminus post quem for the lowest fill of the division line (Battuta 3) of the late 5th c. BC. Angle of rest indicated by throwing striations on the interior, probably a lower wall orientated as shown, rather than a shoulder.

1:2

Tr6-18 BG foot-ring base of open form
Description: Surfaces heavily weathered, but thin semi-matte black slip apparent in recessed areas on the exterior. Interior may have been reserved. Semi-lustrous post-depositional accretions, probably microcrystalline calcite, apparent on surfaces, superficially resembling an underslip or miltos, but apparent also on the fractures. Form unclear, but perhaps from a Hellenistic cup-skyphos (Ch. 6, **Nos. 88–89**) or a dish (**Nos. 129–130**)
Date: Second half of the 4th or early 3rd c. BC?.

1:2

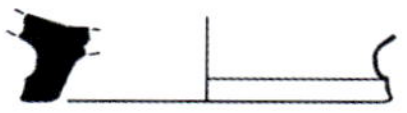

Tr6-19 PW or BW amphoriskos(?) toe
Description: Underside sheared regularly, perhaps where another section of clay was applied to the lower body, with throwing striations visible on the interior. Powdery fabric, unfinished surfaces.

1:2

Tr6-20 PW Plain-ware base of open form.
Description: Shallow disk base with concave underside formed by secondary turning to produce a false foot ring with a carinated external face. Tool-formed junction of foot-ring base and lower body. Finished/washed interior surface suggests an open form (i.e., lekane or medium-sized bowl), with base type paralleled by that of closed form types Cl1 and Cl2 (Ch. 9, **Nos. 302–303**) from the survey.

1:2

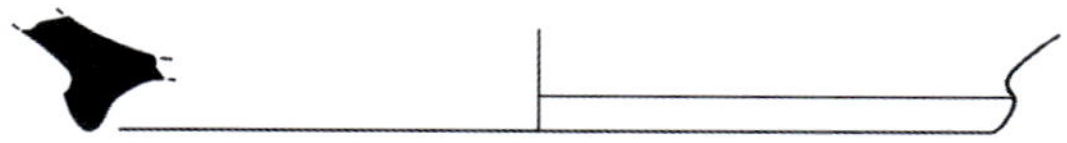

Other:
1 BG fragment of a small handle, from a skyphos or bowl form, ovoid section
3 BG body sherds.
23 PW or BW body sherds.
1 CW bowl or shallow cooking-ware form with slight projecting rim. Usual granular coarse-ware/cooking-ware fabric, heavily weathered surfaces.
1 CW body sherd.
1 amphora body sherd, lower body. Fine powdery, micaceous fabric, perhaps regional. Form indicates a sagging lower body. Pronounced throwing compression ridges on the interior, unfinished exterior.
1 amphora body sherd, upper shoulder. Fine powdery fabric, probably regional.
1 amphora body sherd, upper shoulder. Fine, hard, high-fired fabric with semi-conchoidal fracture. Uncertain form type.
1 amphora body sherd in fabric falling within the range of plain/banded-ware fabrics, though not demonstrably regional.
1 Corinthian body sherd. Small fragment in Corinthian fabric produced from Neogene marl with ARFs (Whitbread 1995). This fabric is associated in particular with Corinthian type A' amphorae, though this piece is not necessarily from a transport amphora.
2 tile fragments.

Non-ceramic:
Iron blade.

Site 736, Trench 6: Cleaning, Level 3 (Lowest)

1:2

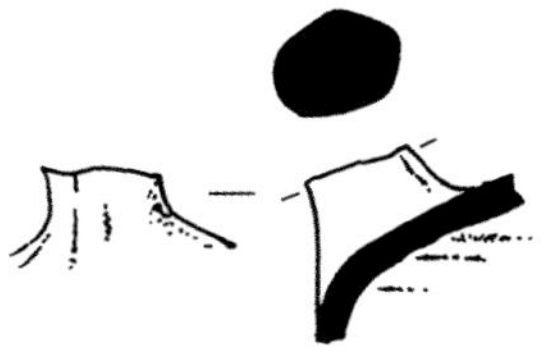

Tr6-21 PW or BW lebes gamikos shoulder and handle
Description: Shoulder of closed vessel with the lower terminus of a vertical shoulder-mounted handle with circular section.
Date: The general form is attested in banded ware (though the handle and shoulder are plain, as with this piece) from the Pantanello necropolis dated ca. 400–370 BC on context (Carter and Toxey 1998a, 700, T 306-10), which is broadly consistent with the dates of the red-figure pottery from Battuta 3.

1:2

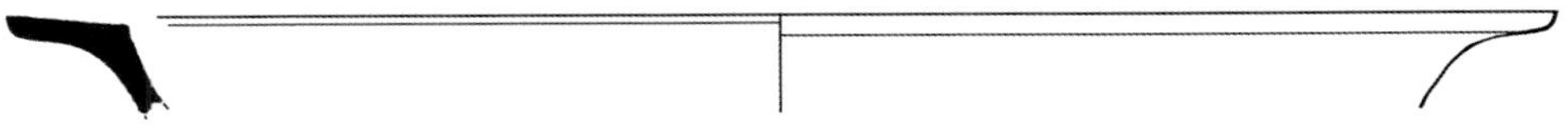

Tr6-22 PWBW lekane rim
Description: There are no close parallels for this rim form from the survey, though the general shape is well-attested.

1:2

Tr6-23 PWBW bowl rim
Description: Type Pl/Lne4a (Ch. 9, **No. 20**). Slight rounded, projecting rim, medium-sized bowl. The form is attested by parallels from the second half of the 5th c. BC onwards.

Site 737: Level 1

1:2

Tr7-01 RF body sherd
Description: Probably dating to the late 5th or early 4th c. BC (Silvestrelli, pers. comm.), but in any case earlier than the black-gloss dish from this deposit (below, Tr7-03).

Tr7-02 BG handle
Description: Uncertain form type, possibly a krater or hydria, which would be consistent in date with the red-figure body sherd from this context.

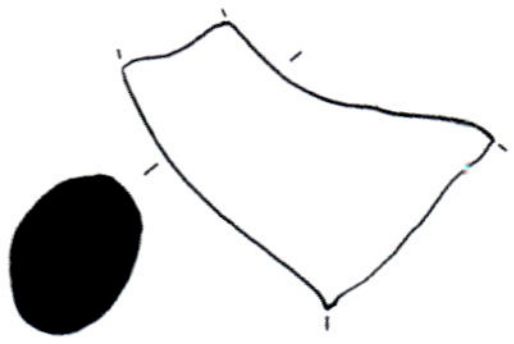

1:2

Tr7-03 BG carinated dish with plain rim
Description: Misfired reddish clay. This form is attested in the Pantanello necropolis, dated by context to ca. 325–300 BC (type D18, Elliott 1998, 653). Similar forms were present in the survey (Ch. 6, **No. 131**, dish rim) and appear to have been produced from ca. 320–280 BC in the kerameikos of Metaponto. This piece provides a terminus post quem for the formation of the deposit of ca. 325 BC.

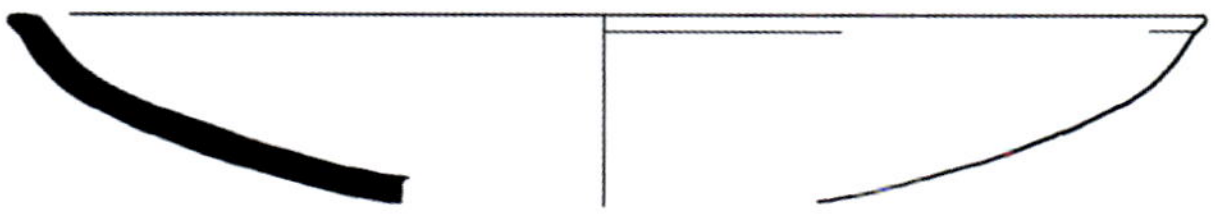

1:2

Tr7-04 BG strap handle
Description: Small fragment of a handle with lenticular section, perhaps from an olpe.

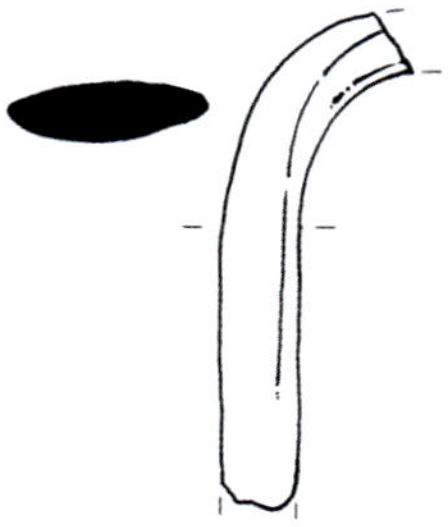

1:2

Tr7-05 BG skyphos rim
Description: Ch. 6, **No. 48**, C-type skyphos rim. Thin, matte bluish-grey slip. Type approx. dated to the late 5th and early 4th c. BC.

1:2

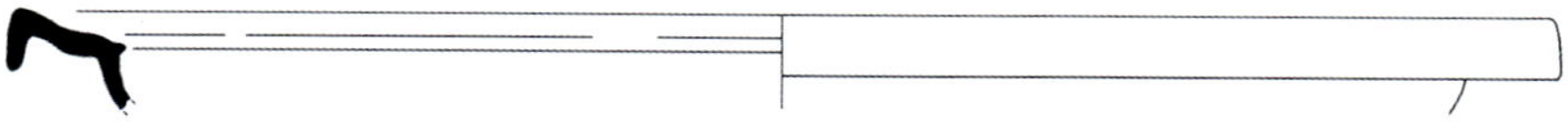

1:4

Tr7-06 PW or BW lekane rim
Description: Type Lne13, Pl. 9.6. Form paralleled by lekanai from the survey, Type Lne13 (Ch. 9, **Nos. 62–63**).

Tr7-07 PW or BW small bowl rim
Description: Type SB/O2 (Ch. 9, **No. 3**). Thin band of reddish-brown slip over the rim. These types have a broad span of production from the 7th c. to the 2nd c. BC.

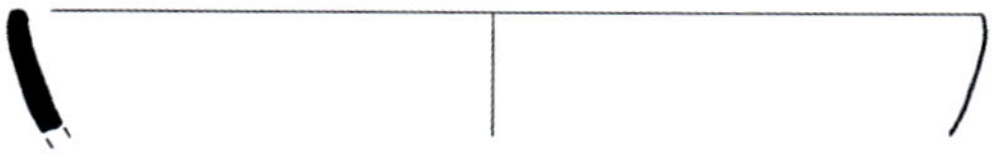

1:2

1:2

Tr7-08 PW or BW disk base
Description: Medium-large vessel, perhaps a lekane or mortar (cf. Mor12c).

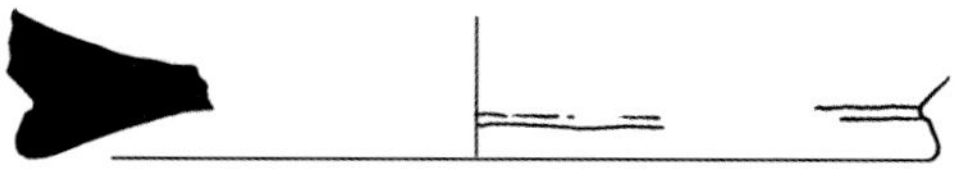

Tr7-09 PW or BW base of open form
Description: Turned foot ring

1:2

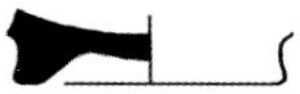

Tr7-10 PW or BW foot-ring base

Tr7-11 CW base of open form(?)
Description: Coarse ware false–foot ring base, with smoothed interior indicative of an open form, with relatively steep extant lower wall.

1:2

1:2

Tr7-12 CW olpe or similar closed form
Description: Restricted neck and plain everted rim.

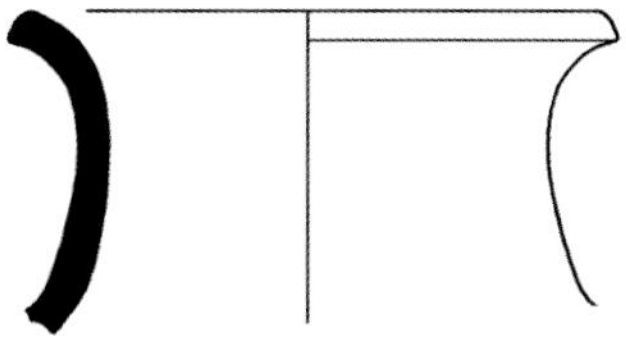

1:2

Tr7-13 AMP Corinthian type A′ amphora rim
Description: Broadly dateable to the 5th c. BC.

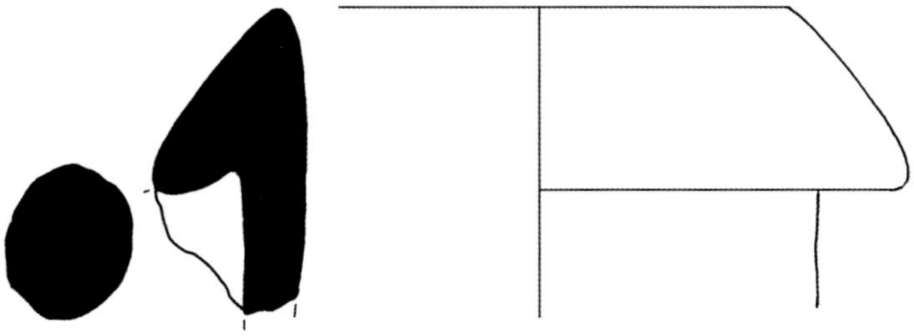

Other Excavated Materials

The pottery recovered from the excavations of trenches 1, 2, 4, 5A, 5B, and 5C, across the transverse feature, L09, to the NW of site 737, across the diagonal feature and across both the longitudinal and transverse features at their intersection (for precise locations, see Figs. 28.6 and 28.14) is recorded below. In contrast to the material from the fill in the longitudinal feature, L09 near the two necropoleis, sites 736 and 737 (where material from destroyed burial or burial markers, as could be expected, is predominant), from these trenches—all at some distance from the necropolis—the material is typical of that recovered in the survey and excavation of farmhouses. Notable is the relatively high proportion of cooking ware and roof tile. None of the dateable fine wares could be dated any later than the end of the 4th century or beginning of the 3rd. This is significant as it indicates that the features were probably not open or frequented after that date. The quantity of material from trenches of very limited size is surprisingly large. Pizzica in its heyday in the 5th and 4th centuries was densely occupied, a fact that is confirmed by the survey (see Vol. II, especially Chs. 23, 24, and 25).

Trench 1

BG
2 body fragments
Impasto
Fragment with lug handle, Prehistoric
BW or PW
1 plate/lekane rim (Pl/Lne4d1b)
1 lekane/basin rim (Lne/Bas1b1)
Ck
5 fragments
Amp
Corinthian B, rim with neck and handle
1 body fragment

Trench 2

Tile
3 fragments
AFW
Ionic cup base
Ionic cup shoulder
BG
Skyphos base
3 fragments of closed vessels
2 fragments of open vessels
BW or PW
2 handles
33 body fragments
Ck
4 body fragments
Amp
Handles and 49 body fragments
Tile
8 Laconian tile fragments

Trench 4

AFW
2 Ionic cup fragments, 6th c.
BG
Skyphos base (**No. 55**, BG 521-114 C-type skyphos base), 430–375 BC
15 body fragments
BW or PW
2 rims (SB/0) fragment
1 vertical handle, unknown shape
4 jug rims J5b, J18, J19a, J1c
3 lekane rims (Lne12, Lne12/13, CLne6c)
Dinos rim Dn6
43 fragments
Ck
Pot/jar rim (PJ1) and body fragment
Casserole body fragment CL

Trench 5A

7 body fragments
Amp
9 body fragments
Tile
5 fragments (one uncertain)

Level 1

BG
Body fragment
BW or PW
5 body fragments
Amp
2 body fragments
Tile
2 fragments

Level 2

BW or PW
1 lekane rim Lne6b
1 handle SB/0
1 handle unknown shape
10 body fragments
Tile
1 fragment

Trench 5B

Level 1

BG
2 handles
BW or PW
1 lekane rim Lne12
1 lekane/basin rim Lne/Bas1b1
8 body fragments
Ck
4 body fragments
Amp
1 body fragment
Tile
6 fragments

Level 2

BG
2 body fragments
BW or PW
1 jug rim J19a
3 body fragments
Amp
5 body fragments
Tile
4 fragments
Lithic
1 obsidian fragment

Trench 5C

BG
1 neck with handle
2 body fragments
BW or PW
1 small bowl rim SB/02
1 carinated lekane rim CLne6a
13 body fragments
Ck
5 body fragments
Tile
8 Laconian tile fragments

30
The Human Remains

Renata J. Henneberg and Maciej Henneberg

The discovery in the summer of 1999 of the two necropoleis in the Pizzica area (sites 736 and 737, see Ch. 18 and 19) presented the authors with a welcome occasion to examine a skeletal sample from a new area of the chora, and to compare the results with those previously documented at Pantanello, Saldone, and Sant'Angelo Vecchio, and from the urban necropolis at Crucinia. It was possible to explore further the observations made in earlier analyses of the human skeletal material at these sites as regards differences in tooth morphology, and to note patterns here that point to residence either in the urban center (Crucinia) or in the chora. The authors are grateful to the excavators, Professor Joseph Carter and Dr. Antonio De Siena, for this opportunity. Apart from the generous support provided by Professor Carter, the authors are most grateful for the Large Australian Research Council grant that also financed parts of this project.

Methods

The analysis of the human skeletal remains from the necropoleis was conducted according to procedures and methods similar to those employed in examinations of skeletal samples previously excavated in the Metapontino.[1] Because of the bones' fragmentary state of preservation, all possible indicators, including some newly described ones, were used in the assessment of sex and age at death.[2] The pelvis and skull took priority in sex assessment, while the age assessment of adults relied heavily on the state of dentition, with pubic symphyses, auricular surfaces of ilia, cranial sutures, and arthritic changes taken into account. All measurements of the long bones were done following Martin's technique.[3] The adult stature was reconstructed using primarily the Trotter and Gleser formulae for whites (1952, 1977) and the formulae presented in Drogman and Iscan (1986). Buccolingual (BL) and mesiodistal (MD) measurements of permanent teeth were also taken to 0.1 mm accuracy using calipers with sharpened ends.[4]

The presence or absence of a few epigenetic traits on skulls and on long bones that could be observed in this sample was recorded following the description of the traits in Berry and Berry 1967, Finnegan 1978, and Buikstra and Ubelaker 1994. The epigenetic traits on teeth were also recorded.[5]

The observations of pathological conditions on bones were made with the unaided eye. Dental caries were recorded on permanent teeth.[6] Tooth loss, abscesses, calculus, and enamel hypoplasia (rings and pitting) were also noted.[7] Root exposure was measured, and changes of alveolar bone (remodeling and pitting) were recorded in order to assess periodontal conditions.[8] Tooth wear was assessed with two scoring systems: Molnar's 8-grade method on all teeth and Scott's 10-grade method on molars.[9]

Results

Only partly preserved skeletons were recovered from most of the tombs at both excavated sites. Each skeleton had many bones missing, and a majority of the preserved bones were in small fragments. Among 15 skeletons excavated from both sites, only one skeleton of a male (site 736, T 22) was nearly complete, which is to say that most parts of the skeleton were represented, but many bones were damaged or fragmented. Because the crania were incomplete and fragmented, only a few standard cranial measurements of one skull were able to be recorded (see individual descriptions of the skeletons below), and only one individual had some upper and lower limb bones preserved, which were measured to attempt the reconstruction of the stature. One individual was represented by only one

1 Henneberg and Henneberg 1998.

2 Acsadi and Nemeskeri 1970; Krogman and Iscan 1986; Loth and Iscan 1994; Buikstra and Ubelaker 1994; Loth and Henneberg 1996a, 1996b, 2001.

3 Martin and Saller 1957.

4 Martin and Saller 1957.

5 Turner et al. 1991; Pinto-Cisternas/Moggi-Cecchi/Paccioni, 1995.

6 Henneberg 1991, 1998; Henneberg and Henneberg 1998.

7 Dobney and Brothwell 1986; Goodman and Rose 1990; Hillson 1996.

8 Goldberg et al. 1976; Clarke and Hirsch 1991.

9 Molnar 1971; Scott 1979.

Pizzica site 736						
No.	**Tomb number**	**Sex**	**Age in years**	**Preservation state**		**Comments on pathological changes**
2	18	F	25–35	Very fragm.	Skull and long bones fragm.	
3	19	F?	25–30	Very fragm.	Skull and long bones fragm., teeth	Dental pathology (see *Table 17.7*)
4	20	F	25–35	Very fragm.	Skull and long bones fragm., teeth	Cribra orbitalia in left orbit; dental pathology; shallow preauricular sulcus
5	21	F	50–x	Very fragm.	Skull and long bone fragm., one tooth M3	
6	22	M	40–50	Nearly complete	Skull and long bones fragm., teeth	Arthritic changes on patellae, foot bones, ulna (olecranon); healed wound to the left side of the mandible

Pizzica site 737						
No.	**Tomb number**	**Sex**	**Age in years**	**Preservation state**		**Comments on pathological changes**
10	18	F?	Adult	Very fragm.	Fragm. of postcranial skeleton	
11	19	F?	40–50	Very fragm.	Skull and long bones fragm., teeth	Skull vault porotic 8.6 mm thick; large diplöic spaces, osteophytes on the left patella, dental pathology
12	20	M?	40–x	Very fragm.	Long bones fragm. only	
13	21	M	50–60	Fragm.	Skull and long bones fragm., teeth	Dental pathologies; arms possibly crossed when buried
14	22	?	10–14	Very fragm.	Long bones fragm. only	
15	23	F	40–45	Very fragm.	Skull and long bones fragm., teeth	Dental pathologies

Table 30.1 Skeletal material excavated at Pizzica, sites 736 and 737. Sex and age estimates, state of preservation, and comments regarding pathological changes on bones. Skeletal remains from tomb 4 (site 736), and tombs 5, 10, and 13 (site 737) not included (see text).

fragment of a long bone, probably a radius. Among the 15 individuals only 7 had some teeth preserved; the number of preserved permanent teeth varied from 28 to only a single tooth. Because most individual skeletons were poorly preserved and lacked many diagnostic parts for determining sex and age, a good assessment of sex could be made for only 6 individuals, a tentative assessment for 5 individuals, and in 4 individuals, sex could not be established at all (*Table 30.1*). The issues regarding preservation of human bones have been discussed in earlier work, including the problems of soils in a heavily agricultural area and other influential factors.[10]

Despite the fragmentary nature of this skeletal sample, some observations are possible in terms of both pathological conditions and morphological characteristics of bones and teeth. The findings are presented in Tables 30.1–4. Systematic recording of some characteristics, other than those in Tables 30.1–5, has been difficult because of the small size of the sample and the fragmentary state of the material. Thus, in addition to the tables, brief descriptions of selected individual remains are included below.

10 Henneberg 1998.

Tooth*/ Indiv.	T 19/736 BL/MD	T 20736 BL/MD	T 21/736 BL/MD	T 22/736 BL/MD	T 19/737 BL/MD	T 21/737 BL/MD	T 23/737 BL/MD
ULM3	10.3/9.3	9.4/7.7					
ULM2	11.2/10.1	10.3/8.6		12.3/10.0			
ULM1	11.1/10.5	10.3/9.3			11.0/–		
ULP2		8.7/6.5		9.3/6.3	9.0/6.3	8.0/6.1	
ULP	8.6/7.0	8.4/7.0		8.7/6.3	8.7/6.2		
ULC	8.0/7.9			9.6/7.6			
ULI2		6.1/6.7		6.5/6.8			
ULI1				7.1/8.3			7.2/–
ULI1				6.8/8.6			6.6/–
URI2		6.0/6.7		6.7/6.8			
URC	8.0/8.3	7.3/7.5		9.5/7.8	7.8/7.0		7.6/6.7
URP	8.7/6.7	8.6/6.7		8.7/6.6			8.9/6.8
URP2		8.8/6.6		9.0/6.2	9.0/6.2		8.5/6.2
URM1	11.2/10.3	10.2/9.4			11.1/9.0		
URM2	11.2/9.8	10.3/8.4			10.7/8.0	12.0/10.0/	11.0/8.8
URM3		10.0/7.7		10.2/7.5	9.5/7.1	11.4/9.2	
LLM3	10.4/11.6	8.9/10.0					9.7/11.0
LLM2	9.9/11.1	9.2/10.4					
LLM	10.2/11.2	9.2/10.3		10.9/11.8	9.3/9.7		
LLP2	8.1/7.4	7.4/6.9		8.4/6.9			
LLP1	7.7/6.7	6.4/6.8		7.6/6.6			7.0/6.8
LLC	7.8/6.6	6.7/6.4		8.3/7.1			
LLI2	6.2/6.0	5.3/5.4		6.2/6.2			6.3/5.4
LLI1	5.7/5.4	5.1/5.0		5.9/5.3			5.7/4.4
LRI1		5.0/5.0		5.8/5.6			
LRI2		5.2/5.4		6.2/6.5			
LRC	7.4/6.5	6.7/6.2		8.4/7.2			
LRP1	7.4/6.6			7.7/6.7			7.1/6.5
LRP2	8.1/7.4	7.5/6.4		8.2/6.6			
LRM	10.2/11.4	9.1/9.8		10.8/11.6			
LRM2	10.2/10.6	9.2/10.2		10.0/11.0		10.3/10.3	
LRM3	–/12.0	9.2/10.1	8.7/9.9	9.7/10.0			

Table 30.2 Dental measurements of permanent teeth. Buccolingual (BL) and mesiodistal (MD) diameters (mm).* The first letter stands for upper or lower; the second stands for left or right; the third letter plus a number stands for a tooth.

Individual no.	T 19/736	T 20/736	T 21/736	T 22/736	T 19/737	T 21/737	T 23/737	Summary
Sex	F?	F	F	M	F?	M	F	
Age (years)	25–30	25–35	50–x	40–50	40–50	50–60	40–45	
No. of teeth	23	28	1	26	10	4	11	103
No. of sockets	27	32	1	31	10	9	14	124
Carious teeth	10	5	–	1?	1	1	4	22
Teeth lost before death (ante mortem loss)	–	–		5	–	5	3	13
Max. number of hypoplastic rings per tooth per person	2	4		4	2	1	4	4
Average wear, Molnar's scoring system (Molnar 1971)	2.2	2.3	3.0	4.4	2.9	4.0	3.7	3.2
Average wear, Scott's scoring system (Scott 1979)	13.3	15.6	24.0	23.5	20.0	27.7	13.5	19.7
Average root exposure (mm)	2.6	2.0		3.1	–	–	–	2.6
Abscesses	2 (possible)				–	3		5

Table 30.3 Permanent teeth of individuals from Pizzica sites 736 and 737: diseases and tooth wear.

Individuals

Site 736, T 18

Female, 25–35 years; no *cribra orbitalia* on the roof of the left orbit; right orbit not preserved. No metopism. No changes in the left glenoid fossa; right glenoid fossa missing. Fragment of the left tibia (circumference 83 mm). Vertebrae not preserved.

Site 736, T 19

Probably a female, 25–30 years; no lipping on the fragments of vertebrae. No parts preserved to observe other characteristics. Upper first molars with double layer of enamel. All upper incisors missing. Other teeth look normal, but extensive caries present. No changes in the left glenoid fossa; right fossa missing.

Site 736, T 20

Female, 25–35 years; no metopism. No supratrochlear foramen on the right humerus; left humerus not preserved. On the roof of the left orbit *cribra orbitalia* grade 2 present; right orbit not preserved. Skull fragments 8.7 mm thick (maximal thickness), and all three tables (external, internal, and diplöe) look normal. Teeth bluish brown in color with a few white spots on enamel, but no pitting present. Shallow preauricular sulcus observed. Partly preserved sacrum has S1, S2, and S3 closed. Sacral segments S4 and S5 damaged.

Site 736, T 22

Male, 40–50 years; this is the best-preserved skeleton excavated from either site. No metopism; parietal foramen present on the right but not on the left side; no supratrochlear foramen on both sides; single mandibular foramina on both sides; mastoid foramina in the suture on both sides; no auditory tori; no foramen of Huschke; no zygomaticofacial foramen; no frontal

	Trait*/individual	T 19/ 736	T 20/ 736	T 21/ 736	T 22/ 736	T 19/ 737	T 21/ 737	T 23/ 737
	Maxilla							
1	Winging				–/–			
2	Shovel shape incisors (grade 0–4): I1				1/1			2/2
	: I2		+(3–4)/+(3–4)		1/1			
3	Canine tubercle	–/–	0/+		+/+	0/+		0/–
4	Canine distal accessory ridge (DAR)	+(2)/+(2)	0/+		+(2)/+(2)			0/0
5	Root number: P1	1/1	2/2		1/1	1/0		0/1
6	Number of cusps: M1	4/4	4/4			4/4		
	: M2	4/4	3/3		4/0	0/4	0/4	0/4
	: M3	4/0	3/3		0/3	0/3		
7	Carabelli cusp grade 5–7	–/–				0/+(5)		
8	Hypocone : M1	4/4	4/4			4/4		
	: M2	4–/4–	3+/3+		4/0	0/4–	0/4	0/4–
	: M3	4–/0	3/3+		0/3–	0/3+	0/4–	
9	Metaconule present (grade 0–4) M1	0/0	0/0					
10	Enamel extension: M1	4/0	2/2					
	: M2	2/2	2/2		2/0		4/4	0/2
	Mandible							
11	Cingular nodule (I2)	–/0	+/+					– /0
12	Canine distal accessory ridge (DAR)	+(2)/0	+(1–2)/+(1–2)		–/+			
13	Lingual cusps number: P1	2/2	1/0		2/2			1/1
	: P2	2/2	2/2		2/2	0/2		
14	Sagittal furrow present (P1)	+/+	–/0					+/+
15	Number of cusps: M1	5/5	4/4		5/5	4/0		
	: M2	4/4	4/4		0/4		0/4	
	: M3	5/6 irr		0/4	0/4			4/0
16	Groove pattern: M1	Y/Y	+/+		Y/Y	+/0	+	
	: M2	+/+	+/X		0/+		0/0	
	: M3	+/+		0/+	0/+			+/0
17	Deflecting wrinkle: M1	–/–	–/–					
18	Protostylid: M1	–/–	–/–					
	: M2							
19	Enamel extension (grade 0–4): M1	4/4	2/2		3/3			
	: M2	4/4	2/2		0/3		0/4–5	2/0

Table 30.4 . Epigenetic traits on permanent teeth (based on Turner et al. 1991).
*0 or empty space = no information;
– = trait absent; + = trait present;
/ = left / right side;
grade in brackets or numbers without brackets correspond to the number of cusps or other gradation of the trait.

Bone/ measurement	Humerus		Radius		Ulna		Tibia	
	Left	Right	Left	Right	Left	Right	Left	Right
Maximum length	–	348	252	252	273	280	381	382
Circumference	69	75	49	48	45	44	87	89
Epicondylar breadth	–	67						
Sagittal head diameter			24	–				
Transverse head diameter			–	24				

Table 30.5. Osteometry of long bones (mm). Male 40–50 year old , T 22, site 736.

notch/foramen—both sides; supraorbital foramen open on the right and closed on the left. No bregmatic bone. Other characteristics cannot be observed.

Craniometry
Units in mm.
"?" indicates measurement taken on slightly damaged bone:

minimum frontal breadth (ft-ft) = 92
face breadth (zm-zm) = 93
upper face height (n-pr) = 76?
nose height (n-ns) = 56?
nose width = 24?
orbit width (mf-ek) = 43
mandibular body length (go-gn) = 91
bigonial width (go-go) = 104
mandibular symphysis height (gn-id) = 37
bicondylar width (kdl-kdl) = 115
external width of the dental arcade (ekm-ekm) = 65
internal width of the dental arcade (enm-enm) = 45
condylar height of the ramus = 56?
coronoid height of the ramus = 68
minimum breadth of the mandibular ramus = 31

Osteometry
See Table 20.5. Epiphyses of both femora had been damaged, but it was possible to measure the circumference of both the left (101 mm) and the right (97 mm) bone.

Stature reconstructed from several bones using formulae of Trotter and Gleser, Manouvrier, and Pearson is 1.70 m.[11]

Pathologies
A healed injury to the left side of the mandible was noted (*Fig. 30.1*). A groove extends posterosuperiorly from just behind the left mental protuberance towards roots of the first molar. It ends about 13 mm below the neck of the M1. The groove is 27 mm long; its width is uneven, approximating 5 mm; the depth varies, reaching in the upper end 3 mm. The surface of the groove is well healed. The lower border of the left mandibular corpus is misshapen by the wound, so that this border extends down about 3 mm immediately behind the mental protuberance. The left second and third molars are missing, and so is the upper first molar. All other mandibular teeth and most maxillary teeth are free of carious lesions, suggesting that the left molars have been lost as a consequence of the injury. There are no signs of extensive inflammation. The wound is likely to have been caused by a light blow of a sharp weapon (possibly a sword) executed by a right-handed opponent. Injuries to the mandible, especially to its body, are rare findings in archaeologically derived skeletal collections. They are almost always associated with combat.[12]

On the right zygapophyseal surface of the axis vertebra, the individual has an oval depression 11 mm long anteroposteriorly, 5 mm wide and up to 3 mm deep with an unevenly pitted bottom. The corresponding left surface is smooth and normal looking, as are the articulating surfaces of the atlas. Apart from the trauma to the mandible, the skeleton displays no pathological changes. No *cribra orbitalia* are seen in the roof of either orbit. Slight lipping (grade 1) around the bodies of cervical vertebrae is related to the age of the individual rather than to any specific reaction to the injury. No carious lesions have been observed, besides one small, dark spot on the enamel, which could be initial caries. Some pitting on alveolar margin is present, indicating mild periodontal disease.

[11] Krogman and Iscan 1986.

[12] For comparison, see Novak 2000.

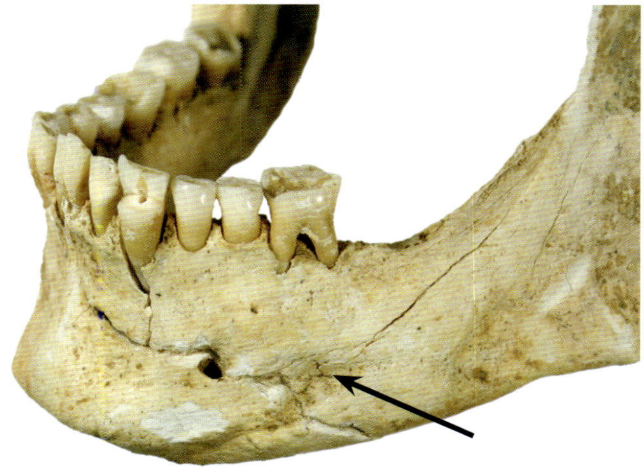

Figure 30.1 Left side of a mandibular corpus in a 40–50 year old male from tomb 22, site 736. Blade injury indicated by arrow. (CR)

Site 737, T 19

Probably female, 40–50 years; skull porotic, with thinner than normal outer table and large diploïc spaces (porotic hyperostosis); fragments 8.6 mm thick.

Site 737, T 21

Male, 50–60 years; no metopism. Skull sclerotic but thin. Caries present and many teeth lost before death. The individual was probably buried with his upper limbs crossed.

Site 737, T 23

Female, 40–45 years; no metopism. In the sacrum, S1, S2, and S3 are closed; the rest of the sacrum is damaged. Slight lipping (1 degree) present on lumbar vertebrae. Extensive caries present.

In general, the sample seems to consist mostly of adults. No bones of small children were identified in this group of individuals, besides the one fragmentary skeleton of the youth 10 to 14 years old. Dental caries and enamel linear hypoplasia were common. Osteoarthritic changes on vertebrae and long bones seem to reflect an aging process. No metopism was observed in this sample.

Discussion and Conclusions

The 15 individuals from the two sites at Pizzica seem to fit well into the larger samples of individuals from rural Pantanello and the two smaller necropoleis at Sant'Angelo Vecchio and Saldone, and also from urban Crucinia. In general, the presence or absence of observable epigenetic traits on skeleton and teeth reflects the frequency of these traits in the entire Metapontino skeletal material.

While most of the common traits in the region are also present in the sample from Pizzica, some findings require particular attention. The canine tubercle on upper canines is present in three out of five individuals (two females and one male) with upper canines preserved. This trait appears in only 13.4% of the urban population (Crucinia) and in 56.5% of individuals from the rural areas outside of Metaponto (Pantanello, Saldone, Sant'Angelo Vecchio).[13] One of the females (T 20, site 736) with the canine tubercle also has pronounced shovel-shaped upper second incisors. Pronounced shoveling of upper second incisors is slightly more frequent in the rural than in the urban necropolis, although the difference is not statistically significant.[14] No "Etruscan trait" is observed in the four individuals with preserved upper second incisors. The "Etruscan" incisor is associated with indigenous Italic peoples and is most frequent among inhabitants of Central Italy.[15] This trait was first described on Etruscan dentition and has not been observed among Greeks from Greece.[16] One of the females (T 19, site 736) and a male (T 22, site 736) have two cusps on lower first premolars. This trait is significantly more frequent statistically in the rural population. This male also has the canine tubercle and slightly shovel-shaped upper incisors. These three traits are more frequently observed among the rural than among the urban population.

The female from T 19, site 736, with two cusps on LP1 has no canine tubercle on upper canines. She has one trait characteristic of rural and one trait characteristic of urban populations. The female with canine tubercles and shovel-shaped incisors (T 20, site 736) also has a reduced number of cusps on the upper second molars, and only four cusps on the lower first molars. The latter two traits are observed with higher frequency in the urban than in the rural population. Two females (T 20, site 736, and T 19, site 737), both with canine tubercles on upper canines, have a reduced number of cusps on lower molars and "+" pattern (see *Table 30.4*), more common—but still rare—in the urban population.

[13] Henneberg 1998.

[14] Henneberg 1998, 273.

[15] Coppa et al. 1998.

[16] Pinto-Cisternas/Moggi-Cecchi/Paccioni, 1995.

The results of the above comparisons suggest that at least some of the people buried in Pizzica at sites 736 and 737 may have come from the city of Metaponto (possibly the females T 20, site 736, and T 19, site 737). The male from T 22, site 736, has more characteristics consistent with the rural pattern. The female from T 19, site 736, besides the lack of canine tubercles, also has more dental characteristics similar to the rural pattern. She has five cusps on the lower first molars and four cusps on all upper molars, which occur more frequently in the rural population. The females from T 20, site 736, and T 19, site 737, have more characteristics consistent with the urban pattern. All three females with well-preserved teeth were buried with grave goods dating to the 5th century BC,[17] thus allowing for the suggestion of mixing of rural and urban populations in this century rather than microevolutionary changes through time leading to reduction in tooth size and cusp numbers (see *Table 30.2* for comparison of tooth size among the three individuals). It should be stressed that the full statistical analysis of the findings has not been attempted at this stage of the study because the sample included very few individuals with observable traits. These suggestions therefore have to be treated with great caution.

The paleopathological observations are also consistent with the findings in both rural and urban samples from Metaponto. One of the findings, porotic hyperostosis in the female in T 19, site 737, suggests thalassemia, common in the area of Metaponto.[18] The best examples of porotic hyperostosis suggesting thalassemia come from the urban cemetery of Crucinia.[19] The double layer of enamel on the first upper molars of the female from T 19, site 736, suggests a sort of systemic disease. According to our previous findings, the specific enamel hypoplasia on the first molars only, while other teeth are not changed, can suggest congenital syphilis.[20] Because no other observations of pathological signs on bones are possible because of very poor preservation of this skeleton, other explanations of changes on enamel have to be considered. Multiple signs of treponematosis, most probably syphilis, have been found in the ancient population of the Metaponto area, suggesting that syphilis was endemic at least in the rural area.[21]

The present study of this skeletal sample, small though it may be, may provide useful information that can clarify and strengthen some of our previous findings in the Metapontino. The importance of the skeletal material, although incomplete in its preservation and insignificant in numbers of individuals, but excavated from a large archaeological area, cannot be overlooked. Each case counts and adds to the larger historical picture.

[17] *Ancient Territories* 1999.

[18] Henneberg and Henneberg 1998.

[19] Henneberg and Henneberg, unpublished data.

[20] Henneberg/Henneberg/Carter 1992; Henneberg and Henneberg 1994, 1998.

[21] Henneberg/Henneberg/Carter 1992; Henneberg and Henneberg 1994, 1998.

31
Aerial Photography and the Organization of the Chora

Alberto Prieto

This chapter describes a new study of the Metapontine "division lines" (hereafter referred to as "lines") using historical aerial photographs that were taken in the early years of the transformation of the landscape by heavy earth-moving machinery in the 1950s.[1] The study was planned and conducted entirely in a geographic information system (GIS) environment with photogrammetric support. It incorporates evidence from archaeological excavations conducted across the territory between 1999 and 2002, which have revealed much new information about these features. The combination of remotely sensed and ground-truthed data represents a significant advance in approach to the subject. The geospatial analysis methods, the evidence for confirming the locations and physical nature of the features, the effects of the methods on the results, and the effects of the local topography on the organization of the chora are described in detail. The results of the study, and the questions it raises, create a more complex and complete picture of the ancient territory than the results of previous studies based on topographic observations and/or photogrammetry.[2]

Aerial Photography and GIS in the Metapontino

With the adoption of techniques and technologies developed in other disciplines, the study of ancient landscapes has evolved substantially in the past six decades from its humble origins in simple exploration and description. In the 19th and early 20th centuries the perspective of topographic research was, with few exceptions, limited to the observer's field of view on the ground, or at best an oblique view from a high location, a perspective that constrained the scope of projects to very small and disconnected areas. The invention and development of aerial photography in the years before and during the First World War finally allowed topographers to study the world with the perspective of Icarus; but the primitive early cameras and aircraft allowed the documentation and mapping of only relatively small areas, such as urban zones and rural archaeological sites. The development of sophisticated high-altitude vertical aerial photography as a source of behind-the-lines intelligence during the Second World War, by both Allied and Axis militaries, paved the way for its rapid maturation and diffusion as a tool for large-scale topographic mapping and planning in the postwar period, particularly in agriculture, forestry, transportation, and urban development.

In Italy, archaeologists, architectural historians, and topographers of ancient landscapes quickly recognized the value of aerial photography in revealing the features and structure of landscapes from prehistory through the Middle Ages. Pioneering researchers demonstrated that value with a series of innovative topographic projects such as the study of the urban plan of ancient Metaponto, which led incidentally to the discovery of the lines in the chora.[3] With these important new contributions these researchers successfully lobbied the Italian government to create a central state archive and photogrammetric service for archaeological and architectonic research located in Rome, the Laboratorio per la Fotointerpretazione e l'Aerofotogrammetria, or Aerofototeca (today part of the Istituto Centrale per il Catalogo e la Documentazione, a section of the Ministero per i Beni e le Attività Culturali). This institution had the authority to acquire important coverages from non-Italian entities such as the Royal Air Force, Luftwaffe, and U.S. Army Air Force and to organize new photographic missions designed to create specialized coverages for topographic research, which continued through the period of Italy's economic recovery (1950s–1970s) and the subsequent mechanical transformation of the landscape.[4] The Aerofototeca's collections remain the

[1] The study was part of the author's doctoral dissertation at The University of Texas at Austin (Prieto 2005), which examined aerial photographs of the major chorai of Magna Grecia. The aerial photos are reproduced with permission of the Aerofototeca, Rome.

[2] Schmiedt and Chevallier 1959; Uggeri 1969; Adamesteanu and Vatin 1976; Guy 1995.

[3] Schmiedt and Chevallier 1959.

[4] On the history of aerial photography and its application to archaeology, see Piccarreta 2003.

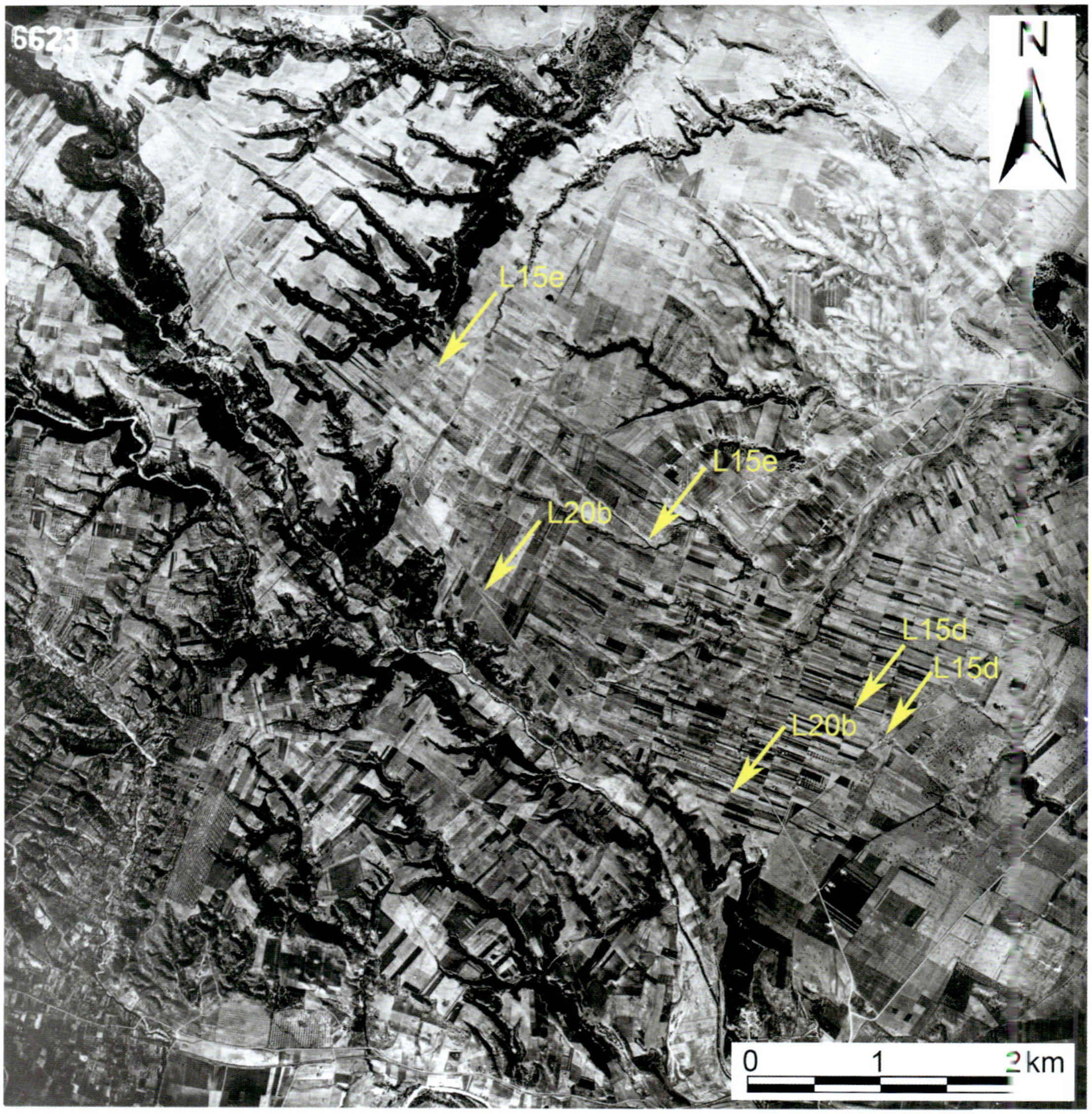

Figure 31.1 Lines visibie in the chora of Metaponto between the Bradano and Basento Rivers. (Aerofototeca Neg. No. 31660, GAI)

foundation of topographic research in Italy and have been employed in every major study of the lines.[5]

In the 1960s and 1970s digital methods for the management of spatially referenced data were developed in the United States for purposes of simple cartographic representation, inquiry, and spatial analysis. In the 1970s photogrammetry began to make the transition to digital techniques, and the introduction of the first generation of digital satellite imagery at the same time prompted the development of software for manipulating remotely sensed data. By the 1990s photogrammetry and remote sensing software, such as GIS, became sufficiently sophisticated to allow nonspecialists to work confidently with digitized aerial photographs and satellite imagery.[6] GIS matured in the 1990s with the development of graphic-based software and a sufficiently wide knowledge base, both of which enabled a broader application of GIS to archaeological research. The availability of these tools enabled ICA researchers to produce an entirely new study of the Metapontine lines using a combination of GIS, photogrammetry, and remote sensing.[7]

[5] Uggeri 1969; Adamesteanu 1973, 1974a; Adamesteanu and Vatin 1976; Guy 1995; Carter/Thompson/Trelogan 2004.

[6] On the evolution and capabilities of GIS, see Wheatley and Gillings 2002, 9–16.

[7] Carter/Thompson/Trelogan 2004.

Sources of aerial photographs

The Metapontine lines appear in several series of black-and-white panchromatic vertical aerial photographs available at the Aerofototeca (*Fig. 31.1*). The only complete photographic coverage consulted for this study consists of twenty-seven photographs in five strips from the Gruppo Aereo Italiano (GAI) mission of 1954–55, a collaboration between the Istituto Geografico Militare (IGM, the Italian national cartographic agency, based in Florence) and the U.S. Army Map Service. These photos cover the entire peninsula at an approximate scale of 1:33,000.[8] Other series of partial coverages include the following:

1. Five photographs in a single strip covering the coastal plain and the Pizzica terrace (ca. 7 x 5 km) between the Bradano and Basento, parallel to the coast, at an approximate scale of 1:18,000 without attribution (unknown series) circa 1970;
2. A series of fourteen photographs along the Bradano valley (mouth to Tavole Palatine, ca. 9 x 3 km) at an approximate scale of 1:10,000 taken by the Aeronautica Militare (AM, the Italian air force) in 1953;
3. Ten photographs along the Basento valley (mouth to Incoronata, ca. 14 x 5 km) at an approximate scale of 1:19,500 taken by the Società Aerofotografica (SAF) in 1965.[9]

Challenges in the interpretation of aerial photos

The linear features visible in the aerial photographs are in general slightly sinuous, shallow depressions. They are distinguishable from the surrounding spaces not only by the elevation change, but also by tonal variations (ranging from light grey to near black) that typically contrast with the prevailing tones of the immediate surroundings. Many features are characterized by alternation of tones along the axis or parallel banding of two or more tones along the axis. The variability in appearance is probably the effect of variable depth of the layers through which the features are cut, including the topsoil, the sand-and-gravel capping layer underlying marine conglomerate left by the receding ancient sea, and the deeper clay deposits formed in older geological epochs.[10] Difficulties such as dense concentrations of field boundaries and roads/paths, numerous random discolorations in the soil (including plow-disturbed ancient sites, abandoned stream courses, and former marshes), and sharp chiaroscuro effects of the sun that can cause the same area to appear very different in two photographs, create an elaborate visual screen through which the viewer must peer cautiously. The screen can obscure the presence of real features and suggest the existence of false ones. Identification of the segments on the ground is further complicated by the massive alterations of the landscape by heavy earth-moving equipment during the past half-century. The subjectivity of aerophotographic interpretation is evident in a comparison of the author's plan of visible segments with the equivalent plans published in previous studies.[11]

The ICA study

The ICA's early attempts to revisit the questions of the extent and nature of the division lines were constrained by spatial inaccuracies in the geographic database, which were derived from a limited GPS survey and a 5-m resolution IRS data, the highest-resolution imagery available at the time.[12]

The scale of the accumulated error is difficult to estimate, but certainly the positions of the lines as measured have absolute errors of at least 30–50 m (across photographs) and relative errors of up to 20 m (within photographs). Overall, the sources of spatial uncertainty in the initial GIS-based study of the chora were large enough to warrant a reassessment of the methods.

In order to reduce inaccuracies, an improved set of base data was required for more consistent and accurate geodetic and spatial control. The 1:10,000-scale topographic map of the Metapontino commissioned by Adamesteanu from ESACTA circa 1970 was used to re-georeference the aerial photos.[13] The map itself was prepared from aerial photographs over a limited area by a company specializing in photogrammetric mapping using precise geodetic controls, so a consistently high level of spatial accuracy can be expected. The spatial accuracy of the map was in fact tested and verified—even in the third, or vertical, dimension—with high-precision GPS equipment in the 2000–

[8] On the GAI, see Boemi 2003a, 31.

[9] On the SAF, see Boemi 2003b, 41.

[10] Most of the photographs are dated only by the year, which hampers any attempt to attribute patterns in the tonality of the features to specific seasonal interactions of climate, land use, and geological structure.

[11] E.g., Schmiedt and Chevallier 1959, 1000–1, Fig. 2 and 1003, Fig. 4; Chevallier 1971, 321, Fig. 8; Adamesteanu 1974a, 79; Guy 1995, 435.

[12] Carter/Thompson/Trelogan 2004.

[13] On ESACTA, see Boemi 2003b, 39.

2003 ICA field survey campaigns.[14] The ESACTA map also has the advantage of large scale, better than that of the photographs to be registered to it, and was close to them in time.

For the current study, the sixteen ESACTA map sheets were digitized with a drum scanner and then accurately georeferenced for display in the GIS.[15] The next step in the process was identification of the minimum set of photographs needed to locate all of the visible lines, determined by careful study: the twelve most central GAI photographs,[16] two of the five photographs of the unknown series over the coastal plain and lower marine terraces,[17] and one photograph from the Aeronautica Militare (AM) mission.[18] Many of the lines are visible in multiple photographs across multiple series, but the amount of time needed to georeference all of them accurately would have been prohibitive, and with questionable benefit. The GAI photographs were digitized at 1200 dpi, and the photographs of the unknown series and the lone AM photograph at 600 dpi, to ensure a uniform spatial resolution of 1.0 m/pixel for the georeferenced versions of all three series.[19]

To reduce the effects of radial distortion in the GAI photographs, each digitized photograph was cropped to leave a minimum of overlap and sidelap with neighboring photographs; in most cases the result was a square or rectangular image confined to the central portion of the photograph, which is the most geometrically faithful part. The photographs at the northern and western edges of the Metapontino cannot be handled this way, because they extend beyond the reach of the ESACTA map, and therefore GCPs (Ground Control Points) cannot be located evenly across their surfaces.

All of the scanned aerial photographs were then georeferenced with ENVI using the provided image-to-map warping process. They were projected to UTM/WGS84 with a spatial resolution of 1.0 m/pixel and exported in TIFF format, a standard format for image processing and display.

The georeferenced TIFF images were displayed in a GIS so that the line segments could be traced and saved in a dedicated data layer. Segments of division lines were identified as linear features oriented between 47.9° and 51.8° west of north and characterized by a uniform dark or light tone contrasting with the surrounding soil, by alternating light and dark tones, or by a combination of light and dark tones in bands parallel to the orientation. A display scale of circa 1:5000 was used consistently; this was found to be an effective scale for resolving the frequent conflict between the need to see most or all of a given segment in order to trace its orientation (especially true of the longer segments), and the need to position each segment's starting and ending points precisely. After initial tracing, each segment was then checked at a closer scale (generally 1:3000–1:4000), and adjustments to their vertices (endpoints) were made if necessary. Midpoints were avoided in the tracing of the segments unless absolutely necessary, since complex linear features (i.e., those with two endpoints plus at least one midpoint) lose their properties of direction/orientation in the GIS.

The GIS layer containing the line segments, examined in detail below, is the product of three successive readings of the georeferenced aerial photographs. The first reading produced a much larger collection of line segments than seen in Figure 31.2; many of these were given a provisional label as a marker for further scrutiny. The second and third readings served to confirm or reject most of the provisional segments. Thus the layer is somewhat subjective: during the study it was necessary to make decisions about the likely reality and likely antiquity of line segments based on the modern organization of the Metapontino and what is known of the ancient organization. Other researchers might agree to the reality of most of the segments so identified, but at the same time might see a different set of segments (see above).

The line layer also incorporates the results of the several excavations involving the lines, starting with Carter's excavation of longitudinal line L21 in the Pantanello necropolis (see Fig. 28.24) and continuing through the excavations for the SNAM petroleum/

[14] See Prieto 2000.

[15] The GIS platform employed in this study was ArcGIS 9.x by Environmental Systems Research Institute Inc. (ESRI). The original high-resolution scans (300 dpi) of the individual 21 x 18" sheets (roughly corresponding to 5.5 x 4.5 km) were downsampled to 150 dpi resolution. They were then georeferenced with ITT's ENVI remote sensing software (version 4.x) via the provided image-to-map warping process. The spatial resolution of the georeferenced sheets is 1.65 m/pixel.

[16] *Aerofototeca* negative Nos. 21314, 30305, 30309, 31660–61, 35092, 35095–96, 35383, 35778–80.

[17] *Aerofototeca* negative Nos. 79422–23.

[18] *Aerofototeca* negative No. 24284.

[19] A photogrammetric scanner would have obtained optimum contrast and tonal range (Ceraudo 2003, 96), but commercial scanners are much less expensive and more readily available.

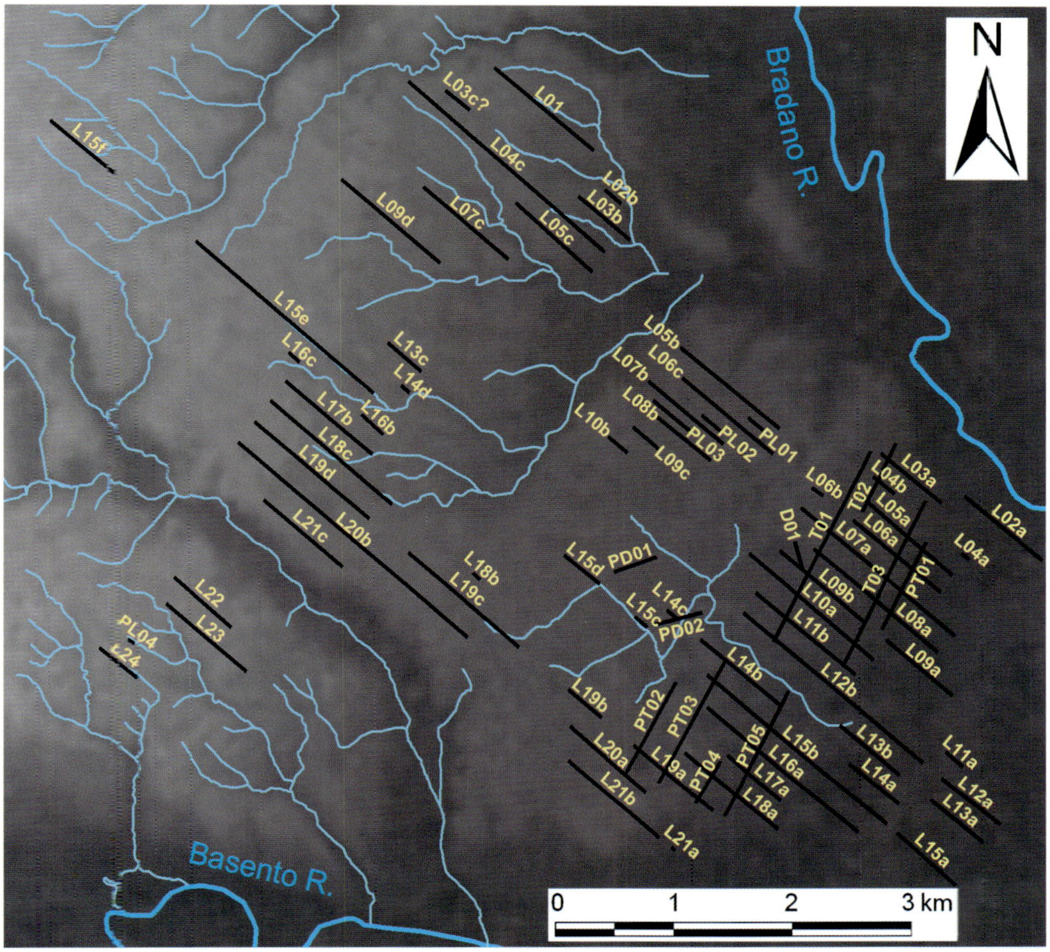

Figure 31.2 Plan of lines between the Bradano and Basento Rivers. L= longitudinal, T = transverse, D= diagonal; P = possible/potential. (AP)

methane pipelines and the Consorzio di Bonifica's irrigation pipelines in 1999–2002 (see Ch. 3 on methodology and Ch. 28 on the excavations). On the one hand, these excavations have confirmed the existence of many segments visible in the aerial photographs; on the other hand, and perhaps more importantly, they have also demonstrated the existence of segments where nothing is visible in the aerial photographs. Both these positive and negative results have been incorporated in the present study.

The maximum spatial error in the aerial photographs used in the study is approximately 20 m. Within the GIS, informal checks on the alignment of identical features in the overlapping portions of the aerial photographs revealed a typical misalignment of 5–25 m, with an average of 10 m; informal checks on feature alignment between map sheets and photographs demonstrated misalignment ranging from 5–20 m, typically 10 m or less.

Description of the Division Lines

Location, size, and orientation

The combination of visual inspection of the aerial photographs and various recent discoveries by archaeologists confirms the existence of 24 longitudinal lines covering approximately 10 x 6 km (6000 ha) of the chora between the Bradano and Basento Rivers (*Fig. 31.2*).[20] None of the lines appear unbroken across the entire breadth of the chora in the aerial photographs.

The longest feature is L15 with 6 segments (L15a–f). The distance between the southeast end of L15a and the northwest end of L15f is approximately 10.1 km, but the combined length of the individual segments comes to 5.9 km, or about 58% of this hypothetical length. At least 17 of the 24 linear features consist of 2 or more discontinuous segments (visible and/or invisible but excavated). All of the remaining features consist of a single segment each. The lon-

[20] *Ancient Territories* 1999; Nava 2000, 2003a, b; Carter/Thompson/Trelogan 2004.

gest visible segment of any feature measures 2.6 km (L20b), while the two shortest visible segments measure approximately 80–85 m (L04a and L18b).

There are 58 segments long enough to determine their orientations, which fall within the range 47.9–51.8° west of north, with a mean of 50.0° and a standard deviation of 0.74°. The aerial photographs reveal distinct traces of at least three transverse linear features (T01–03), all located on the eastern side of the Travaglione/Gravaglione stream in the Pizzica locality at an orientation of 26–27° east of north, including the feature partially excavated by ICA and the Soprintendenza in 1999 (T01; see Ch. 28).[21] The longest visible transverse feature (T01) measures approximately 1.8 km, the shortest (T02) approximately 650 m. The angle formed by the crossing of the longitudinal and transverse lines is 76–78° (typically 77°, as measured from the intersections of T01–03 with L04–12).

One diagonal line (D01) was located in 1999 (Fig. 28.6), a short distance northwest of the intersection of T01 with L09b. It is oriented 16° west of north, forming an angle of approximately 66° with L09b.

There are 15 additional segments that are difficult to discern in the photographs but nevertheless may have been part of the landscape organization: 8 longitudinal fragments (L02a, L03a, L06b, L08b, PL01–04), 5 transverse fragments (PT01–05), and 2 diagonal fragments (PD01–02). Two of the longitudinal fragments, L02a and L03a, are known only through excavation: according to the published plans, L03a lies approximately 200 m northeast of L04,[22] and L02a lies farther to the northeast by approximately another 200 m. Five of the fragments lie in the Saldone locality: two of them can be associated with known lines L06b and L08b, while the other three (PL01–03) fall between confirmed segments, at distances of 70–100 m from them. The eighth longitudinal fragment (PL04) lies roughly 210 m northeast of L24 in the Avinella locality. Its short length (about 70 m) suggests that it is a natural extension of the modern gully into which it flows, rather than part of the ancient landscape organization. The range of orientations for these 6 visible longitudinal segments is 48.9–54.6° west of north.

The 5 possible transverse lines appear on either side of the Travaglione/Gravaglione stream in the Pizzica locality. One (PT01) lies about 165 m southeast of T03, on the east side. The remaining fragments (PT02–05) all lie on the west side, separated by intervals of 260, 280, and 370 m. Their orientations range from 27° to 30° east of north. All are faintly visible in the aerial photographs, with the exception of PT05, which is traced from plans published by Nava.[23]

The two short diagonal features (PD01 and PD02), separated by approximately 565 m and oriented at 70.5° and 76° east of north, lie between L12 and L15 on the northwest side of the Travaglione. They are included in the discussion because they strongly resemble the majority of the visible longitudinal segments in appearance and are unnaturally straight.

Analysis of the Metapontine Lines

The approach taken here to the Metapontine lines—a large set of what appear to be mostly man-made features—is similar to that applied to the landscape divisions at the Greek colony of Pharos on the Croatian island of Hvar.[24] There the orientations of the individual linear features of the division system were analyzed in a GIS environment, an approach that demonstrated the existence of four distinct suborientations within the larger system. These suborientations were in turn traced to a surveying error at the intersection of the two principal axes that caused them to bend slightly over long distances, an error that was then propagated throughout the system. As at Pharos, the structural properties of the Metapontine line segments may preserve traces of the methods and sequence of the ancient division process and the influence of topography on the design.

Most scholars working in the Metapontino have identified L15 as the first longitudinal axis of the organized landscape between Bradano and Basento to have been established by the ancient surveyors because of its unusual length and central position (*Fig. 31.2*), from which all other longitudinal features could be marked in both directions.[25] A DEM (Digital Elevation Model) reveals that the full extent of the organized territory is indeed visible from virtually

[21] Three transverse lines are shown in the Pizzica locality in Nava 2002, 734, Fig. 6, and 741, Fig. 9. The northernmost line is T01. The other two appear to be conjectures (based on aerial photography?) because no excavated sites are marked at any point along them.

[22] Nava 2003a, 665, Fig. 5, and 2002, 741, Fig. 49. Note that L03a is shown in a different position relative to the Bradano in each report.

[23] Nava 2002, 733, Fig. 5.

[24] See Slapsak and Kirigin 2001, 571–73, 578–85.

[25] Schmiedt and Chevallier 1959, 999–1000; Uggeri (1969, 53–54) describes a stone "stele base" that he observed at the southeastern end of L15a; Guy 1995; Carter 1998b, 45.

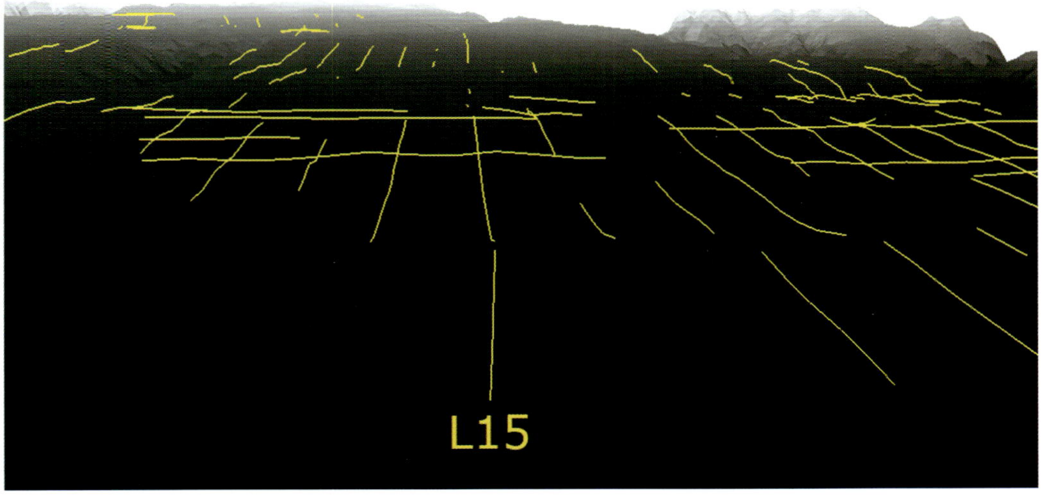

Figure 31.3 Lines between the Bradano and Basento Rivers: view from L15a towards L15f; compare Figure 31.2. (AP)

any location along L15 (*Fig. 31.3*).[26] From the coastal plain one can see, at least in theory, the low Pizzica terrace and the rolling plains of Lago del Lupo and Cogno del Prete behind it, and vice versa. In actuality, vegetation, from scrub to stands of trees, would have created some degree of interference.

The following analysis investigates several aspects of the Metapontine landscape. First, was the territory surveyed all at once, or piecemeal? Second, where did the ancient survey begin and what, if any, were the reference points? Finally, what relationship do the lines have to the traces of ancient roads that have been identified?

Orientation

The most evident characteristics of the Metapontine landscape organization are the orientations of the individual segments. The distribution of orientations across the landscape may be seen to contain artifacts of the survey process, as at Hvar, if one makes two fundamental assumptions: (1) the ancient survey process was founded on at least one primary sighting along the longitudinal axis (a second sighting along the transverse axis cannot be recovered because of the paucity of transverse segments); (2) if the surveyors made errors in the sighting of lines and the marking of subdivisions across the grid because of topographic irregularity, as on Hvar, the errors are likely to have been propagated from the origin in specific directions. (Given the relatively large room for error in ancient surveying techniques, one would not expect perfect alignment to be maintained among all segments in a given orientation.) This means that longitudinal segments visible in any given locality of the Metapontino are likely to have similar orientations, that is, the orientations of the individual segments are likely to preserve at least a small degree of unintended spatial patterning.

The numeric values of the longitudinal orientations (in degrees west of north) in the Bradano-Basento area were calculated in a GIS and added to the lines data layer (*Fig. 31.4*). The subparallelism of the segments is evident also in the distances between the ends of virtually any two neighboring segments (whether consecutive or not).[27] No two neighboring segments of longitudinal lines are perfectly parallel. In the best cases the misalignment is less than 1 m over distances between the measured intervals (i.e., along the axis of the longitudinal lines) ranging from approximately 450 to 1015 m. In the worst cases, by contrast, the difference between intervals ranges from 11.8 to 21.4 m, over approximately the same distances (approximately 540–1050 m).

The orientation values were employed as the sorting criterion of a new color-coding scheme for the

[26] Created from ASTER satellite data (bands 3n and 3b) with the interferometric routines in the Geomatica remote-sensing software by PCI Inc., using 1:50,000 IGM topographic maps for reference elevations; the spatial resolution is 15 m/px, and the vertical accuracy of ASTER-derived DEMs is on the order of 20 m (ASTER DEM 2009, 2, 16).

[27] This procedure was performed by tracing two interval lines between two neighboring longitudinal segments in such a way that the interval lines were approximately perpendicular to the longitudinal segments, and by calculating the difference in length between the two interval lines.

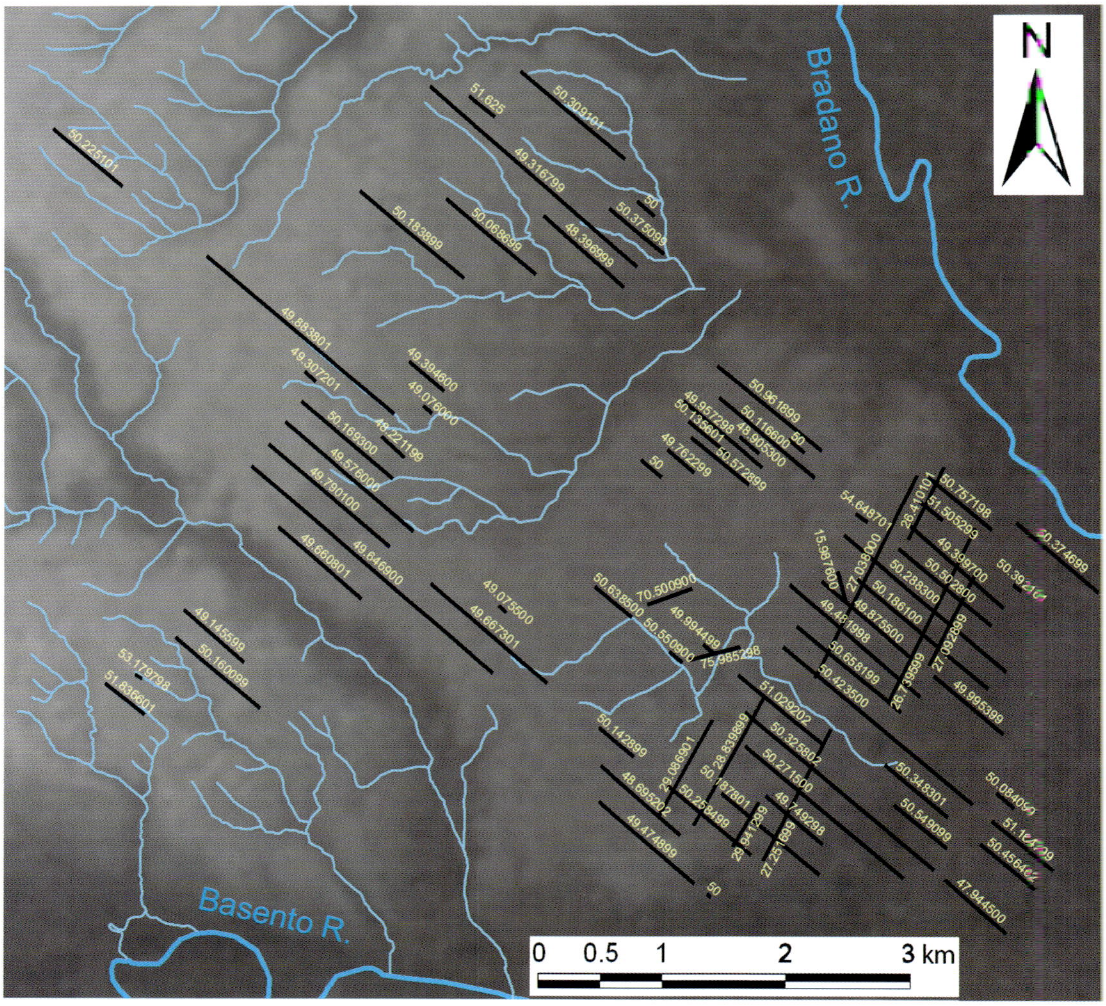

Figure 31.4 Lines between the Bradano and Basento Rivers: segment orientations. (AP)

line layer based on a statistical classification method provided by the GIS software. Several such methods were tested: Jenks/natural breaks, equal interval, defined interval, quartiles, and standard deviation. None of the methods revealed a clear pattern in the distribution of orientations in any part of the territory. There is a consistent but weak distinction (about 1° of difference) between the longitudinal features in the southeast (Pizzica locality) and those in the west (Demanio Campagnolo and Lago del Lupo localities), most apparent in the Jenks/natural-breaks classification (*Fig 31.5*). The differences in orientation are nevertheless so coarse (the color coding remains essentially the same when using any number of orientation groups, from 5 to 30, to represent the approximately four degrees of spread between minimum and maximum values) as to be possible artifacts of the segment-tracing process in the GIS, rather than the ancient surveying process. The largest deviations from the mean orientation—nine examples, all greater than 1° in either direction (i.e., standard deviation greater than ±1.38)—are scattered throughout the length and breadth of the Bradano-Basento area and thus demonstrate no pattern (*Fig. 31.6*). The orientations of the line segments appear to be randomly distributed, not patterned as expected.

The range of orientations of the 6 segments of L15, 47.9° to 50.6° (*Fig. 31.4*), suggests that this particular longitudinal feature was laid out in more than one stage. The most significant variation from the mean is the first segment, L15a (2.06° difference, three times greater than the next nearest value). Its deviation may be due to a shift in the sighting point from the coastal plain to the Pizzica terrace; it is equally possible, however, that the gentle southwestern-facing slope of that location created an optical distortion, making

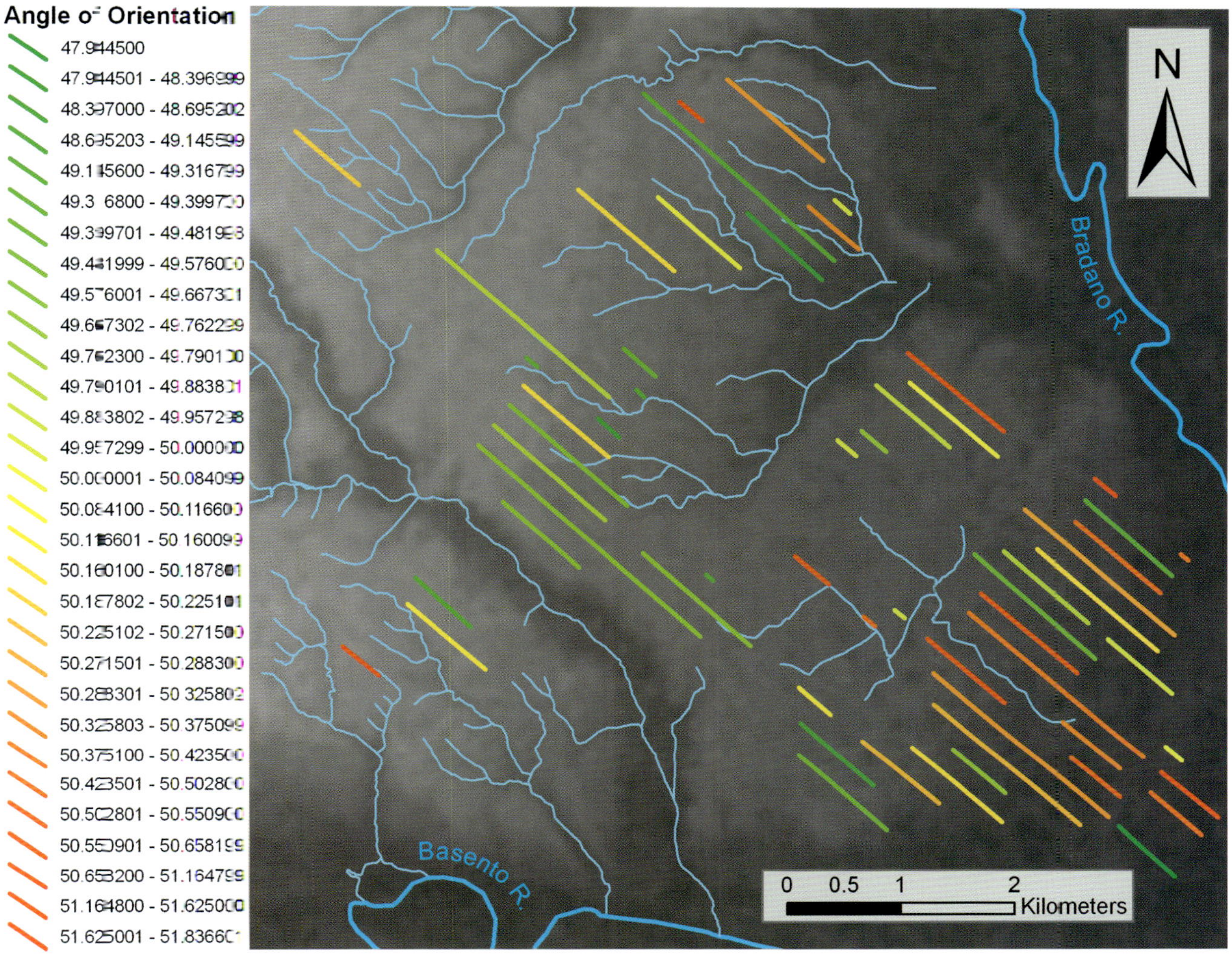

Figure 31.5 Lines between the Bradano and Basento Rivers: classification of segments by orientation (Jenks/natural-breaks method, 30 groups). (AP)

the crooked line appear straight to the surveyors. The same two scenarios (shift in sighting point and optical distortion) are also possible for segments L15c and L15f, which have a similar orientation that is noticeably different from the orientations of the other four segments, but in this case the optical-distortion scenario is less likely because the territory between L15c and L15f is essentially level. Therefore L15 was probably not surveyed in a single operation, but even so it may have played a specific role in the general landscape organization.

In sum, analysis of the segment orientations reveals no trace of a survey origin point or other reference point, or of a survey process as at Pharos. The absence of patterns in the orientations of the lines may indicate that the territory was organized either all at once with poor geometric control or in very small increments involving small (sublocality) areas. It is equally possible that the process of tracing the segments in the GIS is so imprecise as to generate small but significant errors in segment orientation, which confound interpretation. Greater precision is impossible without resorting to ground-based survey using an electronic theodolite (total station).

Interval distance

Another potentially meaningful characteristic of the Metapontine landscape organization is the much-discussed distance between longitudinal segments. Since their discovery, several proposals of a regular, standard interval between the longitudinal lines have been made, typically as part of a projected larger grid-like pattern of plots of identical dimensions, following the known example of such a system at Chersonesos

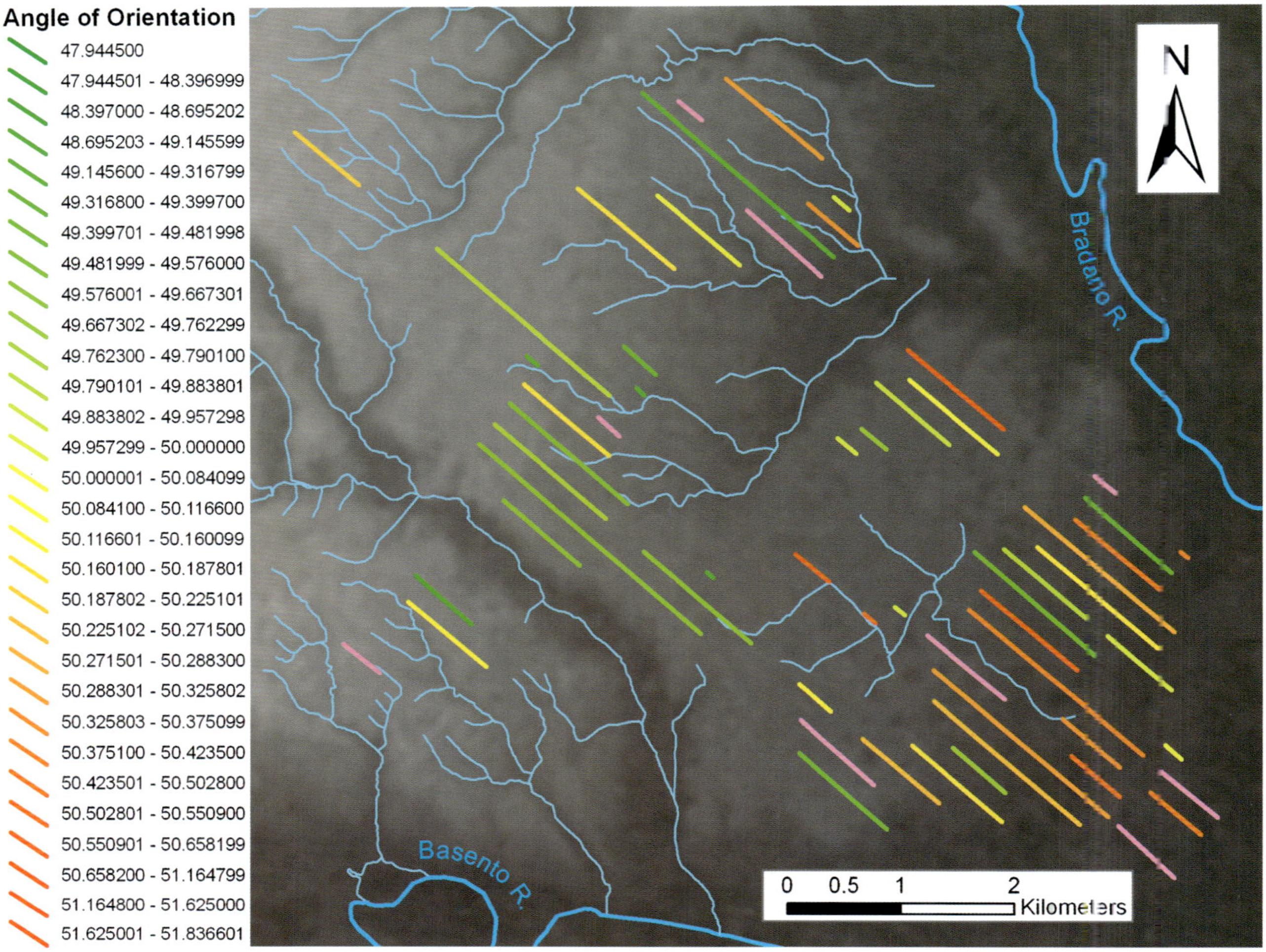

Figure 31.6 Lines between the Bradano and Basento Rivers: classification of segments by orientation (Jenks/natural-breaks method, 30 groups), with nine furthest outlying cases indicated in purple. (AP)

in Crimea.[28] The variety of the proposed standard Metapontine intervals casts doubt on the very idea that the system followed a standard at all. Variability in the interval between longitudinal lines, especially if localized, could be evidence of a locality-by-locality survey process or a piecemeal survey process performed over the course of several years (or even several generations). Localized variations in interval distances might also reveal the locations of survey reference points. For the purposes of this study, the distances between segments were measured on the perpendicular where feasible (i.e., when the neighboring lines were perfectly or nearly parallel), and otherwise as nearly on the perpendicular as possible. In cases of neighboring segments separated by empty spaces of approximately 400 m, the empty spaces were presumed to have been part of the organized landscape, and the measured distances were divided by the appropriate factor (2, 3, etc.) based on the number of missing intermediate features.

The linear features are separated by a broad range of intervals, approximately 179–243 m, with all but two of the 104 intervals measured falling in the range 185–235 m (*Fig. 31.7*). If the single shortest and longest measured intervals are treated as outliers and excluded, the Jenks/natural-breaks method divides the remaining 102 intervals into the groups shown in Table 31.1.

Though the mean interval, 207.8 m, is remarkably close to the most commonly accepted standard intervals for the lines between the Bradano and the Basento (205–210 m), the large standard deviation of the sample (12.4 m) causes approximately 70% of the measured intervals (i.e., 71 of 102) to fall in the wide range of 195.4–220.2 m (reflected in the near total

[28] Schmiedt and Chevallier 1959 (210 m); Uggeri 1969 (205 m); Adamesteanu and Vatin 1976 (210 m); Guy 1995 (200 m). For Chersonesos, see Carter et al. 2000, 713.

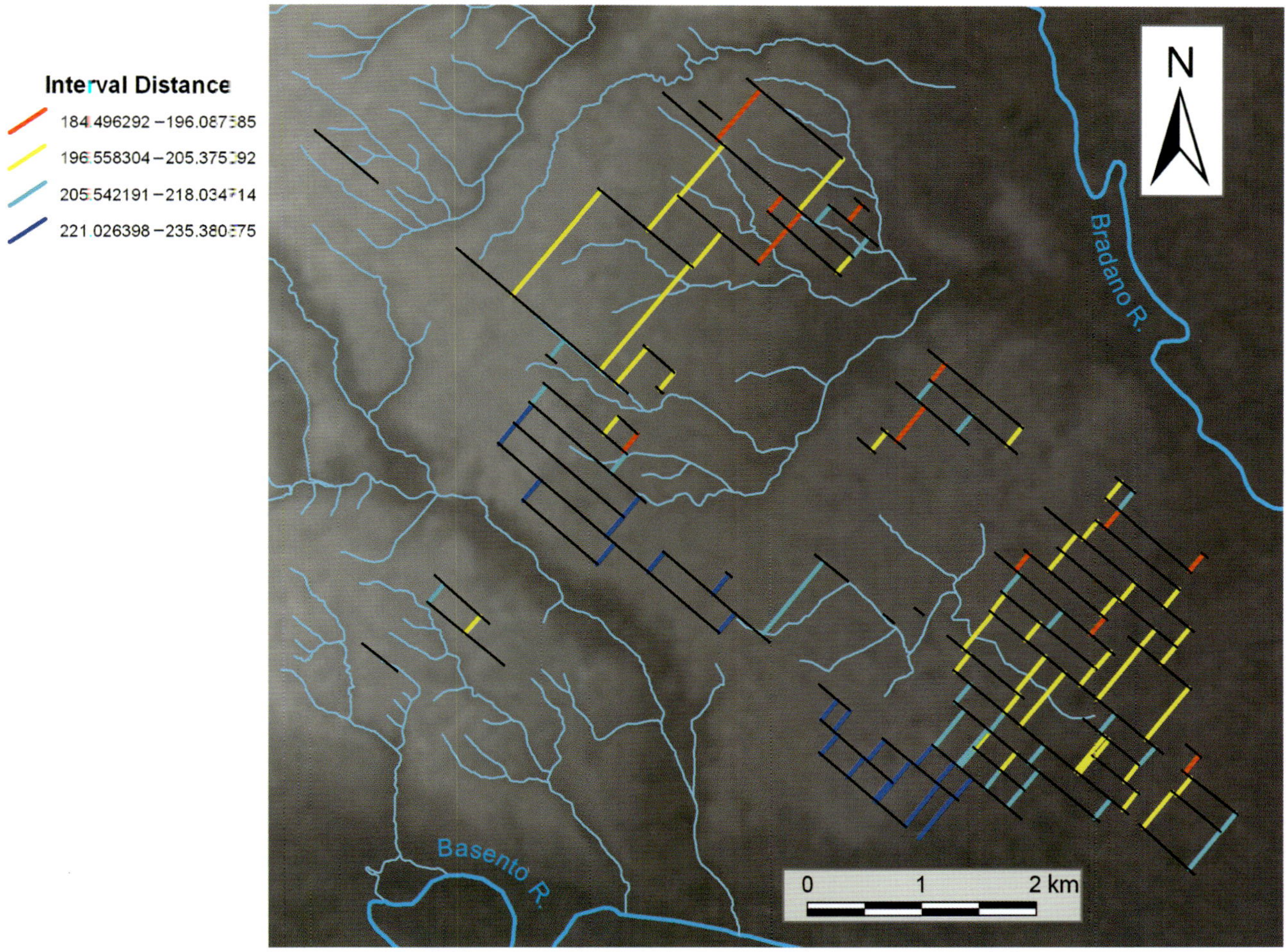

Figure 31.7 Lines between the Bradano and Basento Rivers: classification of interval distances (Jenks/natural-breaks method, 4 groups). (AP)

immobility of the counts in the first and last interval categories between the grouping of five and the grouping of four). The interval groupings, furthermore, are anything but tightly defined, with values spanning ranges from roughly 6 to 14 m. The interval groupings in Table 31.1 suggest that the Metapontine surveyors used several approximate intervals (200 m, 205 m, and 225 m) but could not—or simply did not—strive for perfect regularity.

5 groups	Count	4 groups	Count	3 groups	Count
184.5–193.8 m	9	184.5–196.1 m	12	184.5–203.2 m	48
195.0–201.6 m	32	196.6–205.4 m	43	203.8–218.0 m	33
202.3–208.3 m	24	205.5–218.0 m	26	221.0–235.4 m	21
208.8–218.0 m	16	221.0–235.4 m	21		
221.0–235.4 m	21				

Table 31.1 Jenks/natural-breaks groupings of interval distances.

The distribution of the intervals appears to follow a vague pattern, most evident in the group of four natural breaks (*Table 31.1, Fig. 31.7*). The first two groups (shortest intervals, red) are localized generally on the north side of the central (longest) feature (L15), while the fourth group (longest intervals, dark blue) is limited to the space defined by L15 and the Venella valley (L21). The intervals of the third group (205.5–218.0 m) are scattered on both sides of L15, but concentrate immediately to the south of it. In this arrangement one may see four broadly defined parts of the landscape organization—L01–09, L10–15, L15–18, and L19–21—but clearly one cannot, like Guy, speak of multiple overlapping systems with rigidly maintained intervals.[29]

[29] Guy 1995.

L15 appears to have played a role in the survey that was not shared by the other longitudinal features. The survey may have begun on opposite sides of the chora, at both the Venella and the Bradano, and then converged towards the center; or it may have been performed in two stages (Bradano to L15, L15 to Venella). The approximate positions of the longitudinal features may have been marked over distances of several hundred meters at a time before the main sighting axis was moved to another location.

Analysis of the distances between segments suggests that at least three intervals were used regularly when surveying the lines in the chora, so that the concept of the single canonical interval can be abandoned for good. It is possible that L15, the central line, was the point at which as many as four separate survey processes converged or, perhaps, from which they emanated.

Discontinuity

The Metapontine lines appear discontinuously across the landscape. Many of the features are composed of distinct segments separated by distances ranging from just 230 m (Bradano-Basento L09a–b) to about 4.6 km (Bradano-Basento L13b–c). There are also large gaps in the overall organization, in which most or all lines are completely absent. The largest of these is the 3.0 x 1.5 km gap in the Demanio Campagnolo/Lago del Lupo area, where L01–18 are interrupted.

These gaps in the Metapontine landscape organization pose a significant problem for interpretation of the system as a whole. There are several areas where no segments are visible in the aerial photographs, and therefore one is led to believe that the system did not extend into those areas; yet excavation in the same areas has revealed the presence of buried segments.[30] That parts of the lines may not appear in the aerial photographs should challenge the completeness of the information provided by the aerial photographs.

Still, there are several reasons for which one may have confidence that the aerial photographs, in particular the GAI series, exhibit a complete picture. First, in many areas preserved segments of lines have probably been replaced by modern field boundaries, roads, or channels, so that underlying ancient remains cannot be verified except through excavation. Second, the GAI photographs were taken in 1954–55, when mechanized agriculture had barely been introduced into the Metapontino; since mechanical alteration is the only agent capable of removing all traces of line segments in open, elevated areas, one may reasonably expect that the absence of line segments is not the product of mechanical digging or leveling. Third, according to the published plans and personal communications with the excavators, all of the longitudinal segments excavated by the Italian authorities in 2000–2002 either coincide with segments visible in the aerial photographs (Bradano-Basento L06a, L08a, L09b, L11b, L12b, L14b, L15a–b, L18a; Basento-Cavone L18a, L19a, L23) or lie in relatively isolated or marginal areas (Bradano-Basento L02a, L03a; Basento-Cavone L02c).[31] That none of them occupies an apparently empty space between two visible segments is significant. The absence of line segments in many broad, essentially flat areas with a relatively low percentage of field fragmentation is therefore likely to be a real phenomenon (although exceptions may one day be encountered). Finally, the presence or absence of lines may have a relation to the local topography, a question addressed below.

The last two factors also suggest that the three clearly visible transverse features (T01–03) are unique: no segments following a similar orientation are evident anywhere in the chora northwest of Pizzica. Though they are surely part of the system, they therefore cannot be equated to the longitudinal features in terms of a structuring function in the chora.

It is clear that graphic reconstructions, based on the aerial photos alone, in which the lines stretch continuously across the landscape and divide it into regularly sized plots, are largely fanciful. But the cause of the discontinuity has yet to be discovered, and the possibility remains that even more structure existed than has survived into modern times.

[30] The examples published to date are Bradano-Basento L02a (Nava 2003a, 665, Fig. 5) and L03a (Nava 2003a, 665, Fig. 5, and 2002, 741, Fig. 49). The excavated line in the Pantanello necropolis (see Carter 1998b) also technically falls into this category, since the relevant feature (L21b) disappears in the aerial photographs just a few hundred meters from the site of the necropolis; a token segment (L21a) was inserted at the proper location in the GIS layer.

[31] The matching of segments between the aerial photographs in GIS and the Italian excavation data is a problematic process because of the very small scale of the published plans and the paucity of information included in the summary reports, which are the only sources available so far.

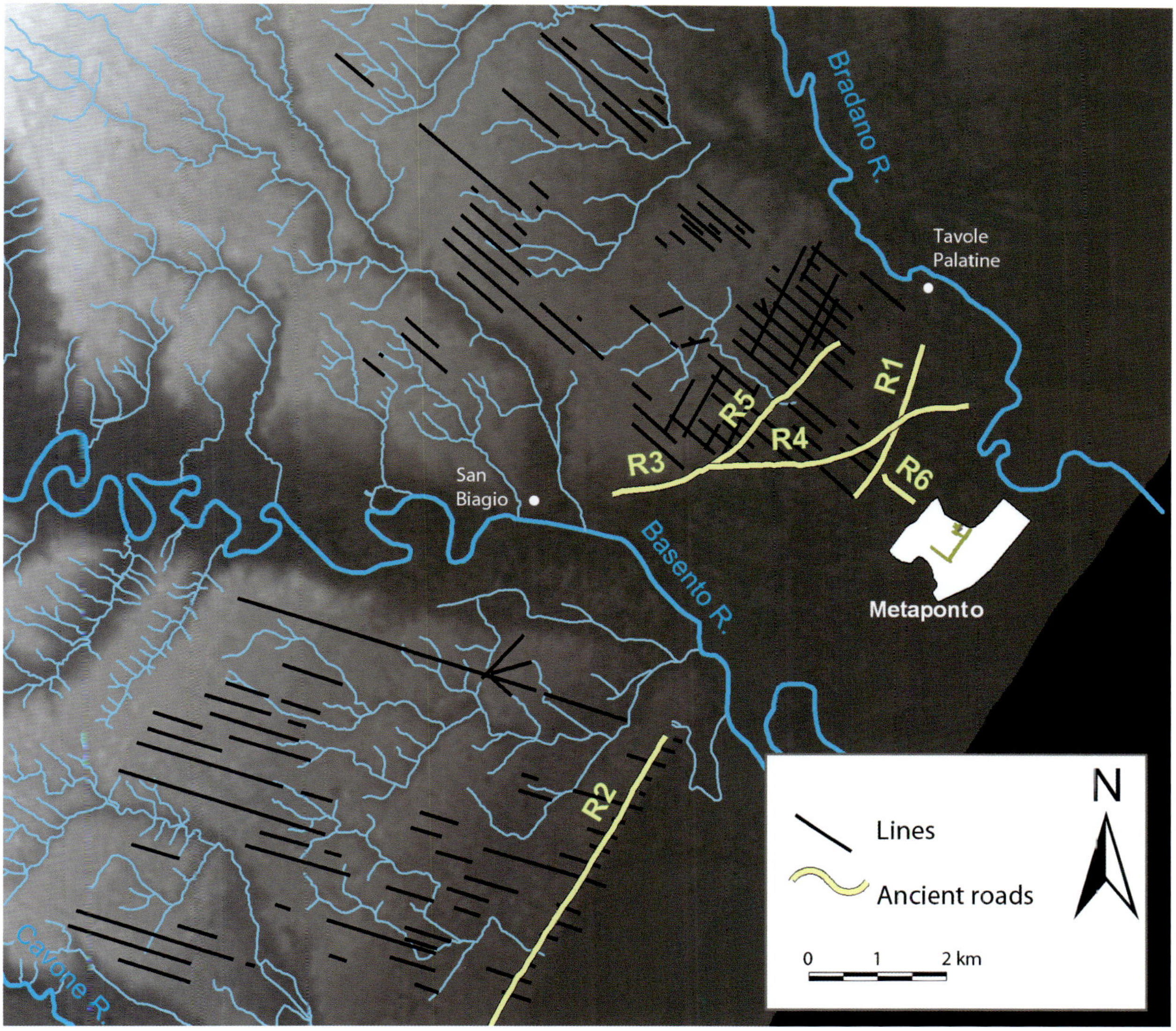

Figure 31.8 Lines and ancient roads in the chora of Metaponto. (AP)

Roads

Six ancient road segments were identified (*Fig. 31.8,* R1–6). Schmiedt and Chevallier noted at least two segments of a road running parallel to the coast, one each on either side of the Basento (R1 and R2). R1 served two functions: (1) it connected the main urban necropolis in the Crucinia locality and the Temple of Hera at Tavole Palatine in the immediate vicinity of the asty,[32] and (2) it gave the asty access to the broader Ionian coastal plain to the northeast. R1 ran across the Bradano, and (along with R2) to the southwest, across the Basento. These segments belonged to a very long road connecting settlements all along the Ionian coast, which Schmiedt and Chevallier dated from prehistoric through historical times. Adamesteanu, however, saw this same segment as belonging only to the historical road linking the Italiote colonies from Taras to Rhegion; he believed that the prehistoric version lay slightly further inland.[33]

Other segments of the ancient roads lack the geometric arrangement characteristic of the lines previously discussed. Their irregular and meandering courses follow the topography rather than override it, with the visible segments connecting features or areas across terrain that is generally elevated and sloping.

[32] Adamesteanu (1973, 53) observed that a 1-km-long stretch of this segment was lined with tombs dating from the 6th c. BC to the 2nd c. AD.

[33] Adamesteanu 1963, 50–52; 1967c, 96. The same feature has continued into recent times, albeit in varying incarnations. In the late 19th c. it was a simple dirt road (Lacava 1891, 120), and most recently it has become the Italian paved state highway SS106 "Jonica," covering the approximately 500 km from Taranto to Reggio di Calabria. See also Lugli 1963, 30.

R3 is the "Basento Road," documented by ICA as part of the Pantanello necropolis excavations.[34] Its continuation to the east (R4) is easily traced in the aerial photographs. In 2003 the Nava reported the discovery of a branch of this same road to the northeast (R5), which crosses the Pizzica terrace towards the extramural Temple of Hera (Tavole Palatine). The existence of this branch has been detected only by excavation, presumably at the very northeastern end, at *sito* M and *sito* X.[35] As Carter has demonstrated at the western end of the Basento Road (R3), the ancient remains of segment R5 probably lie beneath the modern farm road (*tratturo*) visible in the aerial photographs (thus R5 has been traced along the line of the modern road).[36] R3 lay immediately below the monumental sanctuary at San Biagio: via R5 it connected the sanctuary and the lower Basento valley to the Pizzica terrace; via R4 it led to the coastal plain where the urban necropoleis and Temple of Hera (Tavole Palatine) stood. Both R4 and R5 likely fed into a road skirting the south bank of the Bradano, which led into the mountainous zone around modern Matera and gave the colony access to both indigenous communities and the Adriatic seaboard.[37] R3 continued up the Basento valley to the west-northwest, into the mountainous interior of ancient Lucania and beyond the Apennines, to the Tyrrhenian coast. It likely skirted the bottom of the marine and alluvial terraces in the valley, as does the modern SS407 highway to Potenza. R3 was therefore probably an important exchange route linking the indigenous centers of the interior (and possibly the Etruscan and Italiote cities of the Tyrrhenian coast) with Metaponto.[38]

R6 is a short, faint segment immediately northwest of the urban walls that, because of its alignment with Plateia A of the urban grid, should be identified as a road—perhaps the primary road—connecting the asty and the chora. After traversing circa 700 m of land outside the Settembrini (Western) Gate of the asty in an essentially straight line (i.e., the segment follows the orientation of the urban and rural longitudinal features, although it does not quite fall along the line of the nearest longitudinal feature, L13—see *Figs. 31.2* and *31.8*), R6 curves gently to the north in order to feed into R1, the coastal road. (The missing northern intersection of R1 and R6 likely lies beneath the modern SS175 highway and/or the modern channel that abuts it to the east.) The R1 and R6 roads would have allowed the urban residents to reach the extramural sanctuary at Tavole Palatine and cross the Bradano.[39]

The asty and the Basento valley must also have been linked in antiquity. It is tempting to imagine a road parallel to R6 which exited from a gate further to the southwest, at the northwestern end of Plateia K, the longest plateia (*Fig. 31.8*), intersected R1, and probably hugged the base of the lowest terraces to merge with R3 when the latter entered the Basento valley about 4.5 km to the west. No such credible feature is apparent in the aerial photographs.

It is important to note, finally, that the visible fragments of roads are frequently intersected by line segments (*Fig. 31.4*).[40] R2 is intersected by L03, L04, L05a, L07a, L08a, L09a, L10a, L11a, L12a, L14a, L15a, L16a, L17a, L18a, and L19a. R4 is intersected by L12a, L13a, L15b, and L16a. R5 is intersected by L07a, L08a, L09a, L12b, L15b, L16a, L17a, and L18a. The chronological relationships between the roads and the linear features are, however, difficult to establish with certainty at this point. Ceramic evidence indicates that the abandoned streambed R3 was utilized as a road and place of burial by the late 7th or early 6th century BC,[41] but without further evidence it cannot be assumed that all of these roads were in use before the creation of the lines.

Topographical Relationships

All prior studies of the Metapontine landscape organization have proceeded under the assumption that the territory is devoid of relief, as though the local topography had no influence on the condition or function of the various features created in the territory. Consequently the geometric aspect of landscape organization became the principal focus of inquiry. The

[34] See Carter 1998b and Ch. 28.

[35] Nava 2003b, 131, Fig. 249, and 132–33.

[36] Carter 1998b, 34.

[37] No traces of such a road are visible in the aerial photographs. One may presume either that the modern SS175 highway connecting Metaponto and Matera has usurped its lower extent, or that it has been obscured by alluvial accumulation.

[38] On this artery, see Schmiedt and Chevallier 1959, 1006; Buck 1975, 98–102; Mertens 1980a, 37.

[39] See also Mertens 1999, 249.

[40] Schmiedt and Chevallier (1959, 997) first observed this phenomenon. Only one of these intersections has been published in detail, the "crossroads" at Pantanello, where L21a crosses R3 (Carter 1998b).

[41] Carter 1998b, 33–46; De Siena 2001a, 763.

Figure 31.9 Ground-level view of slope of marine terraces south of Cavone valley, looking SW. (AP)

Figure 31.10 Three-dimensional GIS view of slope of marine terraces north of Cavone valley using ASTER-derived DEM (5x vertical exaggeration), looking NE. (AP)

Metapontino is a land of contrasts, however: high, sloping terraces of marine origin loom over essentially flat river valleys and a flat coastal plain. An investigation of the relationship(s) between the topography of the Metapontino and the features of the colonial landscape organization, both planned and unplanned, reveals that topography played an active role in the history of the landscape.

The engineered component of the Metapontine landscape organization consists almost exclusively of the longitudinal line segments. Figure 31.8 shows that their most prominent characteristic is a downward slope from northwest to southeast that follows exactly the natural decline in elevation from the middle marine terraces, at the level of the modern settlement at Bernalda, down to the edge of the coastal plain and the orientation of the subsurface geology (see above). The mean declination, or difference in elevation, is approximately 12 m/km (1–2°).[42] This slope is readily perceptible to an observer standing almost anywhere in the Metapontino (*Fig. 31.9*), but it is virtually invisible to the naked eye in a GIS equipped with a three-dimensional viewer. Slight exaggeration of the vertical scale in the viewer, however, displays the general downward slope clearly (*Fig. 31.10*). The slope is also readily apparent in a

[42] As measured along L15: 128 masl at the western end of L15f, 8 masl at the eastern end of L15a.

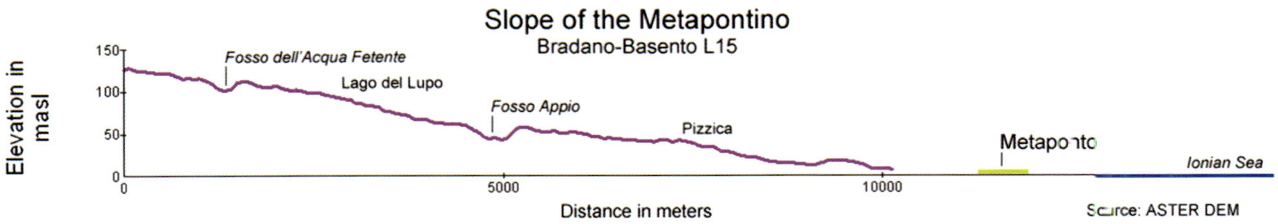

Figure 31.11 Slope of the marine terraces between the Bradano and Basento Rivers: longitudinal (NW-SE) section through ASTER-derived DEM. (AP)

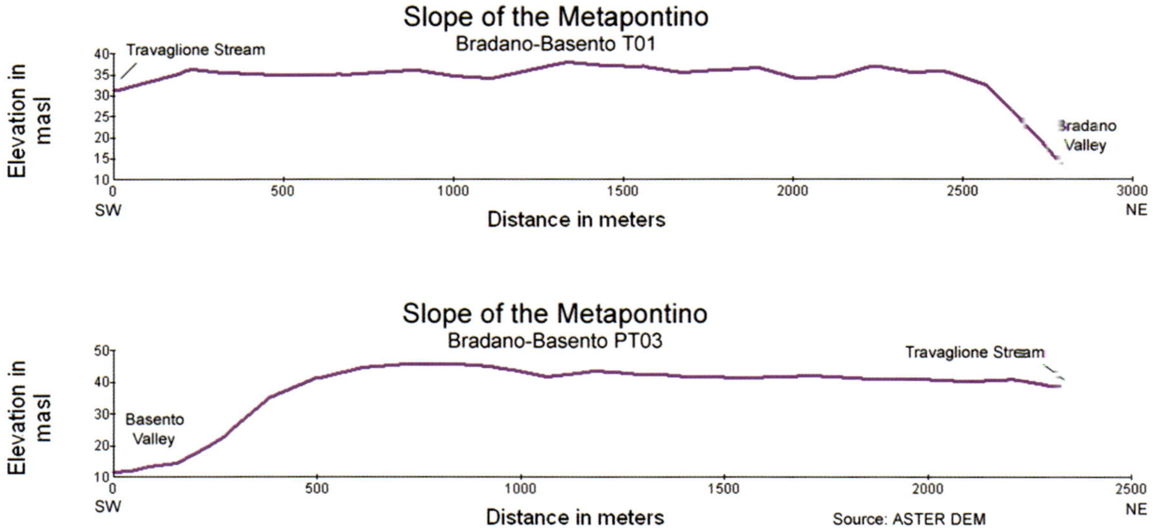

Figure 31.12 Slope of the marine terraces between the Bradano and Basento Rivers: transverse (SW-NE) sections through ASTER-derived DEM. (AP)

section through the ASTER Digital Elevation Model (DEM) of the Metapontino, taken along the total length of L15 (*Fig. 31.11*).

There are no linear features visible in any of the river and major stream valleys, a result that flatly contradicts Guy's assessment.[43] This absence is probably only an apparent one, given the numerous isolated farmhouses and necropoleis discovered in the river valleys and in the Venella valley[44] (attested also in the chora of neighboring Herakleia by the Herakleia Tablets).[45] Roads, paths, ditches, and other typical features of an organized landscape doubtless existed in the Metapontine river valleys, but they have been buried over the centuries by meters of accumulated alluvial fill and are therefore no longer visible from the air or on the ground.[46] Whether or not the same longitudinal and transverse orientations were maintained by the hypothetical features of the river valleys is impossible to determine.

More importantly, there is arguably a relationship between the longitudinal lines and the natural watercourses that crisscross the marine terraces, although it is difficult to describe. At first glance it appears that the lines generally occupy spaces in which streams are absent (*Fig. 31.2*). Closer inspection, however, reveals that many line segments intersect streams or gullies or come very close to doing so: Bradano-Basento L01, L03b–c, L07c, L09d, L13c, L15c–d and e–f, L16b–c, L17a–b, L18a and c, L19a and c–d, L20a–b, L21b–c, L22, L23, L24; Basento-Cavone L02b, L07a and d, L08b, L09b, L10b, L12c–d, L14c, L15d, L16c–d, L17b, L18b, L20a, L22a–b, L23. More significantly, perhaps, several segments of streams or gullies share the northwest–southeast orientation of the lines and even appear almost to serve as line segments: the Travaglione stream coincides neatly with Bradano-Basento L13, the course of the Fosso del Lavandaio next to Marconia matches the orientation and position of Basento-Cavone L16, and segments Bradano-Basento L13c, L15c–d and f, L17a, L22, L24, and Basento-Cavone L09b coincide with smaller streams.

[43] Guy 1995.

[44] Nava 2003a, 668; Abbott 1997, 285–87.

[45] Arangio-Ruiz and Olivieri 1965, 645.

[46] See Abbott 1997 and Ch. 2 for the geomorphological analysis.

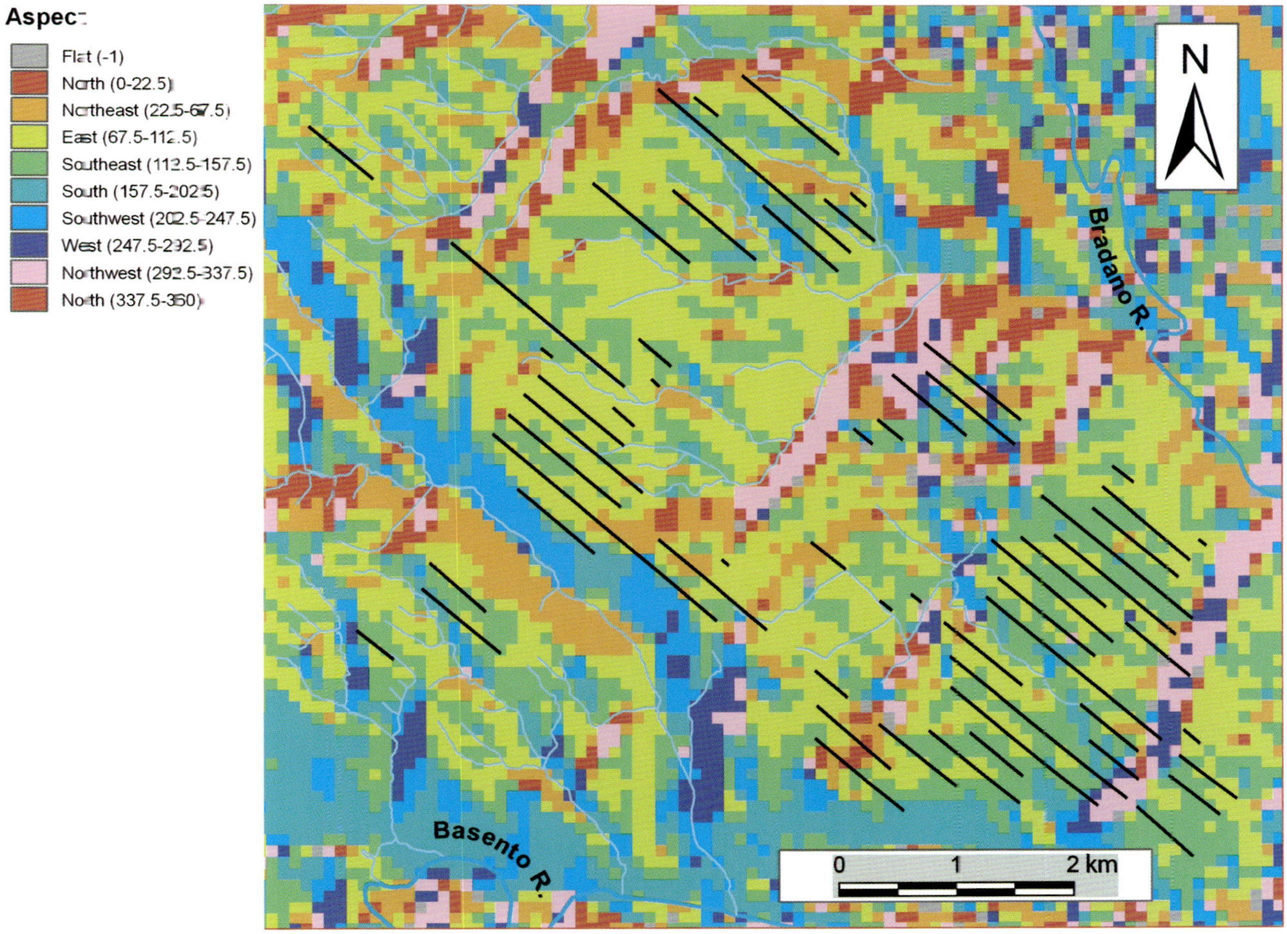

Figure 31.13 Aspect of the Metapontino between the Bradano and Basento Rivers, with line features overlaid. (AP)

Noticing this same phenomenon five decades ago, Schmiedt and Chevallier proposed that the hydrography of the modern Metapontino had been shaped in large part by the lines: many of the carefully excavated ditches were transformed by centuries of erosion into uncontrolled seasonal torrents that carved new paths down the slopes.[47] In this case the linear ditches would not necessarily reflect the hydrographic situation of the ancient Metapontino. The Metapontines, in this view, would have created them as land divisions without considering their ultimate influence on the regional hydrography.

The contrary case—i.e., that the longitudinal division lines were created in a landscape already dissected by fast-flowing watercourses—rests on the assumption that the segments were designed to interact with the regional hydrography by channeling water into the existing streams/gullies, probably to remove it from the terraces. The apparent termination of many segments upon or before intersection with a stream seems to bolster this case. There are many examples, however, of segments that intersect streams and continue beyond them (e.g., L19c). Therefore the first case—that the lines were created in a relatively smooth landscape and subsequently contributed to its dissection—is the more logical.

The transverse lines (T01–03 and PT01–05) also have a clear relationship with the topography of the Pizzica terrace. The landscape in that area varies from essentially flat to a gentle slope down to the northeast, a difference in elevation of approximately 7 m/km, less than one degree (*Fig. 31.12*).[48]

The transverse segments follow the orientation of major changes in aspect (i.e., the direction on the

[47] Schmiedt and Chevallier 1959, 1004–6. J.T. Abbott (pers. comm., 2003) believes that most of the seasonal streams/gullies are of relatively recent origin because the dissection of a landscape by gullying typically occurs over the course of a few generations, rather than centuries or millennia, as a result of intensive settlement.

[48] As measured along T03: 30 masl at the southern end, 20 masl at the northern end.

compass towards which a sloping piece of ground faces) created by large folds in the terraces or gullies cutting through the terraces (*Fig. 31.13*).[49] This last characteristic is another indication that the Metapontine surveyor-planners intentionally designed the lines to coincide with prominent aspects of the topography.

The apparent termination of many line segments upon intersecting with R1, R3, and R4 (*Fig. 31.4*) was probably dictated more by topography than by design: these roads follow the edges of the terrace slopes, beyond which point the ditches would have turned into undisciplined gullies.

The generally parallel arrangement of the longitudinal features indicates that the planned organization of the Metapontine landscape was probably conceived and executed as a unified project. The evidence of the aerial photographs does not, however, support the assumption of Adamesteanu, Uggeri, Vatin, Chevallier, and Guy that the entire chora of Metaponto was crisscrossed by a continuous network of geometrically arranged linear ditches dividing the territory into regularly sized units, as Uggeri reconstructed it.[50] The longitudinal lines appear from the aerial photos to have been created in disconnected segments across the chora. Nevertheless, the close alignment of the segments composing almost any given line suggests that the surveyors at least sighted and plotted the full extent of every longitudinal line; these conceptual linear entities were subsequently given physical form as roads and ditches.

Conclusions

The division lines are a complex archaeological and topographical phenomenon requiring an approach that takes into account more than just their geometric form. The three-dimensional geospatial model of the lines that was created for this study, the first of its kind, combines the most accurate and complete cartographic and photographic coverages available, resulting in a uniform horizontal and vertical accuracy of approximately 20 m. The model indicates that the longitudinal lines were not laid out in a rigid pattern at a regular interval, but rather at a variable interval with minor variations in orientation. This result naturally suggests consideration of ancient surveying techniques and practices, which have similarly been evaluated using geospatial software at Hvar, although the surveying system at Metaponto proves more difficult to reconstruct.

A very small number of transverse features were created only in the northern Pizzica locality. The angle of intersection between these transverse features (T01–03) and the longitudinal features therefore has no particular significance for the landscape organization as a whole. The orientation of the transverse features appears in fact to be a function of the prevailing geological structures in that area, much as Folk has independently concluded (see Ch. 1).

The remarkable straightness of the lines suggests that a sighting instrument—a simple *dioptra* or a device like the Roman *groma*[51]—was employed in the initial survey. The absence of a full set of transverse features, as well as the non-orthogonal orientation of the few visible ones, at first argues against the use of a *groma* at Metaponto. But then one must ask how so many lines, separated by hundreds of Greek feet and extending as far as 10 km from the coastal plain, could have been positioned with such a narrow range of variation in orientation, only four degrees. A *dioptra* alone would probably not have been sufficient, since it was not equipped with a means for determining parallelism. The *groma* is a possible solution to the problem: if a guideline more or less parallel to the coast was marked on the ground, the Metapontine surveyors could have sighted each longitudinal line as a perpendicular to the guideline. The Metapontine situation suggests that a *groma*-like instrument may have existed in the 6th century BC, long before its earliest attestation at Pompeii.[52]

The roads of the ancient territory were positioned primarily in response to topographic concerns. The coastal road (R1–2) runs along the lower edges of the first marine terraces on both sides of the Basento, where the terraces transition into the coastal plain proper.[53] The main portion of the Basento Road

---

[49] The aspect of the Bradano-Basento area was generated from Shuttle Radar Topography Mission (SRTM) digital elevation data distributed by the United States Geological Survey, which has a spatial resolution of ca. 100 m/px. The author wishes to thank Dr. Peter H. Dana of the Department of Geography, The University of Texas at Austin, for his help in obtaining and formatting the SRTM DEM data.

[50] Uggeri 1969, 55.

[51] On the *dioptra*, see Briant 1994 and Goodfield and Toumlin 1965. On the *groma*, see Dilke 1971, 66, 86–87; 1987, I:213–14; Lewis 2001, 15–16, 38, 120–33.

[52] Adam 2005, 10.

[53] The apparent interruption of the coastal road between R1 and R2 is probably due to the same geomorphological activity that has buried many sites in the river valleys (see Ch. 2): the portion of the road crossing the river valley does not show through the accumulated alluvium.

(R3–4) follows the gentlest slope along the upper edge of the first marine terrace before descending into the coastal plain, while segment R5 cuts across the relatively flat terrace below Pizzica to provide a more direct and level route to the Bradano valley. R6 was designed to join the urban center to the coastal road across the gentle slope of the coastal plain.

The diffusion of geospatial technology opens up new paths for future research on the division lines. The quality of the air-photo registration could be improved with a more accurate cartographic base, while the accuracy of the positions and orientations of the surviving segments of the lines would certainly be improved by tracing them on the ground with a GPS receiver and/or an electronic theodolite (total station). The questions raised in this study about the extent and coverage of the division-line features according to the aerial photographs should also be addressed. A brief but intense campaign of targeted trenching and/or geophysical prospection would provide hard evidence about what lies beneath the surface in the apparently sterile areas between the visible segments of a single longitudinal line, information that would in turn clarify the function(s) of the lines and their relationship to the Metapontine chora as a whole.

VII. The Roman Period

32
Imperial and Late Roman Settlement in the Metapontino

Erminia Lapadula

This chapter covers the cultural landscape in the Metapontino beginning roughly with the Roman Imperial period.[1] As noted by Carter (Ch. 27), the settlement patterns in the chora had changed substantially from the peak of the Greek period and the material evidence recovered by the ICA survey is significantly less.

The Early Imperial Period (50 BC–AD 50)

As shown in Table 32.1 and Figure 32.1, there are 11 sites dating to the Early Empire. Because only a few fragments of pottery were recovered from most of these sites, it is difficult to define their exact chronology and function with confidence. Two sites (105 and 428) should likely be considered rural settlements devoted to agriculture and related activities such as ceramic production according to the model outlined by previous research and studies of the Metapontine territory. Site 428 (Pantanello) lies on the lower terraces dominating the Fosso Pantanello, while site 105 (Sant'Angelo Vecchio) sits in the lower Basento valley, not far from the river. The size, density, and typology of the scattered archaeological material indicates that these settlements date to the Late Republican and Augustan periods. Both sites were indentified by heavy scatters of roof-tile fragments. At site 105, some remains of stone walls were found, along with fragments of slag, suggesting there was likely a kiln.

These small to medium-sized sites testify to the fact that the chora continued to be exploited agriculturally after the period of the Greek polis.[2] During the Late Republican period, however, a new pattern of settlement took shape, one substantially different from that of the Greek period. The new rural complexes tended to be built along the major roads (Durante and Bivio Franchi) on commanding high ground (La Cappella), or on sites selected because of their proximity to water sources, for example, site 77, Destra Basento (Ch. 27).[3] These new farmhouses consisted of luxurious buildings or more modest rural dwellings without any public reception room(s). The structures were often mud brick, built on stone foundations with rooms surrounding a central court similar in plan to Hellenistic farmhouses.[4] The new Late Republican (e.g., the farmhouse at Sant'Angelo Grieco, Ch. 27) and Early Imperial complexes had articulated plans developed to function in an economy relying on the cultivation of cereals and on animal husbandry. At several of these new establishments pottery and tile factories produced goods to be used at the farm itself, as well as to be traded in the countryside, nearby centers, and even wider markets. This can be said, for instance, of the Late Republican ceramic production center and farmhouse at Pantanello (Ch. 27), and the farmhouse at La Cappella.[5]

Yet a third type of structure is represented by aristocratic villas of large dimensions, such as that at Termitito (see Ch. 27). Not just one but, essentially, three levels of economic organization were operating simultaneously in the Metapontino in the Early Imperial period. Up to now it has not been possible to define with certainty the relationship between these new rural settlements and the urban center of Metaponto. From the 2nd century BC onward, the city continued to contract, until by the 1st century BC it was restricted exclusively to the area of the castrum. The Early Imperial period saw further dwindling of the city (with modest building activity, limited to the area of the castrum) and reduced settlement of the countryside.[6] The discovery of pottery produced in the surrounding regions, however, especially the wares from Apulia, suggests a revival of the economy and the commerce of the city, as well as of its territory, in

---

1 See Cambi and Terrenato 1994; Carandini and Cambi 2002, 43; and more recently, Cambi 2003 on landscape archaeology. On the geology and climate factors, see Chs. 1–2.

2 Carter 1987, 203–12; 1994a, 182–92.

3 Site 77, on the right bank of the Basento River, was excavated during the SNAM pipeline project in 2002 (Nava 2003a). Bivio Franchi, De Siena and Giardino 1994. Durante, Lissi Caronna 1998–1999.

4 On the study and analysis of the plans and architecture of the rural establishments in the Metapontino, see De Siena and Giardino 1994, 204–09.

5 For the excavation of the Roman Republican farmhouse at Pizzica Pantanello, see Carter 1983a and 1987, 203–05; for the La Cappella farmhouse, see A. Bottini 1983a, 470–71.

6 Giannotta 1980, 73–77; De Siena 2001b, 39.

Site	Location	Elev. masl	Dateable ceramics	Scatter size/ density	Previous occupation	Roman and Medieval occupation
1F	Sant'Angelo	31	1/1	Not recorded	Greek, Late Hellenistic farmhouse	Early Imp. farmhouse
4F	Sant'Angelo	41	1/2	35 x 50/H	Greek farmhouse	Early Imp. farmhouse
95F/K	Demanio C.	54	1/1	80 x 80/H	Late Hellenistic farmhouse	Early Imp. farmhouse + kiln
105F/K	Sant'Angelo Vecchio	16	12/12	90 x 70/H	Neolithic activity; Greek farmhouse + kiln; Late Hellenistic farmhouse	Early Imp. farmhouse + kiln
200F	Giampasquale	75	3/3	150 x 120/L-M	Bronze Age, Greek farmhouse	Early Imp. farmhouse
395F	Avinella	58	1/2	60 x 40/v. H	Prehistoric activity, Greek farmhouse (large)	Early Imp. farmhouse
428F	Pantanello	10	3/18	60 x 40/H	Greek, Late Hellenistic farmhouse	Early Imp. farmhouse Late Roman farmhouse
468F	Giardinetto	67	4/5	25 x 30	Neolithic domestic structure; Greek farmhouse	Early Imp., Imp. farmhouse
627F	Demanio C.	60	1/1	50 x 50	Prehistoric activity; unidentified Greek feature	Early Imp. farmhouse
726F/T	Basento valley	19	6/54	140 x 300	Neolithic activity; Greek farmhouse + tomb	Early Imp., Late Roman farmhouse; Post-Med activity
735F	Pizzica	41	1/1	80 x 70/M	Neolithic domestic structure; Greek farmhouse	Early Imp. farmhouse
Total sites: 11		**Average elev.: 43**				

Table 32.1 Early Imperial sites (25 BC–AD 50). Site 1F was subsequently excavated as Sant'Angelo Grieco, a Greek and Late Hellenistic farmhouse with phases in the 6th, 4th, and 2nd–1st centuries AD. Scatter size in m. H = heavy, M = moderate, L = light, S = sparse, v. = very.

the context of the large and dynamic market-network of the period.[7]

The Imperial Period (AD 50–300)

The Metapontino. Twenty sites can be assigned to the period between the second half of the 1st century and the end of the 3rd century AD (*Table 32.2, Fig. 32.2*). Most of them produced a very low number of fragments, making it difficult to identify the type of settlement.

Nine sites (230, 231, 361, 387, 408, 528, 529, 535, and 726) stand out because of the larger quantities of pottery recovered. Site 230, in particular, because of its high percentage of imported African wares, may have been a large rural establishment—perhaps a relatively important villa—and almost certainly acquired these imported goods via the nearby city of Metaponto.[8] On the surface of the site was a large and dense scatter of materials (ca. 150 x 80 m), consisting both of numerous fine wares chiefly imported from Africa and also quantities of coarse and storage wares (mostly used for wine and oil), which were likely manufactured at the site itself.

All but site 230 can probably be identified as small rural dwellings, characterized by a limited scatter of only a few fine- and coarse-ware sherds, as well as fragments of amphorae and tile. Three of these sites (230, 529, and 726) are distinguished by their location on a height commanding the lower terraces next to the watercourses. Such are the cases at site 230 (San Marco), overlooking the Bradano River, and the Fosso dell'Acqua Fetente, at site 529 (Giardinetto), located beside an ancient branch of the Bradano River, and at site 726, (Cardillo-Azienda Graziadei), near the Basento River.

[7] De Siena 2001b, 39; De Siena and Giardino 1994, 201.

[8] For the goods imported from Africa to Metaponto and their presence throughout the countryside, see Giardino/Lapadula/Auriemma 1999.

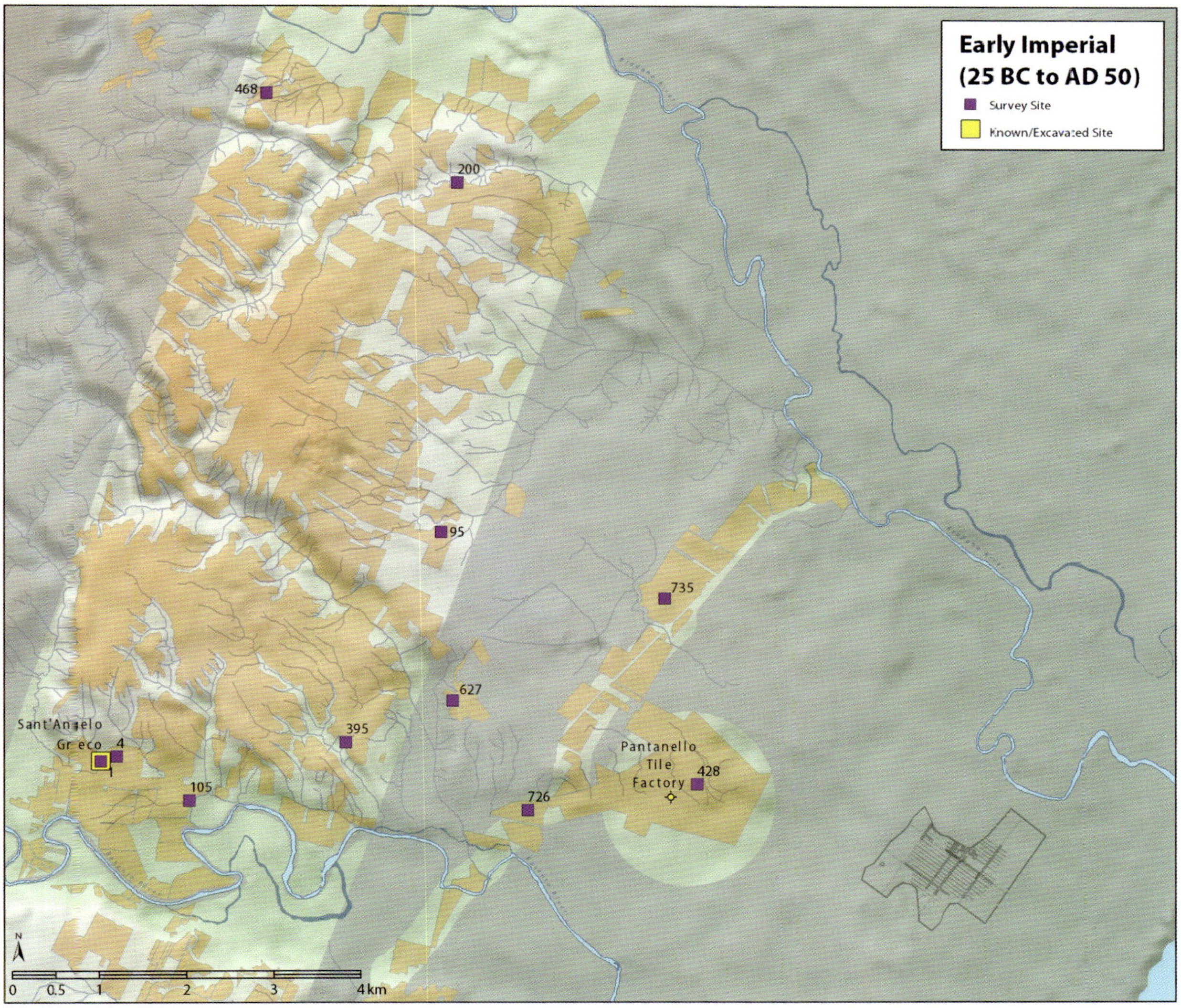

Figure 32.1 Early Imperial period sites. (JT)

These types of settlements may be reflected in the excavated rural settlement at San Biagio (*Fig. 32.3*), located at the head of the Venella valley roughly 5 km from the city of Metaponto. Dating from the mid-2nd to the beginning of the 4th centuries AD, it was first identified as a farmhouse. Study of the excavation materials, however, suggests that it was an important establishment, something more significant than a simple farm, taking into consideration its small bath complex (balneum) together with its rooms furnished with mosaics and frescoes.[9] To this probable dwelling for a vilicus, or farm manager, should be added the villa at La Cappella which stands on the right bank of the Basento River. Together with the villa complex at San Vito, on the right bank of the Bradano River,[10] the three excavated sites provide a fuller picture of settlement in the period of the Roman Empire.

The relationship of these sites to the urban center, which at this time was limited to a small area within the boundaries of the Roman castrum, still needs to be investigated more fully. It seems all but certain that the area of the castrum remained a basic economic reference point for the rural establishments. The archaeological data suggest a sharp contraction in the number of rural establishments in this period, even though some of these survived without interruption into the Late Roman period, develop-

[9] A presentation of the rural settlement of San Biagio can be found in Carter 1987, 205; Carter 1998a, 242–43. The materials from the excavation at San Biagio, carried out by the University of Texas at Austin in 1980 under the direction of J.C. Carter, are being studied by the author and will appear in a future volume in this series.

[10] On the site of La Cappella, see Salvatore 1982, 57. The site of San Vito was excavated during the installation of subterranean irrigation lines by the *Consorzio di Bonifica di Bradano e Metaponto* in 2002 (see Nava 2003a, 667).

Site	Location	Elev. masl	Dateable ceramics	Scatter size/ density	Previous occupation	Roman and Medieval Occupation
230F	San Marco	28	54/192	150 x 80/H	Greek farmhouse + tomb; Late Hellenistic farmhouse	Imp., Late Roman villa
231F	San Marco	35	16/21	50 x 35/v. H	Late Hellenistic farmhouse	Imp., Late Roman farmhouse; Med. farmhouse
320F	Avinella	89	1/1	50 x 40/v. H	Prehistoric activity; Greek farmhouse	Imp., Late Roman farmhouse; Med. farmhouse
361F	Demanio C.	64	3/9	30 x 2/ L		
372F	Demanio C.	62	1/5	60 x 45/v. H	Prehistoric activity; Greek farmhouse	Imp., Late Roman farmhouse; Late Med. farmhouse
386F	Sant'Angelo	38	2/2	20 x 20/M	Greek agricultural structure	Imp. farmhouse
387F	Sant'Angelo	42	3/6	15 x 10/S		Imp. farmhouse
388F	Sant'Angelo	46	4/95	45 x 40/H–M		Imp., Late Roman farmhouse; Early, Late Med farmhouse
408F	San Marco	39	9/11	Not recorded	Greek tomb	Late Roman farmhouse
428F	Pantanello	10	10/18	60 x 40/v. H	Greek, Late Hellenistic farmhouse	Early Imp. farmhouse + kiln
440T	Pantanello	20	1/1	80 x 50/v. H	Greek necropolis	Imp. tomb?
468F	Giardinetto	67	3/5	25 x 35/H	Neolithic domestic structure; Greek farmhouse	Early Imp., Imp., Late Roman farmhouse
528F	Giardinetto	63	6/29	35 x 25/H	Neolithic/Bronze Age domestic structure; Greek farmhouse	Imp., Late Roman farmhouse, Post-Med. activity
529F	Giardinetto	60	32/67	40 x 40/H	Neolithic/Bronze Age nucleated settlement; Greek farmhouse	Imp., Late Roman farmhouse
535F	Giardinetto	55	4/16	40 x 40/M–L	Prehistoric activity; Greek farmhouse	Imp., Late Roman farmhouse
538F	Demanio C.	36	1/1	20 x 15/L		Imp. farmhouse
540F	Lago del Lupo	56	2/2	40 x 20/v. L	Prehistoric activity; unidentifed Greek feature	Imp. farmhouse
726F/T	Basento valley	19	35/54	140 x 300/ H–M	Neolithic activity; Greek farmhouse + tomb	Early Imp., Imp., Late Roman farmhouse; Post-Med activity
727F	Basento valley	19	1/3	250 x 90/M	Neolithic activity; Greek farmhouse	Imp. farmhouse; Post-Med activity
762F	Bradano valley	15	7/15	130 x 100/H	Greek farmhouse + tomb	Imp., Late Roman farmhouse; Med activity
Total sites: 20		**Average elev: 43**				

Table 32.2 Imperial period sites (AD 50–300).

ing sometimes into large complexes with major economic roles.[11]

Lucania. A parallel phenomenon of contraction can be seen in eastern and interior Basilicata, where an archaeological survey has been carried out systematically since the end of the 1980s by the Soprintendenza and the University of Alberta (Canada). The area of the upper Bradano River and of the Basento River valley offer at present a wider context to investigate and understand the Lucanian rural landscape in the Imperial and Late Roman periods. This long-term survey has revealed a shift in settlement pattern: farmhouses appear throughout the river valleys, with larger settlements on the higher terraces, and sites devoted to agricultural production on the lower terraces. Between the end of the 2nd and the 3rd century AD,

[11] Carter 1994b, 187–88; De Siena and Giardino 2001, 161.

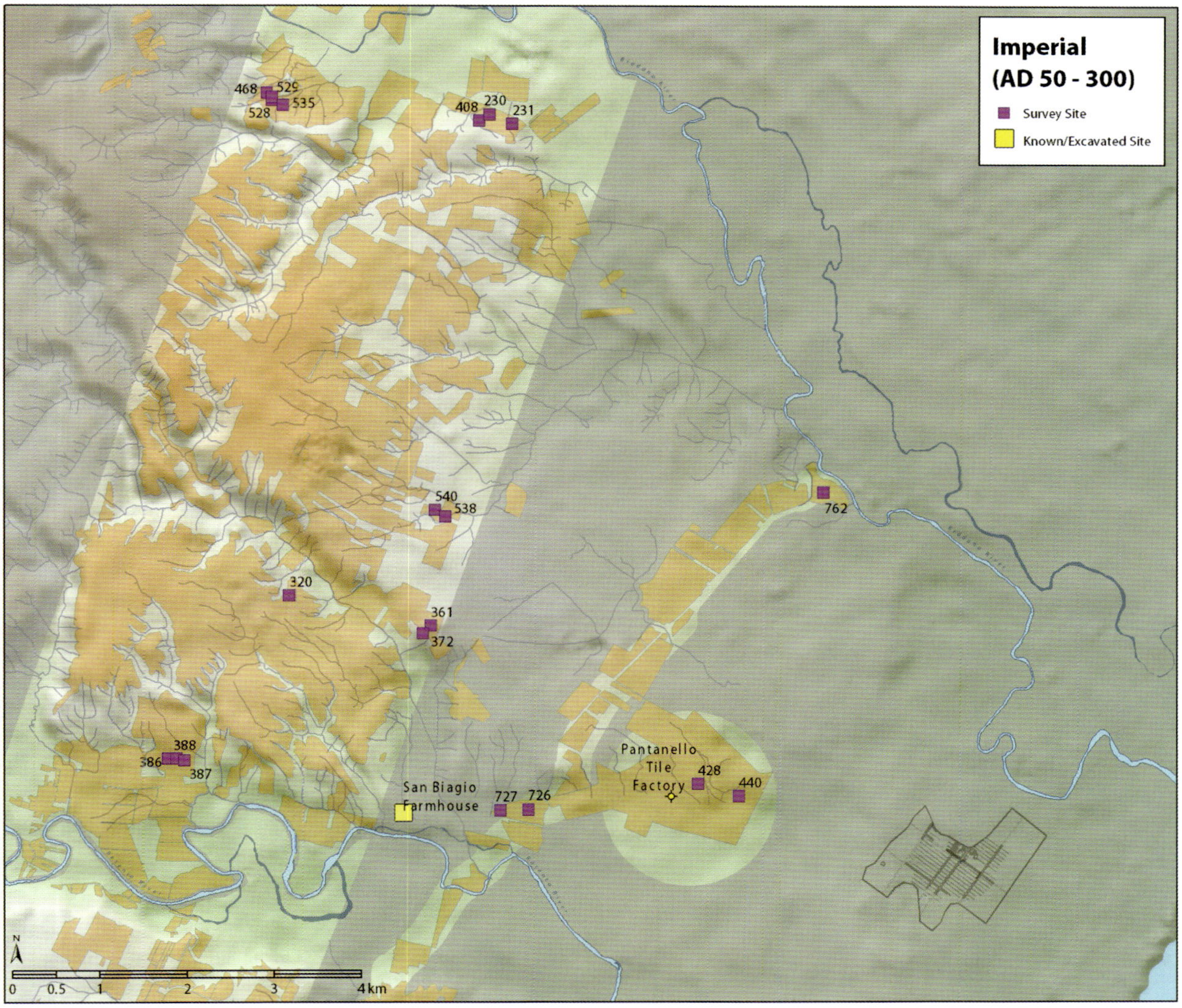

Figure 32.2 Imperial period sites. (JT)

the settlement pattern of the Early Empire developed a new form, characterized by villas that remained active until the middle of the 3rd century AD. During the second half of the 3rd and 4th centuries these villas became increasingly massive and visually impressive, while the small and middle-sized sites disappeared.[12]

The survey record is supported by the evidence from archaeological excavations, especially at the sites of Masseria Cicciotti (Oppido Lucano) and San Pietro (Tolve), which offer much interesting data essential for the understanding of settlement trends. The investigation of the villa, Masseria Cicciotti, which shows the continuous settlement of a site over a long period, is exemplary. This establishment arose in the Late Republican period, then during the Imperial period developed into a complex served by an aqueduct; its *pars urbana* was monumentalized and *pars rustica* expanded.[13] This renewed building activity points to the vitality of the site between the end of the 2nd and the beginning of the 3rd centuries AD, a period characterized by a varied and intense production that reached a high level. The site of San Pietro also shows expanding craft activities, although more modestly. The results of the investigations carried out in the area of the upper Bradano testify to a decline in the number of the settlements in this district around the middle of the 3rd century AD.

12 On the settlement patterns of the upper Bradano, see Fracchia and Gualtieri 1999 and Fracchia 2008.

13 For the new archaeological records from the villa of Masseria Cicciotti, see Gualtieri 1999, and Nava 2001, with related bibliography. The forthcoming study of a context from Masseria Cicciotti is in many ways very interesting, not least for its mostly locally produced pottery (M. Gualtieri, pers. comm.; also Gualtieri 2008a).

Figure 32.3 Reconstruction of the farmhouse at San Biagio (site 320), with the balneum addition at left. (ML)

The same phenomenon can be found during the same historical period in the territory around the old Roman colony of Venusia (modern Venosa), which, according to the epigraphic evidence, was part of the prevailing concentration of rural estates in few hands. The evidence from field surveys suggests a crisis during the 3rd century AD here as well, when roughly 60% of the rural sites were abandoned. In the 4th century AD, however, a revival took place. Large new villas were built, while some old farmhouses continued to operate—the only type of settlement to have survived the crisis of the previous century.[14]

These examples from the area of eastern Basilicata have no parallels in the Metapontino between the Bradano and the Basento Rivers. The results from the topographical surveys do not record any settlement of a size large enough to be identified as a villa-*praetorium* (Masseria Cicciotti) or a *praetorium* proper (San Gilio, San Giovanni di Ruoti).[15]

Beside the villas, the other basic elements of the organization of the rural landscape during the Roman Empire in the interior of Lucania were the *vici*. These can be found discontinuously across the Roman countryside from the Republican to the Late Roman periods. In recent years the archaeological evidence related to the *vicus* settlement in southern Italy, including the interior of Lucania, has grown dramatically.[16] The *vicus* and the large villa-farm, with its grand private estates, are the most prominent elements of the landscape of the Imperial Age. The continuity of settlement is evident in the frequent outgrowth of villages around the structures of a villa (Cugno dei Vagni, Masseria Cicciotti).[17]

Recent archaeological evidence leads us to reconsider the common view of a lack of settlements or of their falling into neglect in Lucania in general in the Imperial period. On the contrary, several signs of vitality and revival often appear in large areas of Lucania, Bruttium, and Apulia. During the 3rd century in particular there is a change in the use of the territory as the number of the villas decreases, and later, the large estates become even larger as they undergo a process of monumentalization.[18]

14 The records from the fieldwork in the area of Venosa were first discussed in *Venosa*; Marchi 2006, provides updated data.

15 On San Giovanni di Ruoti and northeastern Basilicata, see Small 1991, and *San Giovanni di Ruoti* I. Di Giuseppe 1996, 237–40, provides an overview of the villas in the interior of Basilicata during the Empire, with related bibliography; for a recent update, see Russo and Di Giuseppe 2008.

16 An extensive discussion of the question of the Roman rural landscape, specifically of the *ager Brundisinus*, with references to southern Italy, can be found in Aprosio 2008.

17 For an analysis of rural settlement patterns in Roman Lucania, as well as the relationships between the *vicus* and the villa of Masseria Cicciotti, see most recently Gualtieri 2003.

18 Gualtieri 2003, 198–99 and previously mentioned bibliography.

The Late Roman Period (AD 300–600)

For the period from the 4th to the end of the 6th centuries AD, 20 sites have been identified in the area between the Bradano and the Basento (*Table 32.3, Fig. 32.4*). Among these, a group of 7 sites stand out in terms of greater numbers of finds (230, 388, 399, 528, 535, 743, and 762).

In this phase, the settlements continue to occupy the low parts of the valley bottoms. There is continuity of use from preceding occupation at sites 230, 743, and 762. The sites show preference for the flat areas of the first low marine terraces (sites 399, 528, 529, 535, and 726). In all cases the primary necessity appears to have been the presence of water.

The data from the field survey indicate a fairly dispersed occupation of the territory in this period, with scattered settlements of small or medium size which can all be assigned to the category of farmhouse.[19] A small percentage (12%) of these were ex novo foundations. The remainder arose on sites already occupied in the Greek or Late Hellenistic periods, and in certain cases also on pre- or proto-historic settlements.[20] The conspicuous number of Late Roman settlements had an uninterrupted occupation from the Empire (9 sites, or 52% of the total); what has changed is their distribution and the overall pattern of settlement.[21]

Once again, the settlement patterns documented in eastern Basilicata are a source of valuable and interesting comparative information.[22] Beginning in the 4th century AD, the small and moderate-sized sites in the upper Bradano and Basento valleys disappeared or dramatically contracted in number to the benefit of the large sites, which were monumentalized.[23] The contemporary agrarian landscape of Venusia was dominated by large villas and multi-nucleated complexes. These were accompanied by farmhouses, present since the Early Empire, as well as villages (24 attested), some of which developed out of large villas. Among the notable sites of northeastern Basilicata is the villa at San Giovanni di Ruoti. Recent studies and analyses have prolonged the final phase of occupation of this site into the second half of the 6th century AD, contrary to the previous results which dated the abandonment of the villa at ca. AD 540.[24]

The data from the territory of Metaponto fit within the picture created by research on the Late Roman landscape of southern Italy generally, in which crisis and depopulation were accompanied by a transformation of the organizational model, with occupation patterns adapted to the *latifundium* structure and the changed economic and industrial needs.[25] The archaeological documentation from the study area at Metaponto includes no large rural dwellings endowed with luxurious residential quarters alongside the small and medium-sized productive units. The field survey has not encountered concentrations of artifacts or structural remains large enough to indicate the existence of Late Roman dwellings with complex and articulated plans, nor has it recovered indicators of such dwellings in the form of mosaic tesserae or plaster. Considering the profound disturbances to the Metapontino in the wake of the modern *riforma agraria*, only the excavation of a sample of sites, however, could confirm or cause to be rejected the picture obtained from survey. Some useful indication in this regard is provided by the excavation of the settlements in the La Cappella locality and in proprietà Durante, where the recovered materials indicate continuity of occupation there into the 4th and 5th centuries AD.[26]

If the large, sumptuous villa was a form of occupation generally absent from the Metapontino in the 4th–6th centuries,[27] a valid example of the *pagus/vicus* form existed in the castrum at Metapontum.[28] The quantities of imported ceramic materials collected during the survey suggest the existence of a by now dynamic relationship between the rural dwellings of the territory and the village centered on Metapontum and its port, active in receiving merchandise from the Mediterranean and redistributing it in the hin-

19 On the presence of Late Roman farmhouses in the interior of Basilicata, see Di Giuseppe 1996; Russo and Di Giuseppe 2008.

20 On the Late Roman agrarian structures excavated in the Metapontino and their interpretation, see Carter 1994a, 192–95; Giardino/Lapadula/Auriemma 1999, 350; De Siena and Giardino 2001, 161–62.

21 For reflections on these issues in southern Italy, see Carandini 1993.

22 Gualtieri 2008b.

23 On the 4th–6th c. villas of northeastern Lucania and their monumental aspects, see Gualtieri 2003, 259–62.

24 The data from radiocarbon analysis conducted on a porcupine and other faunal remains, as well as carbonized wood, appear in Small 2005.

25 For Apulia, see D'Angela and Volpe 1994, Volpe 1996; for Campania, see Arthur 1991, Peduto 1994. On changes in settlement in southern Italy between the Late Roman and Early Medieval periods, see Delogu 1994.

26 De Siena and Giardino 1994, 204–09; Giardino/Auriemma/Lapadula 1999, 350.

27 On Late Roman villas, see Carandini 1993, 1994; Vera 1994.

28 For definitions of *pagus* and *vicus*, see Volpe 1996, 147–96; 1999. Capogrossi Colognesi (2002) has recently proposed a revised definition of "*pagus/vicus* settlement." Gualtieri (2003, 256) has interpreted Late Roman Metapontum as a *statio*.

Site	Location	Elev, masl	Dateable ceramics	Scatter size/ density	Previous occupation	Roman and Medieval Occupation
161F	Sant'Angelo	26	2/2	40 x 30 L	Greek farmhouse	Late Roman farmhouse
230F	San Marco	28	64/192	150 x 80 H	Greek farmhouse + tomb; Late Hellenistic farmhouse	Imp., Late Roman farmhouse
231F	San Marco	35	1/21	50 x 35 v. H	Late Hellenistic farmhouse	Imp., Late Roman farmhouse
372F	Demanio C.	62	2/5	60 x 45 v. H	Prehistoric activity; Greek, Late Hellenistic farmhouse	Imp., Late Roman farmhouse; Late Med. farmhouse
384F	Sant'Angelo	57	6/6	25 x 15 M		Late Roman farmhouse
388F	Sant'Angelo	46	24/95	45 x 40 H/M		Imp., Late Roman farmhouse; Early, Late Med. farmhouse
399F	Avinella	60	16/16	60 x 40 M/L	Neolithic activity; Greek farmhouse + kiln	Late Roman farmhouse
408F	San Marco	39	2/11	NR	Greek tomb	Imp., Late Roman farmhouse
428F	Pantanello	10	3/18	60 x 40 v. H	Greek, Late Hellenistic farmhouse	Early Imp., Imp., Late Roman farmhouse
465F	Giardinetto	54	2/9	80 x 65 H	Bronze Age domestic structure; Greek farmhouse	Late Roman farmhouse
468F	Giardinetto	67	1/5	25 x 25 H	Neolithic domestic structure; Greek farmhouse	Early Imp., Imp., Late Roman farmhouse
528F	Giardinetto	63	20/29		Neolithic, Bronze Age domestic structure; Greek farmhouse	Imp., Late Roman farmhouse
529F	Giardinetto	60	2/67		Neolithic, Bronze Age nucleated settlement; Greek farmhouse	Imp., Late Roman farmhouse
535F	Giardinetto	55	7/16		Prehistoric activity; Greek farmhouse	Imp., Late Roman farmhouse
536F	Demanio C.	36	2/4	40 x 40 L	Prehistoric activity	Late Roman farmhouse
622F	Demanio C.	55	5/5	45 x 10	Prehistoric activity; Greek farmhouse	Late Roman farmhouse
726F/T	Basento valley	19	11/52	140 x 300 H/M	Neolithic activity; Greek farmhouse + tomb	Early Imp., Late Roman farmhouse; Post-Med activity
743F	San Salvatore	23	1/1	40 x 15 L	Neolithic activity; unidentified Greek feature	Late Roman farmhouse
762F	Bradano valley	15	11/15	130 x 100 H	Greek farmhouse + tomb	Imp., Late Roman farmhouse; Late Med. activity
769F	Basento valley	13	1/1	50 x 50 M	Unidentified Greek feature	Late Roman farmhouse
Total: 20		**Average elev: 38**				

Table 32.3 Late Roman period sites (AD 300–600).

terland.[29] The nucleated settlement at Metapontum, endowed with an early Christian cult building and baptistery in the 4th century AD (hence the suggested interpretation of *vicus*), probably served as a magnet for the scattered rural settlements in terms of both commercial exchange and spiritual care.[30] Characteristic of this rural landscape, where dispersed settlement prevailed over nucleated settlement, is the pervasive presence of empty spaces, interpreted as areas

29 On the Late Roman settlement of the castrum and the port at *zona* Mele, see L. Giardino 1991, and Giardino/Lapadula/Auriemma 1999.

30 For the proposed interpretation of the Late Roman settlement at Metapontum as a *vicus*, see Arthur 2000a, 182. For the basilica and baptistery, see L. Giardino 1991, 842–50. A controversial functional and chronological interpretation of the baptistery has recently been proposed by Sorrenti (2001).

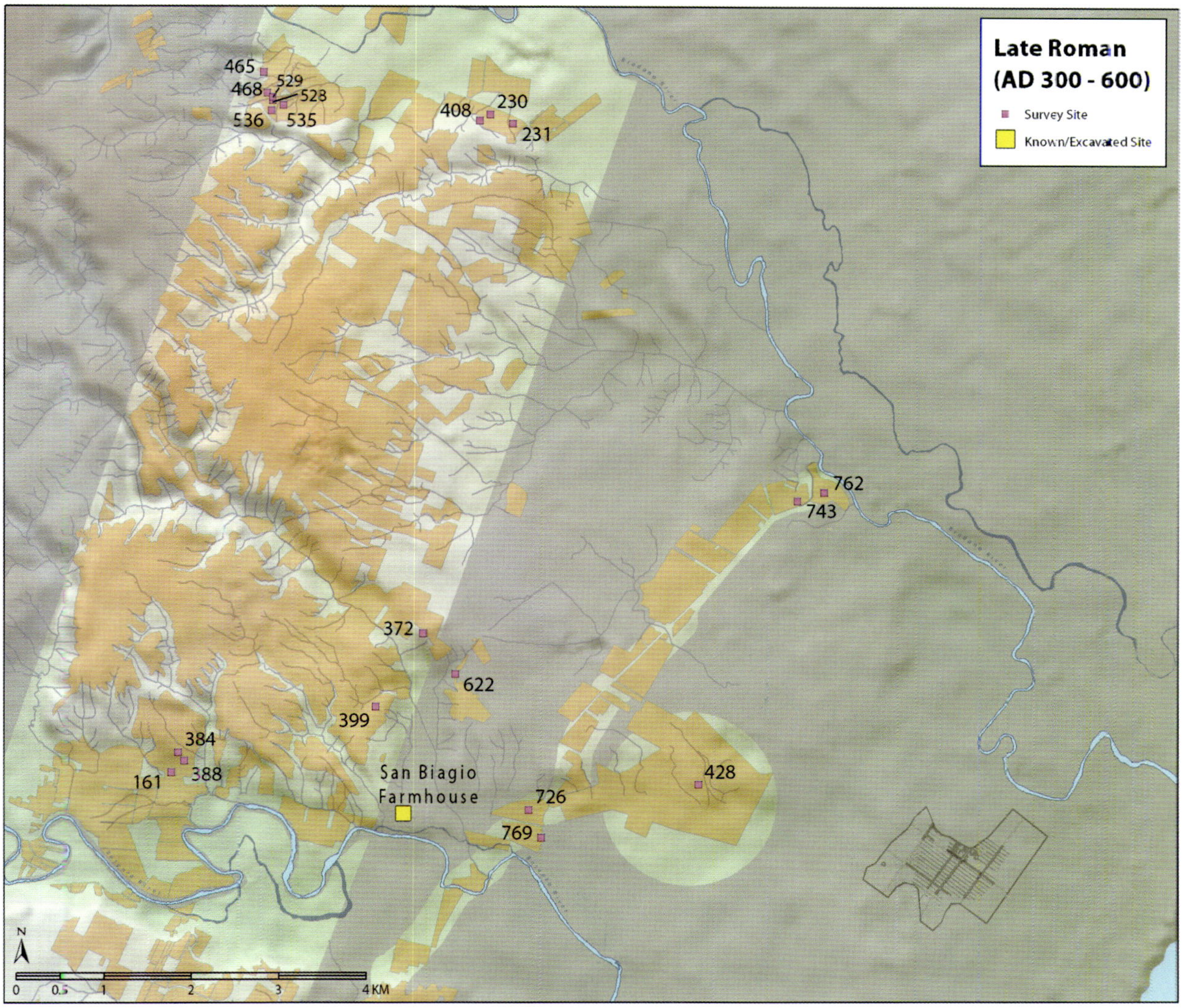

Figure 32.4 Late Roman period sites. (JT)

dedicated to agriculture, rather than abandoned and ignored.[31]

The ceramic evidence recovered in the survey suggests that the majority of the sites were occupied until the end of the 5th century AD. Sites 388 and 762 are exceptions, which appear to have endured through the first half of the 6th century.

The excavations at San Biagio and Pantanello also document the last phase of life of such large facilities in the 4th century AD, while only Pantanello is reoccupied in the Early Medieval period. In any case, the first half of the 6th century AD saw a crisis in settlement of notable proportions throughout the territory under investigation. The causes of this disappearance of settlement are doubtless to be sought in the aftermath of the Greco-Gothic War, but certainly other factors contributed, such as the need to abandon the area in the face of unhealthy conditions provoked by the movement of the Basento and the consequent pervasive stagnation.[32] The factors that favored a change in the landscape are numerous, and the subsequent changes in the economic and institutional orientation could have had a large influence in the later years. This becomes apparent in the 6th century AD throughout southern Italy, with different characteristics and manifestations in the various regions.[33]

31 On the increase of land exploited for agriculture at the expense of pasture, see Gualtieri 2003, 262.

32 Fonseca 2002, 41.

33 For an overview of the various regional situations, see *Atti Taranto* 38 ("L'Italia meridionale in età tardo antica"). Stimulating new observations were offered during the *Seminario sui paesaggi e insediamenti rurali in Italia meridionale fra Tardoantico e Altomedioevo* (published as *Paesaggi e insediamenti rurali*).

33
Imperial and Late Roman Ceramics

Erminia Lapadula

The ceramic artifacts belonging to the Imperial and Late Roman periods (middle of the 1st century BC–6th century AD) form a distinct, minor presence in the enormous quantity of artifacts collected in the survey between the Bradano and Basento Rivers.[1] Only 612 fragments were recovered, with 522 identifiable ceramic pieces, and 90 pieces of pottery and tile that could only be classified generically as Roman.

Most of these artifacts have been reduced to a very small size by abrasion and there are relatively few diagnostic pieces. This situation dictated certain choices of approach and content in the presentation of the materials. A detailed catalog of all materials recovered was not feasible because of their poor state; thus, only the fragments with well-preserved morphological characteristics and sufficient dimensions are presented, along with certain other fragments selected as unique indicators of a ceramic class. The remaining fragments are summarized in tables to show typological and quantitative data, subdivided by chronological periods. For the same reasons, graphic and photographic illustration was restricted to artifacts of sufficient size for valid identification in study and publication.[2]

The catalog follows a chronological sequence within which the fragments are subdivided by functional and/or ware classes, forms, and types. The classes include cooking wares, storage wares, table wares, and transport containers, and are ordered according to the type of surface treatment, from plain to slipped. Each artifact represents a morphological type and is accompanied by a brief description which includes analysis of the decoration and slip, dimensions (where measurement of diameter is possible), and a macroscopic description of the ceramic fabric with color determination according to the Munsell Soil Color Charts. Each entry ends with a general commentary including comparanda from other contexts and a suggested chronology.

The Early Imperial Period (50 BC–AD 50)

The number of ceramic artifacts of this period total 46 and are highly fragmentary, consisting primarily of wall fragments with limited diagnostic value (*Table 33.1*).

The most important group of fragments consists of fine table pottery, represented by Italian terra sigillata and Eastern terra sigillata, with forms generically dateable between the end of the 1st century BC and the 1st century AD. There are only 21 fragments of Italian terra sigillata, very few of recognizable shape or morphological value. Twelve wall and 3 base fragments can be ascribed to open forms; one of the base fragments, though very small, preserves part of the stamp (**No. 4**). Of the 6 rim fragments recovered in the survey, only 4 can be attributed to precise forms. Attested forms include plates with hanging rim and small cups with outward-turned rim dateable between the end of the 1st century BC and the middle of the 1st century AD.

The absence of epigraphic evidence precludes identification of the production centers responsible for the fragments recovered. The forms were generally produced and distributed by workshops in Italy. The Metapontine fragments could have been produced in northern Italian workshops, examples of which are known to have been widely distributed in neighboring Apulia.[3] With its favorable coastal position, Metaponto, along with its territory, probably participated in the commercial currents that distributed Italian terra sigillata in the nearby Apulian centers and likely had an excellent market there.[4]

Though few in number, small in size, and in most cases morphologically uninformative, the fragments of domestic wares and amphorae still fit neatly into

[1] The author wishes to thank Prof. Joseph C. Carter for his confidence and the opportunity to participate in this project. The author also thanks Stephen M. Thompson and Alberto Prieto for their attention to the project and continuous contact with the author, which was indispensable to the success of the study.

[2] The author wishes to thank Cesare Raho and Eloisa Vittoria for executing the illustrations, as well as Giambattista Sassi, who provided the photographic documentation and digitalization. The revision and inking of the illustrations were performed by the author.

[3] See most recently Rizzitelli 2000, 267–69, with relevant bibliography.

[4] Rizzitelli 2000, 271.

Ware	Form	Type	Chronology	References	Site (count)	Subtotal	Total
Coarse ware	Jug				site 105 (2)	2	11
	Bottle				site 105 (1)	1	
	Unknown closed				site 200 (3)	3	
	Unknown open				site 4 (1) site 105 (2)	3	
	Unknown				site 745 (2)	2	
Italian terra sigillata	Plate	Dragendorff 16/ Goudineau 15	20 BC–AD 15	*Conspectus*	site 95 (1)	1	21
	Cup	Goudineau 7/Pucci XVI	20 BC–AD 15	*Conspectus*	site 428 (1)	1	
	Cup	Goudineau 13, 16, 18/Pucci XX, XXI	20 BC–AD 20	*Conspectus*	site 428 (1)	1	
	Cup	Ritterling 5/ Goudineau 27, 37	10 BC–AD 50	*Atlante* II	site 428 (1)	1	
	Unknown open				site 1 (1) site 105 (6) site 395 (1) site 428 (2) site 726 (6) site 735 (1)	17	
Eastern terra sigillata	Open		1st c. BC		site 627 (1)	1	1
Terra sigillata	Unknown closed				site 105 (1)	1	1
Amphorae	Dr. 2/4		1st c. BC–1st c. AD		site 4 (1)	1	12
	Unknown				site 231 (6) site 395 (1) site 468 (2) site 745 (2)	11	
Total							**46**

Table 33.1 Ceramics from the Early Imperial period (50 BC–AD 50).

the chronology created by the fine table wares. Among the unslipped or plain wares, the closed forms (jug, bottle) predominate, while the open forms are much rarer and, consequently, difficult to identify. The few fragments of walls, bases, and handles from amphorae make identification of the forms problematic, with the exception of several fragments of Dressel type 2/4 amphorae.

The Imperial Period (AD 50–AD 300)

There are 262 fragments that can be dated to the Imperial period, but as in the previous period, these too are small in size (*Table 33.2*). The classes most strongly attested are the African imports, both fine table wares and cooking wares. The most common forms, the Hayes type 8, 14, and 50a cups and Hayes type 23 casserole, are broadly dateable between the middle of the 2nd and the middle of the 3rd centuries AD. The five small rim and wall fragments in Eastern terra sigillata (sites 230, 387, and 408) belong to open forms generically dated, via the presence of African fine wares, early in the Imperial period (end of the 1st–2nd centuries AD). The few fragments of locally produced plain and cooking wares, as well as amphorae, also coincide chronologically with the African imports.

The Late Roman Period (AD 300–600)

A total of 214 ceramic fragments can be attributed to the Late Roman period (*Table 33.3*). The range of Late Roman wares and forms is best represented at site 230, where imported and local ceramics were recovered in equal proportions. Numerous open and closed cooking ware forms (pots, pans) appear, as well as mostly open forms in painted coarse wares (bowls, basins) alongside cups and bowls in African red slip ware and Late Roman C (Phocaean) ware. Late

Ware	Form	Type	Chronology	References	Site (count)	Subtotal	Total
Cooking ware	Pot	Ordona VII, 10.5	1st–2nd c.	*Ordona X*	site 230 (1)	1	15
	Pot	Ordona VII,16.1 *similis*	Imperial	*Ordona X*	site 230 (1)	1	
	Pot	Ordona VII,13.2	2nd–3rd c.	*Ordona X*	site 231 (1)	1	
	Lid	Ordona VII,18.1	2nd–3rd c.	*Ordona X*	site 231 (1)	1	
	Unknown closed				site 408 (2) site 726 (2) site 762 (3)	7	
	Unknown				site 387 (1) site 745 (3)	4	
Coarse ware	Bowl				site 727 (2)	2	17
	Unknown open		Imperial		site 408 (1)	1	
	Unknown closed		Imperial		site 200 (3) site 361 (4) site 408 (2) site 538 (1) site 726 (1) site 762 (1)	12	
	Unknown				site 745 (2)	2	
African red slip	Cup	Hayes 8	2nd–beg. 3rd c.	J.W. Hayes 1972	site 33* (1) site 230 (5) site 361 (1) site 372 (1) site 528 (2) site 529 (3) site 726 (3)	16	160
	Cup	Hayes 9	2nd–beg. 3rd c.	J.W. Hayes 1972	site 726 (2)	2	
	Cup	Hayes 14A	ca. 150–200	J.W. Hayes 1972	site 230 (1) site 726 (1)	2	
	Cup	Hayes 14	Mid-2nd–3rd c.	J.W. Hayes 1972	site 529 (10)	10	
	Plate	Hayes 31	ca. 200–250	J.W. Hayes 1972	site 230 (1) site 529 (2) site 535 (1)	6	
	Plate	Hayes 32	ca. 200–250	J.W. Hayes 1972	site 529 (1)	1	
	Plate	Hayes 49	3rd c.	J.W. Hayes 1972	site 726 (1)	1	
	Cup	Hayes 50A	3rd c.	J.W. Hayes 1972	site 230 (3) site 468 (1) site 528 (1) site 529 (3) site 535 (1) site 726 (2)	11	
	Unknown/ *Prod. A*		End 1st–late 3rd c.		site 230 (67) site 320 (1) site 361 (2) site 386 (1) site 387 (3) site 408 (1) site 428 (8) site 468 (2) site 528 (2) site 540 (1) site 726 (21) site 727 (1) site 762 (1)	111	

Table 33.2 Ceramics from the Imperial period (AD 50–300).

Ware	Form	Type	Chronology	References	Site (count)	Subtotal	Total
African cooking ware	Lid	Hayes 22	ca. 100–150	J.W. Hayes 1972	site 726 (2)	2	46
	Casserole	Hayes 23B	Mid-2nd–beg. 3rd c.	J.W. Hayes 1972	site 428 (1) site 528 (1)	2	
	Casserole	Hayes 23	Beg. 2nd–beg. 3rd c.	J.W. Hayes 1972	site 230 (7) site 529 (7) site 726 (2)	16	
	Plate	Hayes 27	Mid-2nd–early 3rd c.	J.W. Hayes 1972	site 230 (1)	1	
	Lid	Hayes 185	ca. 200–250	J.W. Hayes 1972	site 428 (1)	1	
	Lid	Hayes 192	ca. 200–250	J.W. Hayes 1972	site 535 (1)	1	
	Lid	Hayes 195	ca. 150–250	J.W. Hayes 1972	site 230 (2)	2	
	Lid	Hayes 196	End 1st–late 2nd c.	J.W. Hayes 1972	site 230 (2) site 529 (6) site 535 (1) site 540 (1)	10	
	Casserole	Hayes 197	End 2nd–mid-3rd c.	J.W. Hayes 1972	site 230 (1) site 726 (1)	2	
	Lid				site 388 (1)	1	
	Unknown open				site 528 (2) site 529 (2) site 726 (1)	5	
	Unknown closed				site 388 (3)	3	
Eastern terra sigillata	Unknown open		1st–2nd c.		site 230 (2) site 387 (1) site 408 (2)	5	5
Amphorae	Unknown Spanish		Imperial		site 387 (1)	1	19
	Unknown African		Imperial		site 440 (1) site 465 (1)	2	
	Unknown		Imperial		site 361 (2) site 386 (1) site 408 (1) site 465 (8) site 745 (2) site 762 (2)	16	
Total							**262**

Table 33.2 Ceramics from the Imperial period (AD 50–300) cont.

Roman transport amphorae, both African (*spatheion*) and eastern Aegean (LRA 2), are represented in relatively large quantities. Site 388 includes locally produced storage wares (basins and closed forms in painted coarse ware), fine table wares imported from Africa (cups, bowls, plates), and transport amphorae of both African and eastern Aegean (LRA 2) manufacture. This site also produced roof tiles (some decorated with incised motifs) and a few bricks.

Locally produced table, storage, and cooking wares are present at a large number of Late Roman sites. Particularly important among these are the decorated coarse wares (Late Roman painted wares, RPW) designed for use both at the table and in storage, indicators of a lively local manufacture and the material culture of Late Roman southern Italy.[5] The decorated wares, in fact, may have been produced at a local, regional, or sub-regional scale with wide circulation.[6] At the Metapontine sites, the local decorated

[5] On the production and circulation of Late Roman painted coarse wares, see Di Giuseppe 1998, Annese 2000, and D. Leone 2002, with relevant bibliography. Turchiano 2002 re-evaluates the presumed distinction between coastal and inland sites in the distribution of imported and painted coarse pottery.

[6] Manufacturing centers for painted pottery have recently been discovered through archaeological investigations in the upper Bradano valley (Fracchia 2005, 140) and in the territory of Venusia (Marchi 2005, 180). To these can be added the structures for the production of painted coarse wares and cooking wares at San Giusto (province of Foggia) in Apulia for which the results of archaeometric analysis are available (Gliozzo et al., 2005).

Ware	Form	Type	Chronology	References	Site (count)	Subtotal	Total
Cooking ware	Pot	Ordona 6.1 *similis*	4th c.	*Ordona X*	site 231 (1)	1	18
	Pot	Ordona 19.3	5th–6th c.	*Ordona X*	site 230 (1)	1	
	Pot	Ordona 23.7	5th–6th c.	*Ordona X*	site 230 (2)	2	
	Pot	Ordona 22.3	5th–6th c.	*Ordona X*	site 230 (1)	1	
	Pan		4th–5th c.		site 230 (1)	1	
	Unknown open		Late Roman		site 622 (4)	4	
	Unknown closed		Late Roman		site 726 (2) site 762 (3)	5	
	Unknown				site 745 (3)	3	
Coarse ware	Plate/lid		5th c.		site 528 (1)	1	31
	Bowl		5th–6th c.		site 230 (3)	3	
	Jar		5th–6th c.		site 230 (8)	8	
	Domestic amphora		5th c.		site 528 (2)	2	
	Pot		5th–6th c.		site 230 (2)	2	
	Unknown open		Late Roman		site 399 (1)	1	
	Unknown closed		Late Roman		site 528 (8) site 762 (1)	9	
	Unknown				site 745 (2)	2	
Decorated coarse ware	Bowl		Late 4th–5th c.		site 230 (1)	1	13
	Bowl		End 4th–5th c.		site 535 (1)	1	
	Basin		5th–6th c.		site 230 (1) site 388 (1)	2	
	Jar		5th c.		site 384 (1)	1	
	Domestic amphora		5th c.		site 622 (1)	1	
	Unknown open		End 4th–5th c.		site 230 (1) site 408 (1) site 528 (1)	3	
	Unknown closed		End 4th–5th c.		site 230 (1) site 388 (3)	4	
African red slip ware	Cup	Hayes 50B	Mid-4th c.	J.W. Hayes 1972	site 230 (4) site 388 (7) site 428 (2) site 529 (1) site 535 (1) site 726 (1)	16	36
	Bowl	Hayes 61B	ca. 400–450	J.W. Hayes 1972	site 230 (2)	2	
	Bowl	Hayes 61	ca. 350–400	J.W. Hayes 1972	site 230 (1) site 388 (1)	2	
	Plate	Hayes 104a	ca. 500–550	J.W. Hayes 1972	site 388 (1)	1	
	Prod. A				site 320 (1)	1	
	Prod. D		4th–6th c.		site 388 (6) site 399 (1) site 408 (1) site 762 (1)	9	
	Unknown / *Prod. A/D*				site 743 (1)	1	
	Unknown				site 372 (1) site 726(3)	4	

Table 33.3 Ceramics from the Late Roman period (AD 300–600).

Ware	Form	Type	Chronology	References	Site (count)	Subtotal	Total
	Bowl	Hayes 3	5th c.	J.W. Hayes 1972	site 230 (1) site 528 (1) site 535 (1)	3	
Late Roman C (Phocaean) ware	Bowl	Hayes 3E	ca. 450–500	J.W. Hayes 1972	site 230 (1)	1	
	Bowl	Hayes 3F	ca. 500–550	J.W. Hayes 1972	site 762 (1)	1	9
	Bowl	Hayes 5	ca. 450–500	J.W. Hayes 1972	site 428 (1)	1	
	Unknown open				site 230 (1) site 372 (1) site 762 (1)	3	
	African IIC		Second quarter 3rd–beg. 5th c.		site 769 (1)	1	105
	Keay XXVI		5th c.		site 726 (1)	1	
	Spatheion		End 4th–5th c.		site 230 (29) site 384 (1) site 528 (1) site 535 (3) site 536 (1) site 726 (1)	36	
Amphorae	LRA 1		5th c.		site 230 (2) site 399 (3) site 528 (2) site 536 (1) site 726 (1) site 762 (1)	10	
	LRA 2		5th c.		site 230 (9) site 388 (1) site 399 (4) site 528 (2) site 762 (2)	18	
	Mele 3/5		6th c.		site 230 (2)	2	
	Unknown African		5th–6th c.		site 161 (1) site 384 (4) site 388 (4) site 399 (4) site 465 (1) site 528 (1) site 535 (1) site 726 (1) site 762 (1)	18	
Amphorae	Unknown Eastern Mediterranean		5th–6th c.		site 161 (1) site 230 (3) site 399 (1) site 528 (1) site 726 (1)	7	
	Unidentifiable		Late Roman		site 399 (2) site 465 (8) site 745 (2)	12	
Lamps	Unknown African		5th c.		site 230 (1)		2
	Unknown Eastern Mediterranean		4th c.		site 529 (1)		
Total							**214**

Table 33.3 Ceramics from the Late Roman period (AD 300–600) cont.

wares are present in relatively high proportions, albeit lower than the imported wares. The range of forms is limited: open forms (bowls, basins) are more common than closed forms (jars, domestic amphorae), which are often decorated with incised or applied motifs beyond the red or brown paint. The same repertory was discovered in the excavations of the Late Roman settlements at the castrum and the port in Zona Mele.[7] The situation in the Metapontino reflects the general situation of Lucania (Venusia, San Giovanni di Ruoti, Calle) and Apulia (Ordona, San Giusto, Agnuli, Posta Crusta, Bovino, Canosa).[8]

At most Metapontine sites, the fragments of plain coarse and cooking wares belong to morphologically uninformative parts of the vessels and appear in numbers and sizes too small to be identified by type. Only for the cooking wares can it be said that large pots have a greater range of forms than small pots and pans.

Transport amphorae are exclusively imports, of which spatheion amphorae are by far the most common. Among the African imports there is a IIC amphora, the oldest form attested, and a Keay XXVI amphora. Among the less numerous eastern Aegean imports, the LRA 2 amphora is more common than the LRA 1.

There are only two examples of Late Roman lamps in the survey, both imported: the handle of an African product (site 230) dateable generally between the end of the 4th and the 5th centuries AD, and the shoulder of a likely eastern product (site 529) dateable to the beginning of the 4th century AD. Unfortunately, the small size of the fragments makes it impossible to be more precise about their forms or chronologies.

Another important center may be identified in the Salentine peninsula at Cutrofiano in the province of Lecce (see Greco and Lapadula 2004).

7 Giardino/Lapadula/Auriemma 1999.

8 An up-to-date distribution map for Apulia appears in D. Leone 2000, 430, Fig. 382; for Lucania and Campania, see Di Giuseppe 1998, 747, Fig. 10.

Catalog of Roman Ceramics

This catalog includes only fragments with preserved rim or base sufficient for diameter measurement. All dimensions are in cm. An asterisk (*) indicates materials from the 1980 ICA pilot survey (sites numbered 1–44), as well as at the intensive survey sites and plots south of the Basento River through the 2001 season (most of the sites numbered 562 and higher). Additional information about the sites can be found in the Gazetteer. Color was recorded with reference to the Munsell Soil Color Charts.

The Early Imperial Period (50 BC–AD 50)

1:2

1. ITS 095-01 Italian terra sigillata plate, Dragendorff 16/ Goudineau 15
Description: Hanging rim; low and flaring wall accentuated on the interior by two grooves.
Fabric: Pink (5YR 7/4), hard, with clean fractures. Red (2.5YR 4/8), shiny, flaky surface.
Comparisons: *Sibari* II 325, 459; *Sibari* IV 47?; *San Giovanni di Ruoti* I, 52, n. 2; Carter 1983a, 474.
Date: 20 BC–AD 15.

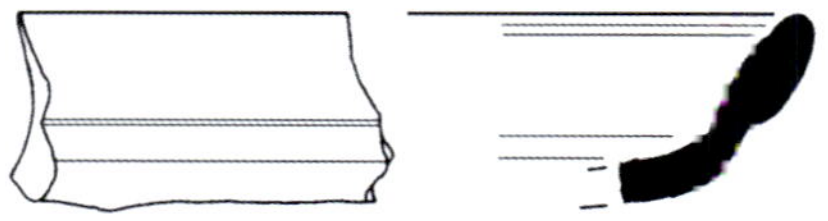

---

2. ITS 428-01 Italian terra sigillata cup, Goudineau 7/ Pucci XVI
Description: Outward-turned, hanging rim.
Fabric: Pink (5YR 8/3), hard, with clean fractures. Red (2.5YR 4/8) surface.
Date: 25/20 BC–AD 15.

---

3. ITS 428-02 Italian terra sigillata cup, Goudineau 13, 16, 18/Pucci XX, XXI
Description: Distinct, outward-turned rim.
Fabric: Pinkish-white (5YR 8/2), hard, with clean fractures. Red (2.5YR 5/6) surface.
Date: 20 BC–AD 20.

1:2

4. ITS 428-03 Italian terra sigillata cup, Ritterling 5/ Goudineau 27, 37
Description: Flaring walls; base bears stamp AUC[. . .] in rectangular frame.
Fabric: Pink (5YR 7/3), hard, with clean fractures. Red (2.5YR 4/8), shiny surface.
Observations: The stamp can reasonably be restored as AUC[TUS], attested in Italian terra sigillata products from northern Italy, especially on form Ritterling 5/ Goudineau 27, 37.
Date: 10 BC–AD 50 (Atlante 197).

---

5. ITS 428-04 Italian terra sigillata cup
Description: Base has molding at the edge, indicating a cup with high foot.
Fabric: Reddish yellow (5YR 7/6), hard, with clean fractures. Red (2.5YR 4/8), shiny surface.
Date: Augustan (30 BC–AD 14).

The Imperial Period (AD 50–300)

6. ROM Ck 230-01 Cooking-ware pan
Description: Rim rounded at tip.
Fabric: Brown (section and surface 7.5YR 5/4), hard, with irregular fractures; abundant small, white, and shiny inclusions.
Comparisons: The form is similar to pans produced at Ordona (type 2.4).
Date: 1st–3rd c. AD.

1:2

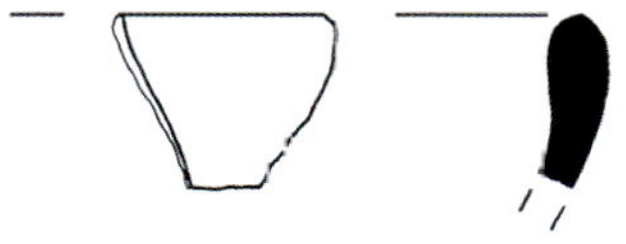

1:2

7. ACk 230-02 African cooking-ware lid, Hayes 195
Description: Raised rim with ashy patina on interior and exterior surfaces.
Fabric: Reddish-yellow (section and surface 5YR 6/6), hard, with irregular fractures.
Date: Second half of 1st–3rd c. AD.

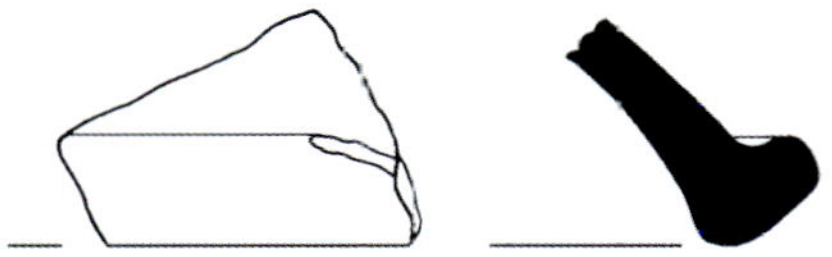

1:2

8. ACk 230-03 African cooking-ware casserole, Hayes 197
Description: Rim with oval section and ashy patina on exterior surface.
Fabric: Light red (section and surface 2.5YR 6/8), hard, with irregular fractures.
Date: End of 2nd–middle of 3rd c. AD.

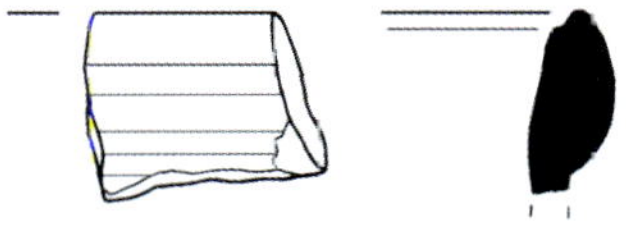

1:2

9. ARS 230-04 African red slip cup, Hayes 8
Description: Outward-turned rim with enlarged exterior fillet/spline decorated with small wheels; a pair of incised lines emphasize the attachment of the fillet on the interior.
Fabric: Light red (section 2.5YR 6/8, surface 2.5YR 5/8), hard, with irregular fractures.
Comparisons: The form is well-attested at many eastern and western Mediterranean sites. See Rizzitelli 2000, with relevant bibliography.
Date: 2nd c. AD.

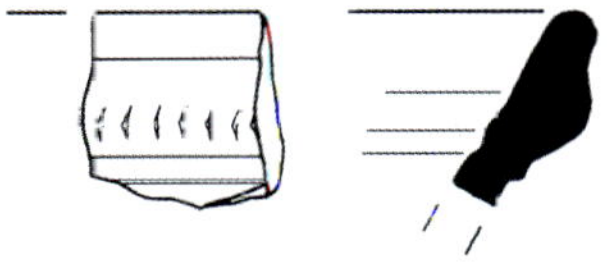

1:2

10. ARS 230-05 African red slip bowl, Hayes 27
Description: Slightly concave rim; the interior attachment of the body is emphasized by a groove.
Fabric: Light red (10R 6/8), hard, with irregular fractures.
Painted decoration: Light red (2.5YR 6/8).
Comparisons: The form is attested at many Mediterranean sites.
Date: End of 2nd–3rd c. AD.

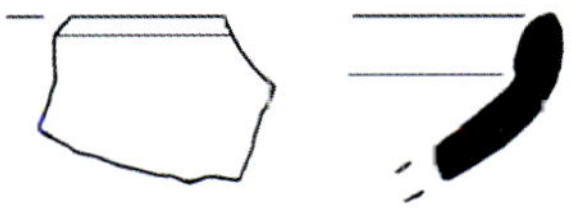

The Late Roman Period (AD 300–600)

1:2

11. ROM Ck 230-06 Cooking-ware pot
Description: Rim with bilobate profile.
Fabric: Reddish-yellow surface (5YR 6/6) and grey core (5Y 6/1), hard, with irregular fractures.
Comparisons: The form, perhaps of Aegean manufacture, is attested at Ordona (Annese 2000, 335, type 23.[illegible]).
Date: 5th–6th c. AD.

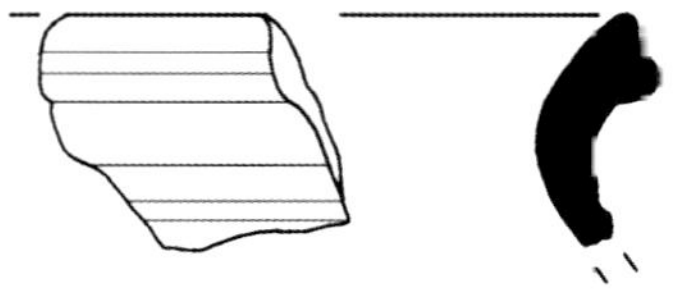

1:2

12. ROM Ck 230-07 Cooking-ware pot
Description: Rim slightly inclined towards exterior; strap handle attached directly to rim; exterior of rim and wall decorated with flattened grooves.
Fabric: Yellowish-red (section and surface 5YR 4/6), hard, with irregular fractures; abundant dark inclusions.
Comparisons: The form is attested at Ordona (Turchiano 2000, 372, type 19.3).
Date: 5th–6th c. AD.

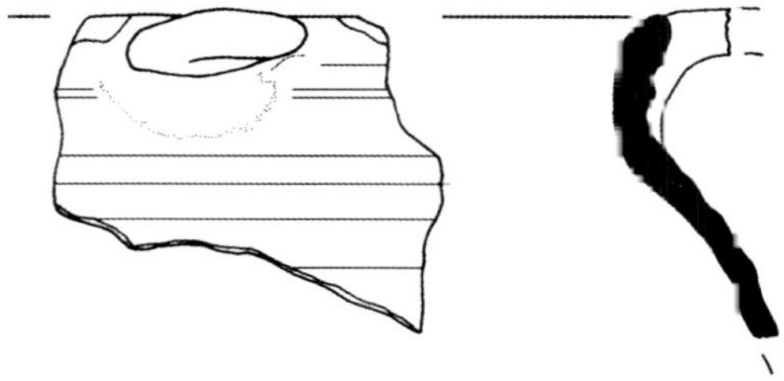

1:2

13. RPW 230-08 Painted coarse-ware bowl
Description: Slightly inward-turning rim with rounded tip; two grooves present on the exterior below the rim.
Fabric: Very pale brown (10YR 7/4), refined, hard, with fairly even fractures.
Painted decoration: Traces of red paint on the interior immediately below the rim.
Comparisons: The form is attested in Lucania and Apulia. Lucania: Calle (Di Giuseppe 1998, 739, Fig. 4.5) and Metaponto (unpublished material from the castrum). Apulia: Ordona (numerous bowls with varying carination of the rim, see Turchiano 2000, 356) and Porta Crusta (D. Leone 2000, 401)
Date: 5th c. AD.

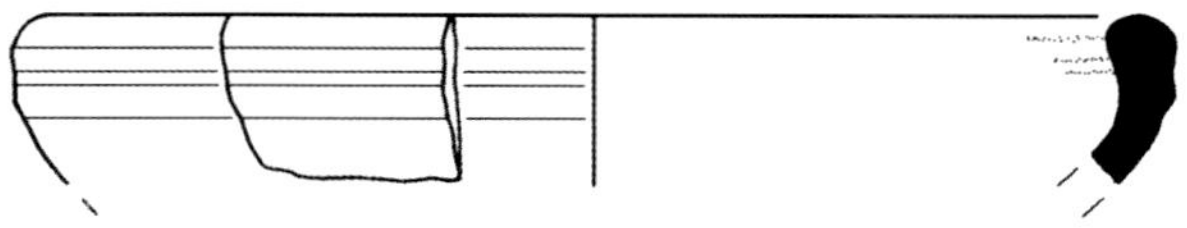

1:2

14. RPW 230-09 Painted coarse-ware basin
Description: Inward-turning rim; the interior surface is decorated with an impressed wave motif.
Fabric: Very pale brown (section and surface 10YR 7/4), purified, hard, with irregular fractures; abundant, very small, shiny inclusions.
Painted decoration: Traces of red paint on the exterior surface.
Comparisons: The form is attested in Lucania at Calle (Di Giuseppe 1998, 740, Fig. 5.6), in Apulia at Ordona (general comparison with numerous basins displaying different articulation of the rim, see Turchiano 2000, 304–305), at Metaponto (unpublished material from the castrum), and at Porta Crusta (D. Leone 2000, 400).
Date: 5th c. AD.

15. ARS 230-10 African red slip bowl, Hayes 61

Description: Base bears decorative motif of concentric circles defined by a series of small straight marks (Hayes style Aiii).

Fabric: Light red surface (2.5YR 6/8) and reddish yellow section (5YR 6/8), hard, with irregular fractures.

Comparisons: The Hayes 61 form with this type of decoration is attested in contexts around the Mediterranean, both urban and rural.

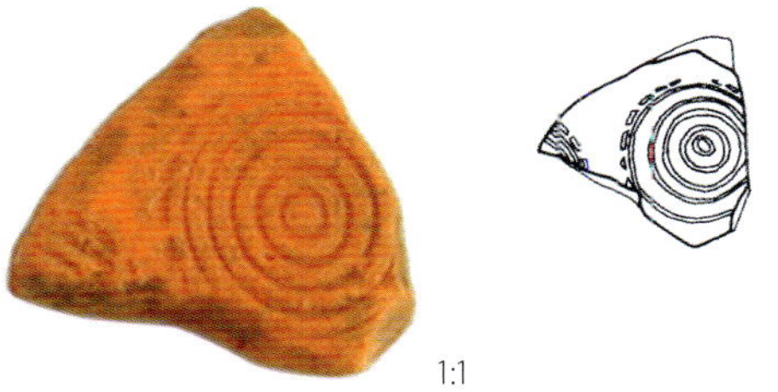

Date: Second half of 4th c. AD.

16. LRC 230-11 Late Roman C (Phocaean) bowl, Hayes 3E

Description: Rim with raised lip and triangular profile.

Fabric: Red surface (10R 5/6) and reddish-brown section (5YR 5/4), very hard, with clean fractures.

Comparisons: The Hayes 3E form is attested in many sites in southern Italy. For a synthesis of the distribution of Late Roman C/Phocaean Ware in Italy, see most recently A. Martin 1998.

Date: Second half of 5th c. AD.

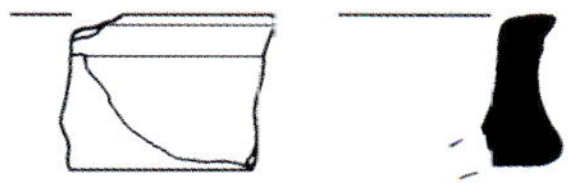

17. LRC 528-01 Late Roman C (Phocaean) bowl

Description: Base bears impressed decoration of individual birds arranged at regular intervals (Hayes pattern 46).

Fabric: Red surface (2.5YR 5/8) and reddish yellow section (5YR 6/6), hard, with clean fractures.

Comparisons: The bird motif, derived from African red slip, is attested in eastern products. See J.W. Hayes 1972, 360, Figs. 77a-d; Carandini 1981 Pl. CXVI.30–33.

Date: 440–490 AD.

18. LRA 769-01 Amphora African IIC/Keay VI

Description: Rim with high, everted lip, convex exterior, essentially straight interior, and pointed tip.

Fabric: Light red surface (2.5YR 6/6) and red section (2.5YR 5/8), soft; abundant white inclusions (calcite).

Comparisons: The form is attested throughout the Mediterranean basin (Villa 1994, 389; Keay 1984, 118–21).

Date: Middle of 3rd–beginning of 5th c. AD.

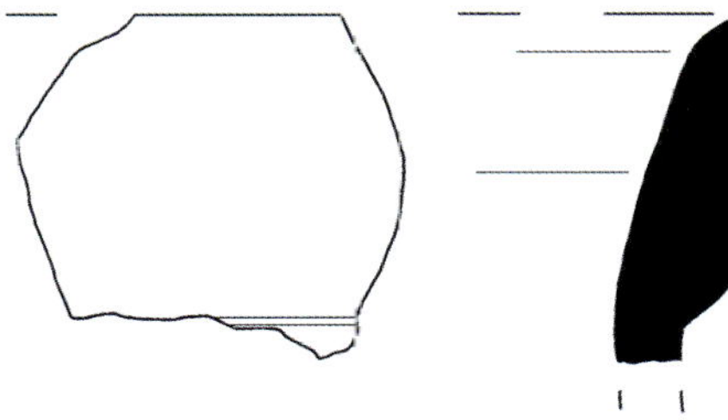

19. LRA 230-12 and **LRA 230-13** Amphora Keay XXVI

Description: Outward-turned rim in the form of a raised ring.

Fabric: Red surface (2.5YR 5/6) and red section (2.5YR 5/8), hard, with irregular fractures.

Comparisons: The form, more commonly known as spatheion, is the most commonly attested type of amphora in the Mediterranean basin in this period. See Panella 1993, and Villa 1994.

Date: Second half of 4th–5th c. AD.

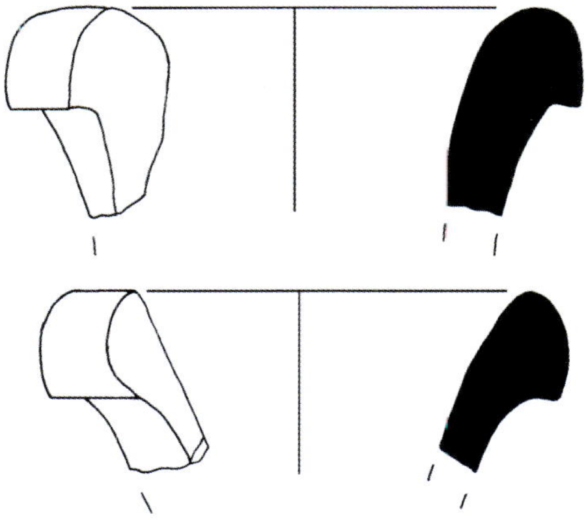

VIII. The Medieval and Post-Medieval Periods

34
Medieval and Post-Medieval Settlement in the Metapontino

Erminia Lapadula

The Early Medieval Landscape (7th–10th centuries)

Three sites only yielded Early Medieval pottery (*Table 34.1*). They are characterized by very small quantities of artifacts that are hard to distinguish, including a few fragments of pottery in the "Broad Line" style, coarse ware, and tiles (see Ch. 35).[1] The general scarcity of artifacts prevents definitive interpretation of the types of settlement, especially in the absence of archaeological excavations, but the assemblages indicate that these were likely farmhouses, and it is surely no coincidence that all three structures were built on the sites of older farmhouses.

Besides the choice of sites previously occupied in Greek and Roman periods (in contrast to Late Medieval settlements), what distinguishes this early period is the concentration of sites on the Basento side of the study area. Occupation was along tributaries (Fosso Sant'Angelo, Conca, and Venella), not at the mouths but secluded upstream. Their elevations, between 46 and 76 masl, varied little and were in the middle range. Each stood on the edge of a valley in a position that commanded these three valleys, taking advantage of the grazing afforded by the valley, the adjacent flat surfaces of the marine terraces, and the presence of springs, as had their predecessors on these sites.

The low number of sites in the Early Medieval Metapontino suggests a depopulation of the territory in this period, the causes of which are no doubt complex and certainly include the critical situation following the Greco-Gothic War (AD 535–553). The general political, administrative, and cultural disintegration, and the advance of an economy based on self-sufficiency were sure features of the central decades of the war. It is reasonable to imagine a severely limited human presence in the countryside, distinguished by poor and modest habitations that were distributed sporadically rather than nucleated. These were likely isolated from larger settlements because of the poor conditions of the abandoned roads. To these problems must be added the shift of the Bradano's bed, the result of the river's poor drainage, and the consequent abandonment of the coastal plain in favor of the surrounding hilltops to escape the unhealthy conditions (see Ch. 1).[2]

The evidence from Lacava's topographic research in the area surrounding the urban center of Metaponto at the end of the 19th century and the excavations at Torre di Mare directed by Noyé at the end of the 1970s demonstrate a very modest, sporadic human presence in the area bounded by the two rivers. This was marked by the existence of a small nucleus of settlement in the place where the castellum/castrum of Torre di Mare would subsequently rise.[3] There are, nevertheless, archaeological traces, however discontinuous, of settlement elsewhere in the landscape in the early centuries of the Medieval period. The finds from the excavation of the castrum at Metaponto suggest an Early Medieval reoccupation of the urban center, site of the Late Roman settlement. This was likely a short-lived occupation, which does not appear to extend beyond the middle of the 7th century. The evidence indicates a funerary rather than residential use of the area by the Early Medieval settlement, which was presumably located at a short distance. Two of the three tombs recovered in the area of the former yielded personal ornaments and clothing accessories dateable between the end of the 6th and middle of the 7th centuries.[4]

The chance discovery of a tomb in the San Salvatore locality produced a domestic amphora with brown bands and a belt buckle dateable to the middle of the 6th century.[5] An excavated context at Pantanello, which included painted pottery, a coin (*follis animo*) dateable to the 6th century, and a commercial amphora bearing the graffito ΣΤΕΦΑΝΟΥ

---

[1] On landscape archaeology, see Cambi and Terrenato 1994; Carandini and Cambi 2002, 43; and Cambi 2003.

[2] J.M. Martin 1993, 86; Fonseca 2002, 41.

[3] Lacava 1891, 325; Noyé 1984; *Torre di Mare* I.

[4] See most recently L. Giardino 1991, 850–51.

[5] D'Andria 1978b, 159–60; Salvatore 1982a, 57.

Site	Location	Elev. masl	Dateable ceramics	Scatter size/ density	Previous occupation	Medieval occupation
326F	Venella	76	2/2	50 x 40/v. H	Prehistoric activity Greek farmhouse (large)	Early Med. farmhouse
388F	Sant'Angelo	46	2/95	40 x 45/H–M	Roman Imperial, Late Roman farmhouse (large)	Early, Late Med. farmhouse
505F	Avinella	62	1/2	20 x 20/M	Prehistoric activity Greek farmhouse	Early Med. farmhouse
Total sites: 3		**Average: 61 masl**				

Table 34.1 Early Medieval sites (7th–10th c.). Scatter size in m. H = heavy, M = moderate, L = light, S = sparse, v. = very.

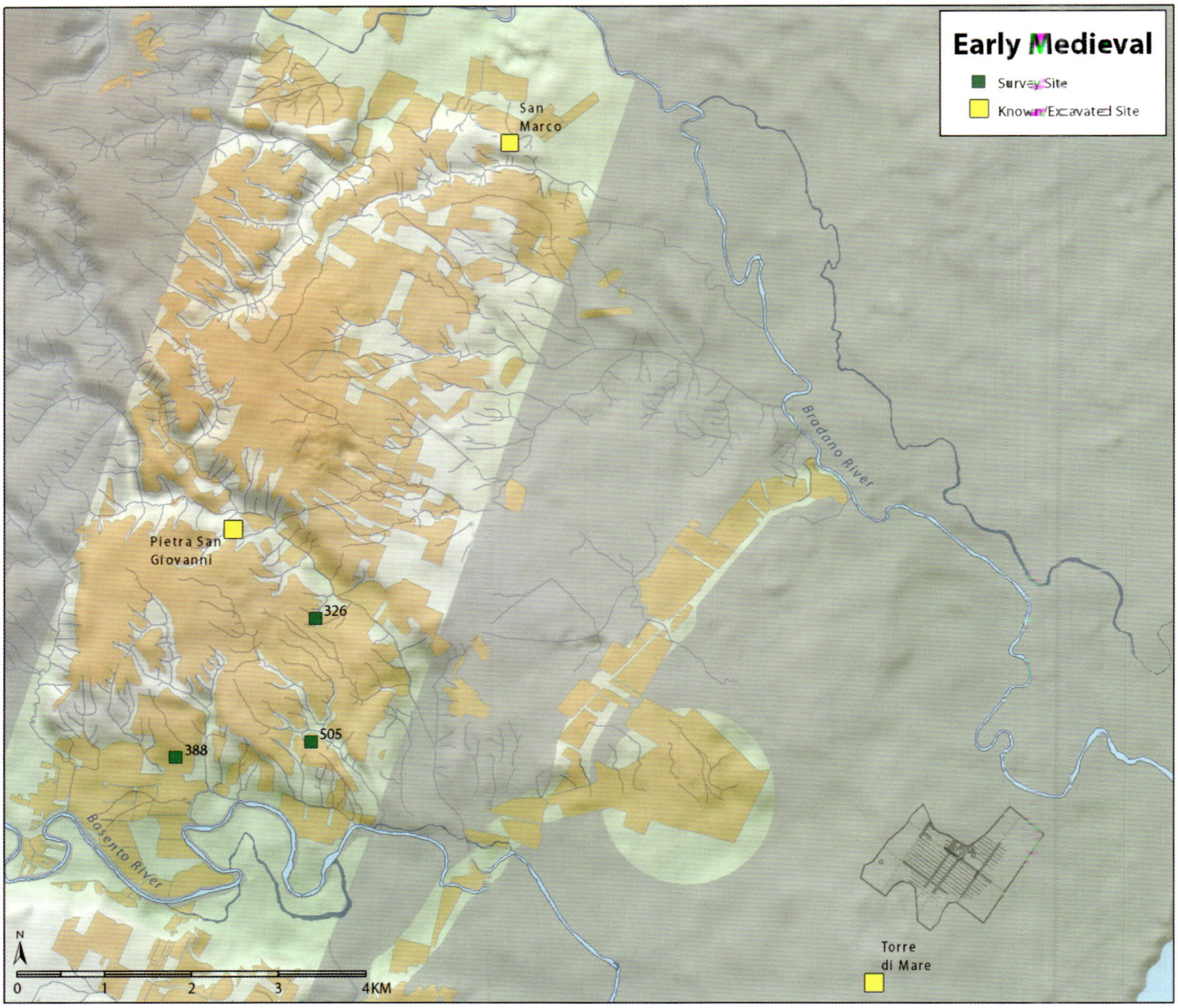

Figure 34.1 Early Medieval sites. (JT)

Site	Location	Elev. masl	Dateable ceramics	Scatter size/ density	Previous occupation	Medieval and Post-Medieval occupation
84F	San Marco	91	7/28	40 x 40/S	Neolithic activity; Late Hellenistic farmhouse	High, Late Med. farmhouse; Post-Med. activity
86F	San Marco	90	7/31	15 x 15/v. S	Prehistoric activity	High, Late Med. farmhouse; Post-Med. activity
90F	San Marco	88	1/18	15 x 15/L		High, Late Med. farmhouse
140NS	San Marco	87	42/135	150 x 100/ M–H	Greek necropolis	High, Late Med. nucleated settlement; Post-Med. activity
300NS	Avinella	105	6/606	200 x 100/ v. H	Neolithic domestic structure; Greek necropolis	High, Late Med nucleated settlement; Post-Med. activity
328F	Avinella	66	1/17	35 x 25/M–L	Greek farmhouse	High Med. farmhouse; Post-Med. farmhouse
381F	Sant'Angelo	45	2/88	30 x 25/H		High, Late Med. farmhouse
507F	Avinella	78	2/29	55 x 45/v. H	Prehistoric activity	High, Late Med. farmhouse; Post-Med. farmhouse – kiln
550F	Venella	49	1/16	25 x 25/M		High, Late Med. farmhouse
551F	Venella	51	5/50	50 x 50/M–L		High, Late Med. farmhouse; Post-Med. activity
Total sites: 10		**Average: 75 masl**				

Table 34.2 High Medieval sites (11th–12th c.). Scatter size in m. H = heavy, M = moderate, L = light, S = sparse, v. = very

ΧΑΡΤΟΛΑΡΗΟΥ, can be dated to the final phase of the Early Medieval period. The morphological characteristics of the amphora (perhaps an Ionian product) resemble those of Salentine types dated between the 9th and 11th centuries. The absence of glazed pottery suggests the 12th century as the terminus ante quem for the closing of the pit. This hypothesis is strengthened by the attestation of a *Stephanos cartularius* in a decree of 1019.[6]

Despite traces of human presence in the Metapontino between the Late Roman and developed Medieval periods, it is at the moment impossible to define the features of this form of settlement. This situation begins to change in the late 11th century with a reorganization of the countryside, the results of which become manifest between the 13th and 14th centuries.

The High Medieval Landscape (11th–12th centuries)

By the end of the 11th century the Metapontino was occupied by 10 sites (*Table 34.2*). Only 2 of them (84 and 328) had previously been occupied by farmhouses in the Greek and Roman periods. The remaining 8 were either on sites where there was only prehistoric activity (sites 86, 507), Greek burials (site 140NS) a combination of these (site 300NS), or which were completely untouched before (sites 90, 381, 550, and 551). All 10 sites continued into the Late Medieval period, except site 328.

The recovered pottery consists of simple, unslipped containers intended for use on the fire, in the pantry, and on the table. There is also a modest quantity of table ware with lead-based monochrome green glaze, as well as a very limited quantity of monochrome green-glazed pottery with decoration in dark green or brown, which can be interpreted as imitations of Sicilian or northwest African pottery. Among the domestic ware, closed forms (jars and jugs) predominate over open forms (bowls), and the table service is composed of open forms for individual use. A 12th-century date for these settlements is suggested by the morphological characteristics of the cooking and coarse wares, and by the particular coating and decoration of the glazed table wares.

The quantity and variety of collected artifacts argue for identifying these structures as small, isolated

[6] Carter 1983a, 463, 466 n. 49; Von Falkenhausen 1978, 122, 194. For a recent study of the amphora, see Arthur 1999, 13.

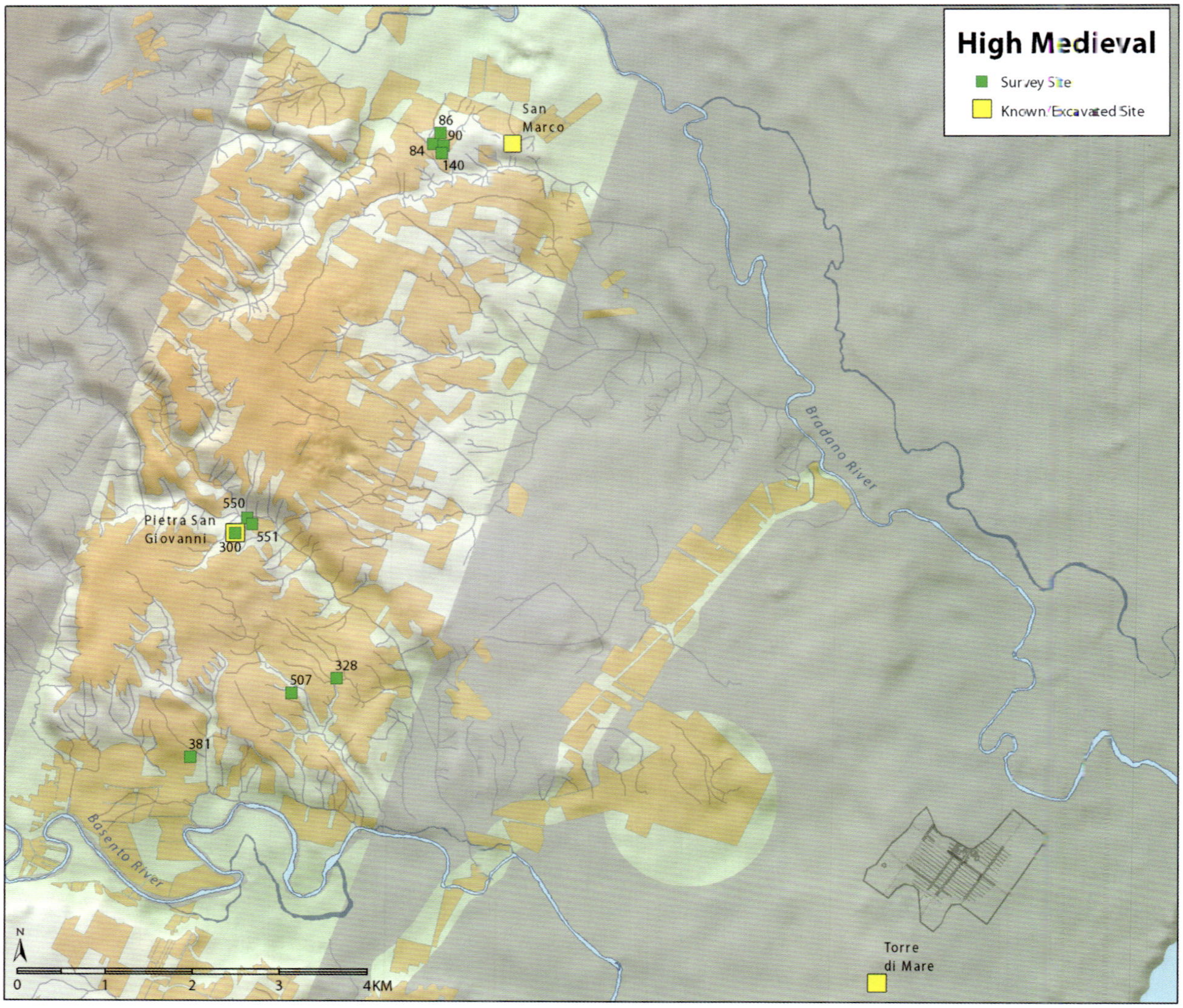

Figure 34.2 High Medieval sites. (JT)

farmhouses, with the exception of site 300 (Pietra San Giovanni), which appears to be a settlement of greater proportions judging from the quantities of artifacts recovered and the architectural remains still visible (*Fig. 34.3*). This survey site was subsequently partially excavated.[7]

The scatter sizes of the previously unoccupied or lightly occupied sites (Neolithic structures, Greek burials) range from 225 m^2 to 2,500 m^2, except in two cases, sites 140 and 300, where the scatter covers 15,000 m^2 and 20,000 m^2 respectively. In most cases, the scatter was of moderate to heavy density, the exceptions being site 84 and two of the smallest sites, 86 and 90. Scatters of betwen 750 m^2 and 2500 m^2 would be appropriate for farmhouses of the Greek and Roman periods. Site 551 was entirely similar to farmhouse scatter sizes documented in the Greek and subsequent periods.[8] The very large ones imply a different kind of settlement (see below). The presence of slipped table wares places these sites in a commercial and economic circuit that was perhaps limited in range, but in any event also beyond mere internal consumption and perhaps not exclusively agrarian.

This period saw a considerable geographic expansion of settlement with respect to the Early Medieval period, though the diffusion of sites was still limited to a few areas. Isolated farms continued in Sant'Angelo and Avinella. The lower Venella was abandoned and new clusters sprang up in the upper Venella, and, on the Bradano side for the first time, at San Marco. Two sites (381 and 507) remained in Avinella and in the Basento valley. Seven sites arose in the Venella valley

[7] Noyé 1976.

[8] See Ch. 18 on types of structures and farmhouse sizes in earlier periods.

Figure 34.3 Site 300, Pietra San Giovanni, overlooking the Venella valley, with remains of the church of St. John the Evangelist.

and above the Bradano valley, arranged in tight clusters, indicating an inability—or reluctance—to live in isolation. Sites 550 and 551 stood on the Venella valley floor at the confluence of several small tributaries; above them loomed site 300 at the edge of the marine terrace to the southwest. The inhabitants of the cluster on the Bradano, which consisted of 4 sites (84, 86, 90, and 140), occupied the gently sloping marine-terrace edge leading down into the valley between Giardinetto and San Marco; these sites were presumably interested in the nearest part of the floodplain, hard by the marine terraces, doubtless the driest part. In all cases there was a clear desire to avoid low-lying areas: the lowest elevation was 45 masl, well above the lowest elevations (ca. 15 masl) occupied by the Greek colonists. Conditions in the Bradano and Venella valleys obviously had improved, but not so much that the inhabitants felt secure in living closer to them. Site 34 stood alone on the eroded hills west of the main transect, and one may reasonably interpret its isolation as a sign of the poverty of the soil.

Within a territory marked by limited occupation and a pattern of small habitations, there were three conspicuous settlements at San Marco, Pietra San Giovanni, and Torre di Mare. The complex at Torre di Mare appears in documentary sources beginning in the 11th c. as a castellum/castrum under the name *Civitas Sanctae Trinitatis* (thereafter, by the early 13th century at the latest, also under the name *Turris Maris*).[9] Recent excavations of several features have added little to its archaeological history.[10] It began in the 10th century,[11] but there is precious little data for the Norman phase. The site of Pietra San Giovanni (*Fig. 34.3*) is commonly described at the end of the 11th century as a fortified settlement (castrum), equipped with walls and extending over the highest part of the promontory which dominates the valley of the Venella at its upper end.[12] The ceramic artifacts recovered during the survey suggest that occupation of the promontory began in the middle of the 12th century and continued for another two centuries, thus confirming the evidence supplied by Noyé's excavations of the late 1970s.[13] The materials dating to the 12th century belong to several forms designed for kitchen use, jars in particular, and for table use, such as several types of bowl with lead-based monochrome green glaze. There are also several fragments of glazed monochrome green ware with decoration in dark green and brown on the rim, which may be interpreted as a local imitation of Sicilian/northwest African pottery (see Ch. 35). The overall typology of the wares present at this site suggests an economic

[9] Many of the diplomi (original Medieval grants of privileges) are reproduced in Lacava 1891, 243–68.

[10] Bertelli 2002, 159–60.

[11] On the Byzantine and Norman phases of Torre di Mare, see Noyé 1987 and 1988; J.M. Martin 1993, 86 n. 103; and the work of Bertelli and Roubis, *Torre di Mare* I, with relevant bibliography.

[12] Lacava 1891, 27; Noyé 1976, 719.

[13] Noyé 1976, 718–19.

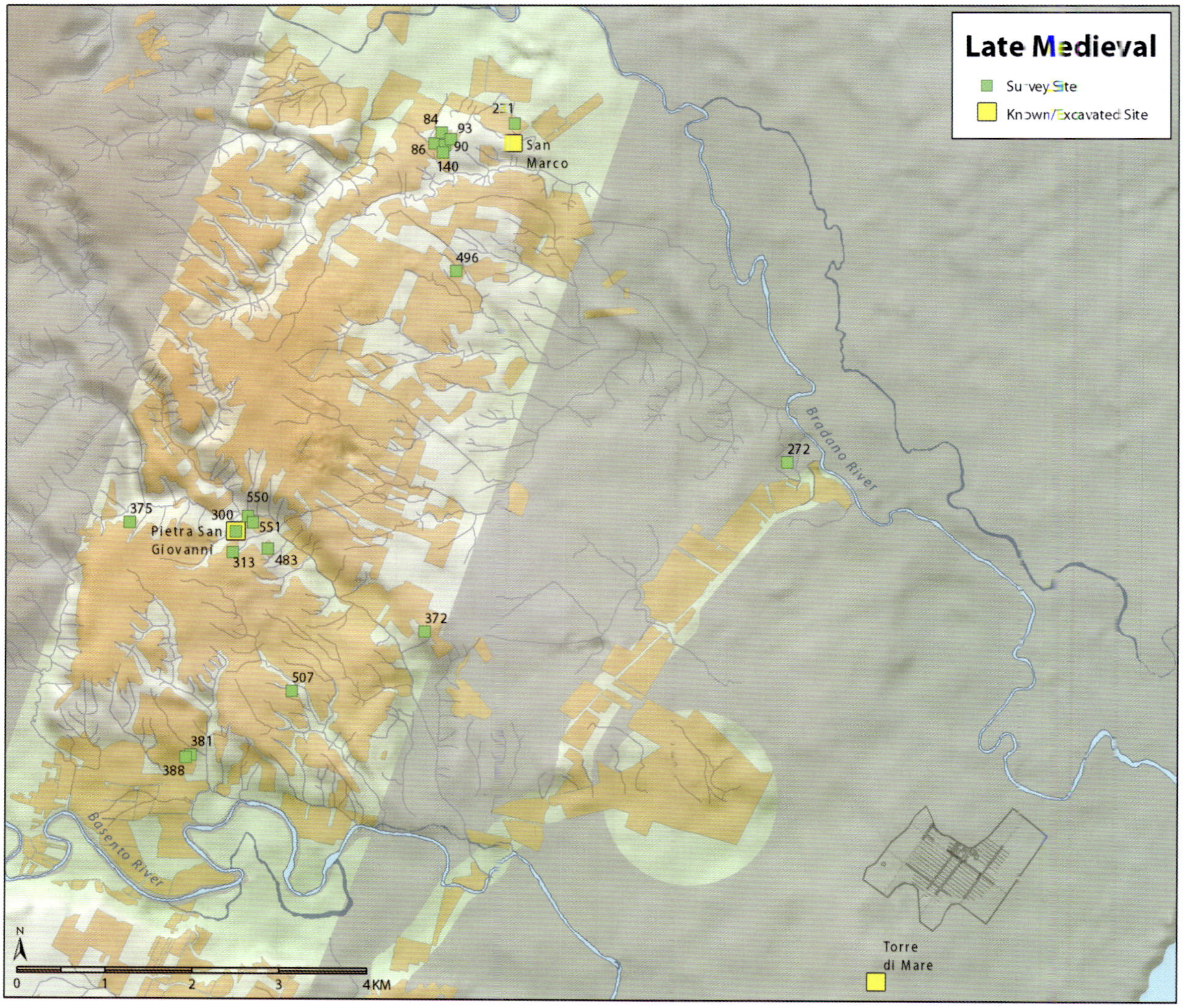

Figure 34.4 Late Medieval sites (13th–15th c.). (JT)

outlook of a certain level. The glazed table wares attest a purchasing power connected to a lively economy characterized probably by exchanges and commercial relations with the most important of the neighboring population centers.

The scatter at site 140 on the Bradano side of the study area is of the same order of magnitude as that of Pietra San Giovanni. This was likely an estate farm or masseria, antedating the later Masseria San Marco approximately 800 m to the east. Nearby are 2 small possible farmhouse sites (86 and 90) and one larger one. An additional 5 sites, with little pottery and none dateable (85, 89, 90, 91, and 92), were very likely active in this period, and were perhaps sheds, rather than residences. This agglomeration of 9 sites plausibly formed a type of agricultural village with a larger central structure.

The average elevation of the sites dateable to the High Medieval period is 72 masl, which is notably higher than in the preceding period (61 masl). This average is also significantly higher than that of farmhouses in the Greek period, which ranged from 57 masl to 83 masl (in 200 BC), and in the Roman, with a maximum of 52 masl in the 2nd century BC and a minimum of 38 masl in the Late Roman period. All across the chora in the Medieval period there was an evident desire to avoid low-lying areas: conditions in the Bradano and Venella valleys were such that inhabitants did not feel safe living near them.

The Late Medieval Landscape (13th–15th centuries)

Only one site (328) from the High Medieval period failed to prolong its existence into the Late Medieval,

Site	Location	Elev. masl	Dateable ceramics	Scatter size/ density	Previous occupation	Medieval and Post-Medieval occupation
84F	San Marco	91	5/28	40 x 40/S	Neolithic activity; Late Hellenistic farmhouse	High–Late Med. farmhouse
86F	San Marco	90	14/31	15 x 15/v. S	Prehistoric activity	High–Late Med. farmhouse
90F	San Marco	88	5/18	15 x 15/L		High–Late Med. farmhouse
93F	San Marco	86	1/5	25 x 25/M		Late Med. farmhouse
140NS	San Marco	87	41/135	150 x 100/M–H	Greek necropolis	High–Late Med. nucleated settlement
231F	San Marco	35	5/21	50 x 35/v. H	Late Hellenistic, Early Imperial, Late Roman farmhouse	Late Med. farmhouse
272F/T	San Salvatore	32	13/13	NR	Greek farmhouse + tomb (large)	Late Med. farmhouse
300NS	Avinella (Pietra San Giovanni)	105	386/606	200 x 100/v. H	Neolithic domestic structure; Greek necropolis	High–Late Med. nucleated settlement; Post-Med. activity
313F	Avinella	103	7/7	30 x 25/v. H	Prehistoric activity; Greek necropolis	Late Med. farmhouse
372F	Demanio Venella	62	1/5	60 x 45	Prehistoric activity; Late Hellenistic, Roman Imperial, Late Roman farmhouse	Late Med. farmhouse
375	Scarciullo	114	1/5	30 x 30/H	Neolithic activity; Greek farmhouse	Late Med. farmhouse
381F	Sant'Angelo	45	37/88	30 x 25/H		High Med. farmhouse
388F	Sant'Angelo	46	13/95	45 x 40/H–M	Roman Imperial, Late Roman farmhouse	Early, Late Med. farmhouse
483F	Venella	63	1/2	40 x 30/H		Late Med. farmhouse
496F	Cogno del Prete	71	1/2	25 x 20/M–L		Late Med. farmhouse; Post–Med. activity
507F	Avinella	78	10/32	40 x 40/H	Prehistoric activity	High–Late Med. farmhouse
550F	Venella	49	7/16	25 x 25/M		High–Late Med. farmhouse; Post-Med. activity
551F	Venella	51	27/50	50 x 50/M–L		High–Late Med. farmhouse; Post-Med. activity
Total sites: 18		**Average: 72 masl**				

Table 34.3 Late Medieval sites (13th–15th c.). Scatter size in m. H = heavy, M = moderate, L = light, S = sparse, v = very.

but there was generally a great increase in life during this period (*Table 34.3*). Of the 18 sites active in the Late Medieval phase, 8 (sites 93, 231, 272, 313, 372, 375, 483, and 496) were new foundations with respect to the preceding phase; 4 of them (sites 231, 272, 372, and 375) were constructed on the remains of farmhouses dating from the Greek and Roman periods. None of these new sites had scatters of dimension exceeding what would be expected of a farmhouse.

The Late Medieval farmhouses are present on the surface as scatters of domestic pottery and tile fragments, as well as rare examples of table pottery bearing lead- and tin-based coatings. Several new farm sites, like 313 and 496, had extents of 30 x 25 m^2 and 25 x 20 m^2 respectively, suggesting that some Late Medieval farmhouses may have been significantly smaller than their ancient predecessors. Nucleated settlements (sites 140 and 300) continued to be a significant presence in this period, distinguished primarily by their greater dimensions (average ca. 175,00 m²), and also by the variety of ware classes present (high-quality table ware, domestic wares for table and pantry, and cooking wares), and occasionally by indications of activities associated with agriculture (querns), metal-working, and ceramic manufacture (metal and clay slag).

The economy of the Late Medieval Metapontino was relatively lively, as demonstrated by the recovered pottery belonging to ware classes of a certain level of distinction. Particularly noteworthy is the recovery at site 381 of a bowl fragment in graffito ware, which can probably be ascribed to a northern Adriatic producer of prestige wares (see Ch. 35, **No. 72**). This presence could be described as an important indicator of the socioeconomic power of the site in question, capable of purchasing valuable products originating in distant manufacturing centers and, as such, fully integrated into the Adriatic commercial circuit, probably via Torre di Mare.

The overall pattern of settlement remained a combination of nucleated and isolated farm sites, as it had been just one or two centuries before, but the occupants of the countryside now spread themselves out over a greater proportion of the landscape, moving further inland and closer to the stream valleys than in the previous period. In the northern section of the study area, for example, the agglomerate of farmhouses near San Marco was increased by the addition of site 93. At the very edge of the Bradano valley, approximately 800 m to the east, site 231 was reoccupied, as was site 272, 5 km to the southeast; the rehabilitation of two positions so close to the river after so many centuries of abandonment indicates a renewed interest in the natural resources offered by the Bradano floodplain. Yet the high plains of the northern and central chora remained empty, with the exception of site 496 in Cogno del Prete, which had uncontested access to all of the fertile land around it.

In the southern third of the chora, the loose nucleated group (sites 300, 550, and 551) at the boundary between Scarciullo and Avinella grew to 5 sites with the addition of sites 313 and 483, located on the marine terrace edge and in the Venella valley, respectively. One new farmhouse (site 375) arose at the northern and northwestern edges of Scarciullo. The new interest in the Scarciullo area may have something to do with the town (borgo) that came into existence sometime around AD 1200 just to the west, which later became Bernalda. Southeastern Avinella and the southern end of the Venella valley hosted dispersed farmhouses positioned on the marine terraces and slopes (sites 372 and 507). In the Basento valley, a cluster of two farmhouses (381 and 388) stood near Masseria Sant'Angelo Vecchio on the eastern edge of the alluvial terrace.

The important nucleated complex on the coastal plain, Torre di Mare, was matched in the territory by the sites at Pietra San Giovanni on the Venella (with which site 300 is related) and San Marco on the Bradano (with which site 140 is likely connected). Torre di Mare, identified as a castrum of Frederick II, was the principal settlement in the Metapontino in this phase, occupied continuously until the second half

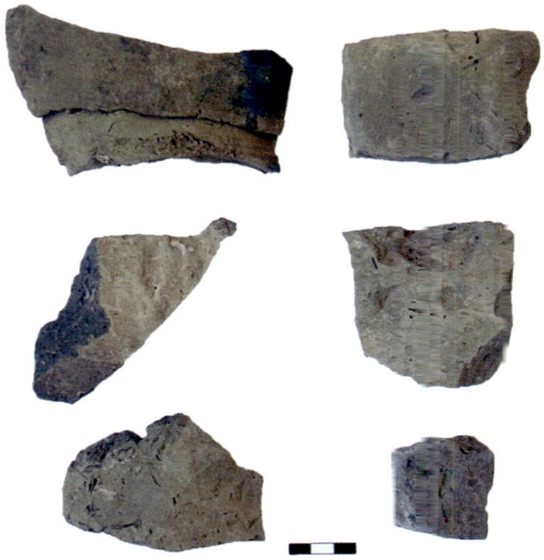

Figure 34.5 Kiln wasters, site 300.

Figure 34.6 Fragment of stone quern, site 300.

Figure 34.7 Decorated wares, site 300

Figure 34.8 Metal slag, site 300.

of the 17th century, when it was almost completely abandoned.[14] At this time it was configured as a castle or military fortress, mentioned in the *Statutum de reparatione castrorum* of Frederick II pertaining to the years 1241–1246; it was associated with a settlement consisting of several small residential clusters, a harbor/port, and a salt works.[15]

Evidence from the excavations at Pietra San Giovanni suggests that in the 12th–14th centuries this fortified nucleus also became an open settlement of considerable size, a village, probably belonging to the monastery of San Michele Arcangelo at Montescaglioso, from which depended its cult building dedicated to St. John the Evangelist (*Fig. 34.3*).[16] This settlement, commonly known as "casale dell'Avinella," was depopulated and abandoned starting in 1488 according to several Aragonese documents.[17] Its position was naturally fortified, granting control over the Basento and Venella valleys, and it was associated with an important transport route connecting Taranto and Salerno, as well as another route in the direction of northern Calabria. In the Late Medieval phase the defensive structure was replaced by an open habitation better suited to the exploitation of the soil.

The ceramic evidence collected during the survey from site 300 is consistent with this hypothesis. A large number of tile fragments were recovered, as well as considerable quantities of coarse pottery and cooking ware for use in the kitchen and pantry, and slipped and decorated wares for table use. Among the kitchen wares pots predominated, which were used for slow cooking of meats; among the coarse wares, the closed forms for the pantry predominated. These were large jars equipped with lids, used for the storage of solid and liquid foodstuffs or for the transport of water over short distances; there were also bowls and cups for individual use and basins and bottles for collective use. Particularly interesting is the presence of the cauldron among the unslipped wares, suggesting room-temperature use of a container generally attested in refractory fabric designed for exposure to heat. Site 300, moreover, is the only site in the period to have produced commercial amphorae, albeit in modest quantities (*Fig. 34.5*), indicating the integration of the settlement into a commercial circuit probably based on agricultural exploitation of the area. The evidence from the morpho-typological analysis of the pottery, as it pertains to dietary habits, is consistent with the consumption of agricultural products such as cereals and meats, the latter perhaps associated with small-scale pastoral activity.[18] The recovery of several stone querns (*Fig. 34.6*) on the site confirms the prevalence of an agrarian economy based on cereals.

The large quantities of impermeable ceramics with lead-based glaze (*Fig. 34.7*) and, in two cases, the presence of containers with tin-based glaze (proto-maiolica) attest the vivacity of this rural nucleus's exchanges and its commercial relations with nearby civitates (Taranto) or larger centers (such as Torre di Mare), from which it likely acquired its high-quality pottery. Nevertheless the possibility of ceramic production in loco cannot be excluded, even if indications

[14] Bertelli 2002, 163.

[15] Sthamer 1995, 116; Bertelli 2002, 161, 166.

[16] Noyé 1976, 719. On the concept of the village (casale), see Poso 1988, 191–202; and J.M. Martin 1993, 282–89.

[17] Bubbico 1996, 165.

[18] On later Medieval dietary habits, see Bertelli 2002, 166.

of manufacture have not yet been verified.[19] Also significant are the ten fragments of metal slag (*Fig. 34.8*) recovered on the surface and attributable probably to a metallurgic production facility present on the site.[20] The presence of such activity suggests the possibility of commercial relations with nearby metal-working settlements—Torre di Mare and perhaps Taranto—for the instruction and technical knowledge, and with Calabria and perhaps even the Balkan area for the supply of the raw materials.[21]

The settlement at Pietra San Giovanni was abandoned in the second half of the 15th century, a date provided by not only the documentary sources but also the archaeological evidence, which records very few ceramic artifacts dating beyond the 15th century.[22]

In addition to those at Torre di Mare and Pietra San Giovanni, one must also consider the nucleated settlement at San Salvatore, to which site 272 is surely related. It was recently identified by excavation, but has been known for some time from documentary sources, occasional surface ceramic finds, and architectural remains still visible on the surface.[23] This was probably a fortified village of Norman foundation, later extensively renovated beginning in the 17th century. Its abandonment is attested in an Aragonese document of 1488, which lists it among deserted settlements along with Pietra San Giovanni.[24] The nucleated settlement at San Salvatore, which included a cult building, was therefore a third important pole playing a notable political, religious, and economic role in the territory in the later Medieval period.

The Post-Medieval Landscape (16th–19th centuries)

Evidence of settlement in the Metapontino after the 15th century comes from 29 sites (*Table 34.4*) but the relative meagerness of the ceramic evidence suggests a frequentation of the area more than a stable occupation, with the exception of sites 61, 328, and 800. Fifteen, or more than half, of these sites were constructed on the remains of earlier farmhouses dating from the Greek and Roman periods and one on a farm of the Medieval period. These still may have provided useful building materials. Some Post-Medieval sites may reasonably be identified as farmhouses. Site 61, a new foundation where previous occupation consisted of a single Greek tomb, had a scatter of 30 x 30 m^2 extent after shallow plowing. This would be indicative of a modest farmer's residence. In general, among the attested ware classes, coarse pottery is predominant (by number of fragments), in closed forms (jugs, jars) and open forms (bowls, basins). The cooking ware and coated wares, both slipped and glazed-enameled, are present in reduced quantities of examples. An important result is the complete absence of the polychrome graffito ware typical of the Terra d'Otranto in the Renaissance, which circulated intensively in the Salento between the 15th and 17th centuries.[25]

The distribution of settlement changed radically between the end of the Medieval and the Post-Medieval periods. All activity in the Bradano valley was once again concentrated in Giardinetto (abandoned since the Greek period), where three farm sites (465, 469, and 793) cultivated the rich soils of the mass movement deposit and part of the floodplain (see Ch. 2 for discussion of the geomorphology). Site 496 remained the lone settlement on the "high plains" of Cogno del Prete. As in the Medieval period, the majority of settlement was concentrated in the southern third of the territory between the Venella and the Basento, but the sites were now more clearly arranged in clusters spread over a greater extent. Northern Scarciullo hosted a community of two farm sites (368 and 800), as well as a new farm site (61) about 500 m to the southeast. Northern Avinella was occupied by a neighborhood consisting of the reduced settlement at Pietra San Giovanni (site 300) and farmhouse site 551 in the valley below, while in the southeastern section stood two farm sites (328 and 507) separated

[19] On the production of glazed polychrome ceramics in Puglia and Basilicata, see P. Tagliente 2001 and 2002, with relevant bibliography.

[20] The current lack of archaeometric analysis of the slag precludes speculation as to the type of metal-working practiced in the settlement.

[21] Evidence of iron-working activity was recovered at Torre di Mare, for which the raw material appears to have been supplied from Calabria (Cuteri 2002a, b). On commercial relations for the supply of raw materials between the Salento and the Balkans in the Medieval period, see Arthur 2003.

[22] For the documentary sources, see Bubbico 1996, 165; for the archaeological evidence, see Noyé 1976, 718–19.

[23] The settlement at San Salvatore is mentioned in numerous documents beginning in the Norman period. For an historical synthesis, see Bubbico 1996, 163–64, with relevant bibliography. An excavation project in September 2001 was carried out during the expansion of the irrigation network by the *Consorzio di Bonifica di Bradano e Basento*. The investigation (currently being prepared for publication by the author) identified an inhabited context dated by ceramic finds between the 13th and 14th c. The author wishes to thank A. De Siena for the preliminary data and the opportunity to study the site.

[24] De Coronei 1880, 92.

[25] For Salentine graffito ware see Ciminale 2000, 93–99, and P. Tagliente 2002, 551–54.

Site	Location	Elev. masl	Dateable ceramics	Scatter size/ density	Previous occupation	Post-Medieval occupation
61	Scarciullo	116	6/6	30 x 30/S	Greek tomb	Farmhouse, 16th–18th c.; 18th–20th c.
84	San Marco	91	2/28	40 x 40/S	Neolithic activity; Greek farmhouse; Late Hellenistic farmhouse; High–Late Med. farmhouse	Activity, 16th–17th c.; 18th–20th c.
140	San Marco	37	7/135	150 x 100/H–M	Greek tomb; High–Late Med. nucleated settlement	Activity, 19th–20th c.
154	Basento valley	19	10/10	40 x 40/M	Greek farmhouse	Activity, end 15th–18th c.; 19th–20th c.
300	Avinella	105	6/606	200 x 100/H	Neolithic domestic structure; Greek necropolis; High–Late Med. nucleated settlement	Activity, end 15th–18th c.; 16th–18th c.
328	Avinella	66	7/17	35 x 25/L–M	Greek farmhouse	Farmhouse, 15th–18th c.
368	Scarciullo	112	4/4	45 x 30/M	Greek agricultural structure	Farmhouse, 19th–20th c.; 16th–18th c.
413	Pantanello	15	2/2	30 x 30/H	Prehistoric activity; Greek farmhouse	Activity, 19th–20th c.
464	Giardinetto	43	2/2	15 x 10/L	Greek necropolis	Activity, 19th–20th c.
465	Giardinetto	54	7/9	80 x 65/H	Bronze Age domestic structure; Greek farmhouse; Imperial–Late Roman farmhouse	Activity, 18th–20th c.; 16th–18th c.
469	Giardinetto	35	1/1	40 x 25/M–L	Prehistoric activity; Greek farmhouse	Activity?
475	Demanio C.	63	1/1	30 x 30/L	Neolithic domestic structure; unidentified Greek structure	Activity, 19th c
478	Venella	35	1/1	40 x 25/L	Prehistoric activity; Greek agricultural structure	Activity
491	Venella	55	1/1	25 x 20/H	Greek farmhouse	Activity, 18th–20th c.
492	Venella	57	1/1	30 x 25/H	Greek farmhouse	Activity, 18th–20th c.
493	Venella	63	1/1	30 x 20/H	Greek farmhouse	Activity, 18th–20th c.
496	Cogno Prete	71	22/22	25 x 20/H	Greek farmhouse	Activity, 19th–20th c.
503	Lago Lupo	68	2/2	60 x 50/H	Prehistoric activity; Greek, Late Hellenistic farmhouse	Activity, 19th–20th c.
507	Avinella	78	9/29	40 x 40/H	High–Late Med. farmhouse	Farmhouse and kiln, end 15th–18th c.
528	Giardinetto	63	1/29	35 x 25/H	Neolithic/Bronze Age domestic structure; Greek farmhouse; Imperial, Late Roman farmhouse	Activity, 18th–20th c.; 19th c.
551	Venella	51	7/50	50 x 50/M–L	High–Late Med. farmhouse	Activity, 16th–18th c.
557	Sant'Angelo	40	2/2	15 x 10/L	Greek tomb	Activity, 18th–20th c.; 16th–18th c.
632	Demanio C.	58	1/1	35 x 25/M–L	Prehistoric activity; unidentified Greek structure	Activity, 18th–20th c.
637	Pizzica	46	1/1	80 x 45/H–M	Prehistoric activity; Greek farmhouse	Activity, 18th–20th c.
639	Pizzica	45	2/2	40 x 40/L	Unidentified Greek structure	Activity, 19th–20th c.
726	Basento valley	19	1/54	140 x 300/H–M	Neolithic activity; Greek farmhouse + tomb; Early Imperial, Late Roman farmhouse	Activity
727	Basento valley	19	3/4	250 x 90/M	Neolithic activity; Greek farmhouse; Imperial farmhouse	Activity, 18th–20th c.; 19th–20th c.
793	Giardinetto	53	1/6	30 x 10/v. L	Prehistoric activity; unidentified Greek structure	Activity, 18th–20th c.
800	Scarciullo	114	6/12	30 x 25/M–L	Unidentified Greek structure	Activity, 18th–20th c.; 19th–20th c.
Total sites: 29		**Average: 59 masl**				

Table 34.4 Post-Medieval sites (16th–20th c.). Scatter size in m. H = heavy, M = moderate, L = light. S = sparse, v. = very.

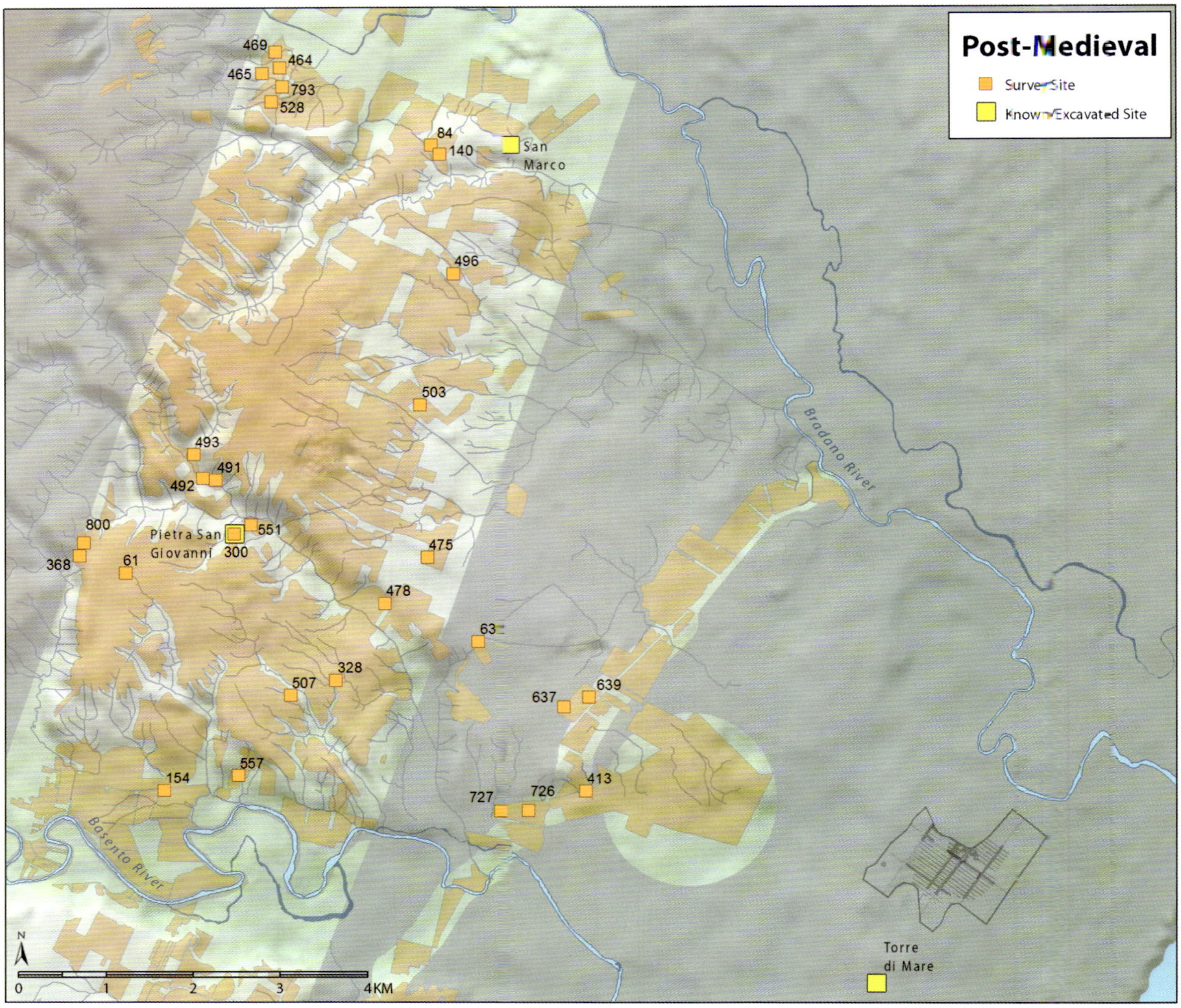

Figure 34.9 Post-Medieval sites (16th–19th c.). (JT)

by deeply incised streams. Finally, the alluvial terrace remnant at the base of the Pizzica/Cardillo marine terrace was frequented for the first time since the Late Roman period with modest reoccupation at sites 726 and 727.

The most substantial settlement in the vicinity remained Torre di Mare, where archaeological excavations document a strengthening of the defensive structures performed between the 15th and 16th centuries, visible in the construction of an isolated tower and a wall circuit with scarp wall.[26] The site has undergone extensive conservation and restoration, and now serves as a public venue for cultural functions and as a tourist attraction.

[26] Bertelli 2002, 162–63.

35
Medieval and Post-Medieval Ceramics

Erminia Lapadula

The pottery fragments recovered by the ICA survey for the Medieval and Post-Medieval periods are considerably fewer in number than for the preceding Greek and Roman eras, as noted in Chapter 34. The purpose of this chapter is to describe the range of wares and illustrate the significant diagnostic pieces. During this time new forms of decoration/glazing emerged and some wares continued over long stretches of time, and for these reasons the ceramics of the High and Late Medieval periods are treated together. Table 35.1 shows the wares, forms, and types that can be dated between the 7th and 15th centuries.

The Early Medieval Period

Only 12 of the fragments recovered could be attributed to the Early Medieval pottery. Of these, 9 belong to coarse ware and 3 to painted vessels used in the kitchen and pantry (*Table 35.1*)

Coarse ware

The morphologically significant fragments of coarse pottery are a rim of a jug and a rim and handle of a cauldron (paiolo), both from site 388. The latter form is characterized by a globular body with a flat base and a round handle for hanging that arches over the rim.[1] Typically produced in very coarse fabric, the paiolo was intended for direct use over the fire to cook soups, stews, and meats, and boil water; it is possible that it was used for storage.[2] This vessel type was fairly diffuse in the Early Medieval period, particularly between the 6th and 7th centuries, and saw increased distribution in the 13th and 14th centuries as attested by discoveries in central Italy (Tuscany, Emilia Romagna, Umbria, Marches, Lazio).[3] The examples recovered in the excavations at Otranto, perhaps brought to the Salento via an Adriatic route, are in large part dateable between the 9th and 11th centuries.[4] After the 14th century, the paiolo continued to be produced in ceramic, albeit in very reduced quantities, having been replaced by the more costly but certainly more robust and lasting metal version.

The Metapontine paiolo fits into the overall picture of coarse ceramic vessels for daily use.[5] The Lucanian example differs in the articulation of the rim, as well as the section of the handle, which is circular instead of flat. The relatively refined fabric, enriched with frequent white inclusions (probably calcite), suggests it was used for heating water. This container was widespread both inside and outside Italy in the later Medieval period.

Painted ware

The few fragments with painted decoration, characterized by simple wide bands in red and brown, belong to jars/domestic amphorae (sites 326 and 505) and, in a single case, an open form that cannot be precisely identified (site 326). Ceramics decorated with wide painted bands (Broad Line style) have been found at many excavated Early Medieval (8th–11th c.) sites of central-southern Italy, with attestations in Campania, Molise, Puglia, Basilicata, Calabria, and Sicily.[6] The evidence consists predominantly of closed forms (amphorae, jugs), while the open forms (bowls, basins) are present in very limited quantities. A considerable corpus of forms with Broad Line decoration was recovered at Otran-

[1] For a more detailed description of the form, see Cuteri 1993, 329.

[2] On its use as a container for the storage of foodstuffs, see Cuteri 1993, 331, with relevant bibliography.

[3] Cuteri 1993, 335.

[4] Patterson and Whitehouse 1992, 96–98.

[5] On the same form in cooking ware, see Mercando 1970, 409–10; Francovich and Vannini 1977, 221; De Marinis 1979, 287; Brogiolo and Gelichi 1986, 313; Gelichi 1986, 122–24; Bernardi et al. 1991, 464–66; and Cuteri 1993.

[6] For a definition of Broad Line ware, see D. Whitehouse 1966a, and Patterson and Whitehouse 1992, 107–10, for a wide-ranging analysis of the ware found in many 9th–12th-c. contexts at Otranto. For a synthesis of painted Medieval ceramics, see Arthur and Patterson 1998, 526, and Patterson 2001, 315–17, with relevant bibliography. For the attestations in Campania (Capaccio Vecchia, Altavilla Silentina), see Maetzke 1976 and Peduto 1979 and 1980; for Molise (San Vincenzo al Volturno), see Hodges and Patterson 1986; for Basilicata (territory of Matera), see D. Whitehouse 1969 and D'Andria 1978b; for Puglia (Ordona, Lucera, San Severo, Otranto, and other Salentine settlements), see D. Whitehouse 1988 and 1966b, as well as Patterson and Whitehouse 1992.

Ware	Form	Type	Date	Sites and counts
Cooking ware (unglazed)	Pot		12th–14th c.	84(1) 140(26) 300(17) 381(1) 550(1)
	Pot			300(6)
	Pot			300(53)
	Pot			300(1)
	Unknown closed			34*(1) 84(5) 86(7) 90(2) 300(27) 313(2) 381(5) [illegible]3(1) 507(4) 550(3) 551(3)
	Unknown			34*(1) 368(1)
Glazed cooking ware	Pot		15th c.	84(2) 86(1) 140(1)
	Unknown open			84(1)
Coarse ware	Bowl		End 13th–14th c.	140(1) 300(2) 375(1) 381(1) 550(1)
	Bowl		13th–14th c.	34*(2) 140(1) 300(1) 375(1)
	Cup			300(3)
	Basin		14th c.	140(2) 300(12) 381(1)
	Paiolo		10th–11th c.	388(1)
	Paiolo		13th–14th c.	381(1)
	Paiolo			300(13) 381(3)
	Paiolo		10th–11th c.	388(1)
	Jug			272(1) 300(2) 381(1) 551(1)
	Trilobate jug			381(1) 550(1)
	Bottle?		13th–14th c.	300(4) 550(1)
	Flask			300(1)
	Jar		13th–14th c.	33*(5) 86(3) 140(6) 300(233) 381(31) 388(13) 507(6) 550(5) 551(25)
	Lid of paiolo		13th–14th c.	381(1)
	Unknown open			140(2) 550(2)
	Unknown closed			34*(4) 84(4) 90(2) 121(5) 140(33) 231(5) 272([illegible]) 313(3) 381(40) 388(32) 505(1) 507(12) 550(2) 551(6)
	Unknown			368(3) 372(2) 496(2)
Painted ware (Broad Line)	Jar		9th–10th c.	326(1) 505(1)
	Unknown open		Early Medieval	326(1)
Painted ware (Narrow Line)	Unknown closed		14th–16th c.	34*(1) 86(1) 140(1)
Monochrome green glaze	Small bowl	*Otranto* 6:16 570	13th c.	86(2) 300(1)
	Bowl	*Otranto* 6:16 571		140(1)
	Bowl	*Otranto* 6:16 568	13th c.	300(1)
	Bowl	*Otranto* 6:17 580	13th–14th c.	300(1)
	Bowl	*Otranto* 6:16 576		551(1)
	Bowl	*Policoro* CXCII 43	13th–14th c.	140(2) 300(1)
	Bowl	*Policoro* CXCIII 47	13th–early 14th c.	300(1)
	Bowl		13th–14th c.	86(2) 300(32) 381(2) 483(1)
	Bowl	*Policoro* CXCII 33	13th–early 14th c.	300(1)
	Bowl			551(1)599(1)724(1) 762(1)
	Bottle/jug		13th c.	300(2) 90(1)
	Jug			300(1)
	Unknown open		12th–14th c.	34*(1) 84(6) 86(7) 90(1) 140(16) 300(43) 507([illegible]) 551(3)
	Unknown closed			84(1) 140(4) 300(14)

Table 35.1 Medieval wares and forms (7th–15th c.).

Ware	Form	Type	Date	Sites and counts
Monochrome brown glaze	Bowl		13th–15th c.	86(2) 300(2)
	Unknown open			86(1) 140(1) 300(2)
	Unknown closed			84(1) 300(2)
Monochrome yellow glaze	Unknown open			84(1) 300(1)
Green-brown decoration	Bowl		12th c.	300(1) 551(2)
Glazed pottery with brown decoration	Bowl	*Montescaglioso* II 16	13th–14th c.	90(1) 300(1)
	Bowl	*Montescaglioso* II 17	13th–14th c.	90(1)
	Bowl	*Otranto* 6:25 693	13th–14th c.	300(1)
	Bowl	*Policoro* CXCII 31	13th–14th c.	300(1)
	Bowl	*Policoro* CXCIII 46	13th–14th c.	300(3)
	Bowl	*Policoro* CXCIII 48	13th–14th c.	300(1)
	Bowl	*Policoro* CXCII 35	13th–14th c.	140(1) 551(1)
	Bowl	*Policoro* CXCIII 54	13th–14th c.	300(1)
	Bowl	*Policoro* CXCVII 8a	13th–14th c.	300(1)
	Bowl		13th–14th c.	34*(1) 140(2) 300(9)
	Bowl		13th c.	140(1)
	Domestic amphora		13th–14th c.	313(1)
	Unknown open		13th–14th c.	84(2) 140(6) 300(1) 507(1) 551(1)
	Unknown closed			90(1) 140(2) 300(2)
Glazed pottery with painted decoration including red	Plate		13th–14th c.	300(1)
	Plate	*Policoro* CXCII 32	13th–14th c.	90(1)
	Bowl	*Policoro* CXCII 39	End 13th c.	140(3)
	Bowl	*Policoro* CXCIII 53	13th–14th c.	90(1)
	Bowl	*Policoro* CXCII 41 *similis*	13th–14th c.	86(1)
	Bowl	*Policoro* CXCII 31	13th–14th c.	300(1)
	Bowl	*Policoro* CXCIII 50	13th–14th c.	300(2)
	Bowl	*Policoro* CXCII 34	Late 13th–14th c.	300(1)
	Bowl	*Policoro* CIII 48		300(1)
	Bowl	*Policoro* CXCII 21		300(1)
	Bowl	*Policoro* CXCII 35		551(1)
	Bowl	*Otranto* 6:24 674	Late 13th–14th c.	300(4)
	Bowl	*Otranto* 6:24 686	Late 13th–14th c.	300(3) 507(1)
	Bowl		End 13th c.	33*(1) 34*(1) 300(1) 507(1)
	Bowl		12th–13th c.	300(1)
	Unknown open			86(1) 140(2) 300(4)
	Unknown closed			90(1) 300(3) 496(1)
Glazed pottery with painted decoration including brown, yellow, and green	Bowl		13th–14th c.	140(1)

Table 35.1 Medieval wares and forms (7th–15th c.) cont.

Ware	Form	Type	Date	Sites and counts
Glazed pottery with painted decoration including brown and green	Bowl	*Policoro* CXCII 29	13th–14th c.	300(1)
	Bowl	*Otranto* 6:24 685	13th–14th c.	300(1)
	Bowl	*Otranto* 6:24 675	13th–14th c.	300(1)
	Bowl		13th–14th c.	300(2) 507(1)
	Bowl		13th–14th c.	300(2)
	Unknown open		13th–14th c.	34*(1) 140(2) 300(5)
	Unknown closed			300(2)
Glazed pottery with brown, red, and green decoration ("RMR")	Bowl			33*(1) 34*(2)
	Bowl	*Montescaglioso* III 23	Mid-13th–14th c.	93(1)
	Bowl	*Montescaglioso* I 1	Mid-13th–14th c.	140(2)
	Bowl	*Policoro* CXCII 35/38	Mid-13th–14th c.	140(1)
	Bowl	*Policoro* CXCII 35	Mid-13th–14th c.	140(1) 300(1)
	Bowl	*Montescaglioso* I.5	Mid-13th–14th c.	140(1)
	Bowl	*Policoro* CCII c	Mid-13th–14th c.	300(1)
	Bowl			300(4)
	Basin		13th–early 14th c.	300(1)
	Unknown open			84(1) 90(3) 140(2) 300(16) 313(1)
	Unknown closed			84(1) 86(1) 140(1) 300(2)
Glazed pottery with green decoration	Unknown open			300(1)
Proto-maiolica	Bowl		Early 13th c.	300(1)
Glazed pottery with graffito decoration	Bowl		End 12th–13th c.	381(1)
	Unknown open		14th–15th	84(1)
Double-dipped ware	Bowl		End 14th–15th c.	36(2) 140(6)
Glazed pottery with green decoration	Bowl		Mid-15th c.	300(1)
Amphorae	Otranto-type		End 11th–12th c.	328(1)
	Unknown			300(3)

Table 35.1 Medieval wares and forms (7th–15th c.) *cont.*

to in 8th–11th-century contexts,[7] but the discovery of several products at Brindisi in contexts dated to the 13th century indicates a modest continuation of production or use beyond the 11th century, contemporaneous with the new forms bearing a decoration of narrow bands (Narrow Line style) which began to spread in the late 12th century.[8]

The High to Late Medieval Periods

A total of 1049 fragments can be dated to the High–Late Medieval periods. They are represented by domestic vessels—coarse table, storage, cooking, and painted wares—as well as fine glazed tablewares (monochrome, polychrome, double-dipped, proto-maiolica, graffito). Transport containers are extremely rare, attested by only 5 amphora fragments. Though the specific proportions may vary, the combination of wares reflects the general distribution of ceramics in High–Late Medieval southern Italy, as confirmed by numerous comparisons with materials from other contemporary contexts (*Table 35.1*).

[7] Patterson and Whitehouse 1992, 105–10.

[8] Patitucci Uggeri 1976, 159. For new ideas on the transition from Broad Line to Narrow Line style based on the painted Medieval ceramics recovered at Quattro Macine (province of Lecce, Puglia), see Tinelli 2000–2001.

Unglazed cooking ware

Domestic pottery employed in the kitchen and for use over fire is attested by a modest number of fragments. Of these 97% are coarse-ware containers (136 fragments), which overwhelmingly outnumber those covered with transparent glaze (4 fragments).

The repertory of forms is restricted. The most prevalent is the jar, a form widely utilized for the cooking of foods. It enjoyed a considerable increase in diffusion beginning in the 10th–11th centuries, when it was produced also in partially refined fabrics to reheat foods rather than cook them. The Metapontine examples have a generally triangular rim with variable edge, ovoid body, and flat base, as well as an occasional handle. Pots, on the contrary, are present in low proportions, even if, in many cases, the reduced size of the fragments prevents secure attribution to one or the other form. Closed forms predominate, with open forms only attested by a single glazed fragment.

Analogous forms are found throughout southern Italy, since these products appear to have circulated very widely. Nevertheless the closest comparisons are found in Lucania and Puglia.

Glazed cooking ware

The 4 recovered fragments of glazed cooking ware belonged to pots. Cooking vessels with impermeable slips, successors of the Early Medieval cooking wares, began to be produced in the 13th century, but spread only in the Post-Medieval period, gradually replacing coarse-ware products.[9]

Coarse ware

Coarse ware for table and pantry use is the most common class (541 fragments, approximately 51% of all Medieval ceramics). Closed forms heavily outnumber open forms. Amphorae lead among the closed forms, followed by jars and a few examples of jugs and bottles, as well as a flask. There is also an amphora lid. The storage amphora/domestic amphora (*Fig. 35.1*) is the form most widely attested in Medieval contexts of central Italy beginning in the 11th century. It is moderate to large in size and characterized by a globular or trapezoidal body with flat base, a simply articulated rim with small variations in the inclination, and grooved ribbon handles that can reach a considerable width (up to ca. 10 cm). The form was completed by a cover, as suggested by the recovery of a trapezoidal example with internal grip.[10] They were certainly used to store solid foodstuffs (grains); it also seems reasonable to propose their use in the transport and storage of water.[11]

The Metapontine amphora, with its painted red and brown decoration on the body and handles, does not have any parallels in terms of shape and decoration among the ceramics found at the numerous contemporary settlements of the Salento, in the southeastern extremity of Puglia.[12] Parallels among the painted amphorae of Puglia are very widespread, but they are limited to morphology and do not involve decoration. This anomaly could be explained by assuming different intended functions for the same form of container: painted products for the table, coarse products only for the pantry and domestic use.

Jugs and bottles complete the range of shapes designed for the storage of liquids. These are receptacles of modest capacity, both probably employed in the pantry and on the table for pouring. This is suggested by the absence of closed forms among the glazed ceramics. Jars in undecorated coarse ware, documented by a small number of examples, were containers used in the kitchen to store semi-liquid foods or, when placed near a source of heat, to reheat water.

Among the open forms the large containers predominate, represented only by basins, used to serve the courses and generally for non-liquid foods. Forms for individual use (bowls) are present in lesser quantities. There are also few examples of paioli: 2 rim fragments with an overarching rim that is circular in section.[13] The presence of these items shows that the form endured in both coarse and slipped or glazed versions, for a fairly long time span stretching from the final centuries of the Early Medieval to the High Medieval period.[14]

The typical fabrics of the open and closed vessels are refined and mostly light in color, varying from beige to pale yellow, with some orange and pink examples. In both cases the consistency is hard, and the

[9] On Medieval glazed cooking ware, see Patitucci Uggeri 1977, 124–27; on Post-Medieval glazed cooking ware in Puglia, see Castronovi 1999, with relevant bibliography. The Medieval cooking ware from Lecce is treated in P. Tagliente 2003a.

[10] On the use of lids with storage amphorae, see Romei 1986, 529.

[11] On the use of this form, see Maccari Poisson 1984, 258.

[12] Patterson and Whitehouse 1992; Blattmann D'Amelj 1996.

[13] On the problems associated with this form, see the section on Early Medieval coarse ware in this chapter.

[14] For functional comparisons with examples produced in enameled ware, see most recently Deferrari/Palazzi/Parodi 2001, 712.

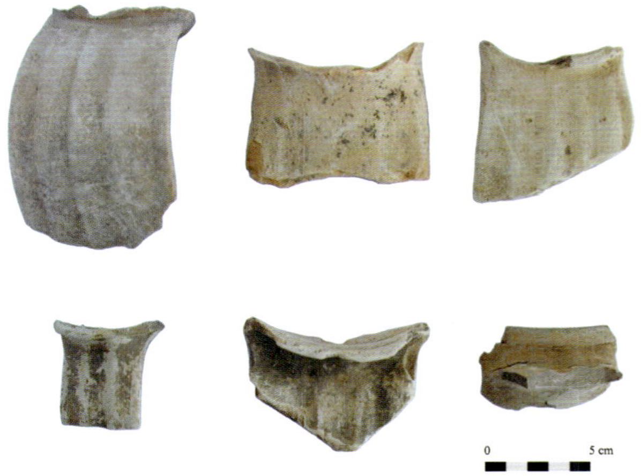

Figure 35.1 Grooved ribbon handles from Medieval amphorae.

inclusions, when present, are small and either white or shiny. The limited presence of small or medium-sized containers among both the closed and open forms could be attributed to the increase of this type of vessel among the glazed table products. It is therefore likely that the coarse wares were used for the preparation and storage of foods, while the coated wares were used on an individual basis for the consumption of foods.

The production of coarse wares for daily use and storage could have been on a local, regional, or sub-regional scale, but the lack of analyses and archaeometric studies does not allow more than speculation.[15]

Painted ware

Three wall fragments that belonged to a closed form with Narrow Line decoration attest to the use of painted ceramics. Beginning in the 12th century, this type of decoration characterized ceramic production in southern Italy, particularly in southern Puglia (the provinces of Brindisi, Taranto, and Lecce). It has a very limited presence also in excavated contexts of the late 11th and 12th centuries, and the examples have often been recovered in association with Broad Line forms. Such findings indicate that this period was a transitional phase between the Broad Line and Narrow Line productions. Instances of Narrow Line products increase considerably in the 14th century and persist at least until the 17th century. The forms frequently have morphological characteristics similar to contemporary products in coarse pottery, and belong to both closed containers (amphorae, jugs), prevalent in the initial phase of production, and to open containers (bowls, cups, basins), widespread and numerous beginning in the 14th century. Production was standardized, particularly in southern Puglia, and the vessels were intended primarily for storage.[16]

Glazed pottery: monochrome green glaze

The 153 fragments of this pottery with monochrome green glaze (approximately 14% of all Medieval artifacts), made exclusively for the table, are the most fully attested class after the coarse pottery. The morphological repertory of the monochrome green production has a greater range of open forms, represented primarily by bowls and, to a lesser extent, cups. The bowls, intended for individual use (serving of soups and semi-liquid foods), have a carinated profile with a rim characterized by a short, straight and outward-turned flange. The lead-based coating, generally shiny and of good quality, is applied to the interior surface and the rim (both interior and exterior).

The fabric of these products is refined, light in color (pale yellow or light brown), and quite hard, with rare white and shiny inclusions; there is also a second fabric, present in lesser proportion, very hard, refined, red-orange or pale brown in color, often containing white or red and shiny inclusions. There are only a few examples of bowls and cups, the latter likely used to hold spices, seasonings, and condiments for meats. Less numerous are the closed forms, attested only by a few rim fragments from bottles and jugs. These containers, intended to hold a modest quantity of liquid at the table for a short time, have simple, outward-turned rims that are not accentuated, and fairly high cylindrical necks. The glaze completely covers the exterior surface, while the interior is covered only in part with green glaze or with a diluted yellow glaze.

The monochrome green glaze represents the first High Medieval ceramic production with lead-based coating in southern Italy, and is typical of the Norman table service. The beginning of monochrome green glaze production is still the subject of research and discussion;[17] its diffusion in Campania and southern Puglia, however, certainly seems to have started in the first half of the 12th century, along with local produc-

[15] One case of local production of domestic ceramics in the Late Medieval period is Castel Porciano in Lazio; see D. Whitehouse 1967, 135–36.

[16] For a typological analysis and aspects of manufacture and circulation of Narrow Line ceramics, see Patterson and Whitehouse 1992 and Blattmann D'Amelj 1996, with relevant bibliography.

[17] On the earliest appearance of monochrome glazed ware in the Otranto area, see Arthur 2000b.

tion likely scattered throughout southern Italy.[18] The examples from the Metapontino suggest that, besides imports from southern manufacturing centers, there was probably local production in the neighboring Tarantine area and important centers of the Lucanian coast.[19]

Glazed pottery: monochrome brown glaze

Pottery with lead-based monochrome brown glaze is attested by only 11 fragments (1% of all Medieval artifacts), of which 8 belong to open forms. The only morphologically recognizable shape is the bowl. This form was intended for the table and resembles the products in green and yellow glaze with respect to manufacture and circulation. The chronological range of this material spans the 13th–15th centuries, although its existence in prior periods is possible.[20]

Pottery with green-brown decoration

Among the High–Late Medieval ceramic artifacts recovered in the survey there are 3 fragments of bowls bearing on the rim a decoration in dark green and brown beneath a monochrome green coating. The current interpretation of this class of pottery as an imitation of *ceramica siculo-magrebina* appears correct on the basis of both morphological and decorative criteria. Glazed *ceramica siculo-magrebina* is widely attested in Sicily and, to a lesser extent, in several areas of southern Italy (Campania, Calabria, and Puglia). To the southern Italian discoveries must be added the numerous imported basins utilized in the façades of Pisan churches.[21] Importation of this pottery from Sicily and northern Africa was certainly launched in the first half of the 12th century, but its diffusion in the Salento possibly began in the 11th century.[22] The circulation of these products was part of the general pattern of exchange accompanying Norman politics in southern Italy, a pattern which also included an increase in exports from Sicily to the Tyrrhenian area.[23]

Glazed pottery with polychrome decoration

The most widespread type of tableware in southern Italy and Sicily during the Swabian and Angevin periods was polychrome glazed pottery.[24] The ceramic bears a transparent slip, either colorless or pale yellow, beneath which appear painted motifs in brown, red, and green, used individually or in various combinations of two or three colors ("RMR").[25] It should be emphasized, however, that often the small dimensions of the fragments prevent certain identification of a precise chromatic combination. The morphological and decorative repertories, much as the fabrics employed, are essentially homogeneous with respect to the varying color combinations used.

In the Metaponto survey corpus, polychrome glazed ceramics are represented by 150 fragments, approximately 14% of all Medieval pottery; this is the highest proportion after the coarse and monochrome green glazed pottery. The fabric used in the Metapontine samples—beige or pale yellow in color, predominantly hard, with rare white inclusions—appears traceable to a single type and presumably to a single production source.

Open forms predominate: small bowls in particular are heavily represented, while plates and basins appear in a very limited number of examples. Closed forms are conspicuously rare, the jug being the only recognizable form.

The bowls are characterized by a carinated profile, rim with short and variably articulated flange, and ring foot. They display recurrent decorative motifs found also on basins, most commonly the central medallion and the geometric pattern. The medallion motif, painted at the center of the interior, appears in the form of the Taranto type (**No. 61**) consisting of a circle divided in half by one or more parallel lines, with a wavy line inside each section; the motif is completed by a series of dots arranged radially outside of the circle and by a gridiron design (**Nos. 59 and 63**) consisting of a circle filled with a dense reticulate pattern.[26]

[18] Patitucci Uggeri 1977, 100, 251; Patterson and Whitehouse 1992, 130.

[19] Archaeometric analyses have not yet been performed on the monochrome glazed ceramics from the excavations at Pietra San Giovanni and Torre di Mare.

[20] The proposed chronologies for monochrome brown pottery are reviewed in Arthur 2000b.

[21] For Sicily, see Fiorilla 1996 and Molinari 1997, 139; for Campania, see Romei 1992; for Calabria, see Di Gangi and Lebole 1998 as well as Capelli and Di Gangi 2000; for Puglia, see Patterson and Whitehouse 1992, 173–75, also Arthur 2000b; for Pisa, see Berti and Gelichi 1995, and Berti 1997.

[22] On the problems associated with the Sicilian/North African pottery, see Arthur 2000b, with relevant bibliography.

[23] Capelli and Di Gangi 2000, 198, Arthur 2000b, 164.

[24] Fundamental studies of polychrome glazed ware include the materials from the Salento (Patitucci Uggeri 1977), Otranto (Patterson and Whitehouse 1992), and several Lucanian, Calabrian, and Apulian sites (Dufournier/Flambard/Noyé 1986). For Basilicata, see Fiorillo 2005.

[25] For an analysis of the technological aspects of this pottery, see Dufournier/Flambard/Noyé 1986.

[26] The Taranto-type designation stems from the discovery of wasters in this style in the old city of Taranto. See Dufournier/Flambard/Noyé 1986, 272, and P. Tagliente 2001, 463.

The geometric pattern, known as the Torre di Mare type, is also widely attested. It consists of parallel bands alternating with wavy lines painted on the inside of the vessel beneath the rim, completed by parallel small arcs and strokes on the rim.

Naturalistic designs of plants and animals are extremely rare. One base of a bowl bears a design of elongated leaves converging at the center (**No. 62**), while on a wall fragment from an open form it is possible to recognize part of the body of a bird (**No. 64**).[27] Heraldic and anthropomorphic designs are entirely absent.[28]

The circulation of products bearing Taranto-type and Torre di Mare–type decorations between the second half of the 13th and the 14th centuries involved the Ionian coastal zone, in particular the territory of Taranto, northern Calabria, the territory of Matera in Basilicata, and the Salentine coast.[29] These wares were produced in several workshops, likely distributed around the known area of circulation. The current archaeological data positively identify several manufacturing locations in northern Puglia (Salpi) and in the Terra d'Otranto (Taranto, Brindisi, Ugento, and Cutrofiano).[30] Furthermore, the products from the Terra d'Otranto were objects of trade and reached the opposite shores of the Adriatic, the coastal zones of Dalmatia, Albania, and northwest Greece.[31] The consistent presence of glazed polychrome ceramics on the eastern side of the Adriatic bears witness to the complexity and intensity of commercial contacts between this area and the Terra d'Otranto in a particularly favorable and advantageous political climate.

Glazed polychrome pottery is widely attested in numerous urban and rural contexts of southern Italy (Campania, Puglia, Basilicata, Calabria) and Sicily between the mid-13th and 14th centuries, revealing a change in culture (more than in dietary habits) initiated previously in the Norman period with the spread of a monochrome glazed table service. The process was completed in the Swabian and Angevin periods with the production of richly decorated glazed table vessels.[32]

Proto-maiolica

There are only 2 fragments of proto-maiolica in the survey corpus, both bases belonging to open forms (bowls). **No. 70** is decorated with a geometric pattern consisting of concentric circles and dots in the same blue-green; **No. 71** has a reticulate pattern in a diluted blue-green. The fabric is very refined, consistently hard, and diluted in color (beige or pale yellow). The enamel is fairly thin and covers the surface poorly.

This type of table pottery, very similar in its decorative and morphological characteristics to the glazed polychrome products, is characterized by an opaque tin oxide slip. The technique, borrowed from the Islamic world, reached southern Italy via Sicily.[33] The ware is typical of the Swabian and Angevin periods and appears to have originated in Puglia between the end of the 12th and the beginning of the 13th centuries in two distinct areas: the Tavoliere, with centers at Lucera, Fiorentino, Torrione del Casone, and perhaps Ordona; and the Salento, with a workshop at Brindisi. Other manufacturing centers have been documented in Sicily (Gela) and in Campania (Naples), and there may have been yet more in southern Italy.[34] This ware was certainly a prestige product, likely used by only the dominant classes of the most important settlements. The distribution of finds with stronger presences in the coastal centers of southern Italy and attestations in both northern Italy and the Aegeo-Balkan region, reflects the traffic of this ware as a luxury item.[35]

Glazed pottery with graffito decoration

Among the ceramic artifacts recovered in the course of the field survey between Bradano and Basento there are two fragments of glazed pottery with graffito decoration. One fragment, a bowl rim preserving

[27] The converging leaves motif is common in the area around Lecce and in many Lucanian sites on glazed and painted products in brown and green or in brown, green, and red. Maetzke 1977, Pl. IV, No. 11; Salvatore 1984, 444, Pl. CXCVI, No. 6c; P. Tagliente 2003b.

[28] A classification of these decorative motifs is found in Dufournier/Flambard/Noyé 1986, 263–72.

[29] P. Tagliente 2001, 464; 2004, 91; De Crescenzo 2002, 178. On the presence of glazed products with the Taranto-type and Torre di Mare-type motifs among surface materials recovered at Motta di Cirò Marina and among casual finds collected during construction work in Croton, see Corrado 2003.

[30] Dufournier/Flambard/Noyé 1986, 257–59; P. Tagliente 2001, 465.

[31] On diffusion of these wares outside of Italy, see P. Tagliente 2001, 471–79; 2004, 89, 91–92.

[32] On diffusion in southern Italy, see Ebanista and Fusaro 2000, 120–27.

[33] For the definition of proto-maiolica and its identification, see Waagé 1934 and *Corinth* XI. For the problems associated with this ware in general, see D. Whitehouse 1966a; Patterson and Whitehouse 1992, 154–60; and *La protomaiolica*, with relevant bibliography.

[34] On production centers and circulation, see *La protomaiolica* 16–54. For Basilicata, see the materials recovered at Lagopesole (Fiorillo 2005).

[35] P. Tagliente 2001; 2004, 86–87.

part of the wall, is particularly interesting for its morphological and technological characteristics as well as its fabric. The fabric resembles the Byzantine Incised and Painted graffito ware, albeit very approximately; still, the red color, the abundance of both white and dark inclusions, the decoration, and the coating preclude an eastern origin. At the same time one can also exclude a local production, defined as that part of Puglia towards which Lucania gravitated culturally and economically. An Adriatic origin, however, can be reasonably proposed, though the possibility of confirming such a hypothesis is limited by the absence of exact parallels, and above all the current lack of archaeometric analyses.[36]

Double-dipped ware

Pottery dipped twice in yellowish brown or green lead-based glaze is represented by only 8 rim fragments (less than 1% of all Medieval artifacts), all belonging to bowls.[37] The Metapontine examples are morphologically indistinguishable from the Puglian types, although in several cases the Lucanian rims display variations in the height of the vertical part of the wall, which is usually not very tall. The interior and exterior surfaces are characterized by the twofold coloration (although never with dark tones); the fairly refined fabrics are predominantly soft and beige or pale yellow in color. These containers, intended for use at the table, generally have a trapezoidal profile, with flanged rim and low ring foot.[38] The Metapontine examples, too, have flanged rims, with the exception of one fragment (**No. 74**), in which the rim has a very short flange.

Double-dipped ware circulated principally between the end of the 14th and the 15th centuries in Basilicata, north-central Puglia, and the Salento. The latter, comprising the administrative area of the Terra d'Otranto, has yielded the greatest number of examples. In the Metapontino, the castrum of Torre di Mare has produced a collection of double-dipped ceramic fragments.[39] The ware also circulated along the northern Adriatic coast (Venice, Ferrara), the Dalmatian coast (Spalato/Split), and Sicily (Brucato), albeit with lesser intensity.[40] The current state of research indicates production of this ware in the Terra d'Otranto, as attested by the recent discovery of manufacturing waste deposits at Lecce.[41] Other workshops were likely active in the lower Salento and also in neighboring Lucania, as the variable macroscopic qualities of the fabrics appear to suggest.[42] Still unclear are the commercial relations between the Salento and the Adriatic area, and, above all, the role of Venice in the exchange and circulation of both the double-dipping techniques and the finished products.[43]

Ware	Chronology	Sites and counts
Cooking ware		61(1) 328(1) 368(1) 469(1)
Coarse ware	16th–18th c.	61(4) 300(5) 328(3) 368(1) 465(1) 507(1) 551(7) 557(1)
Coarse ware	19th–20th c.	140(7) 154(7) 368(2) 413(2) 496(22) 503(1) 727(1) 800(4)
Pottery with monochrome green glaze		61(1) 478(1) 726(1)
Pottery with monochrome brown glaze	End 15th–18th c.	154(3) 300(1) 507(1)
Pottery with monochrome brown glaze	19th–20th c.	464(2) 503(1) 639(1)
Pottery with yellow glaze	19th c.	475(1) 528(1)
Glazed pottery		507(1)
Graffito ware	16th–17th c.	84(1)
Enameled ware	18th–20th c.	61(1) 84(1) 465(1) 491(1) 492(1) 493(1) 507(6) 557(1) 632(1) 637(1) 639(1) 727(1) 766(1) 793(1) 800(2)

Table 35.2 Post-Medieval pottery (16th–19th c.).

36 The author wishes to thank Prof. S. Gelichi for useful suggestions as to the attribution of this artifact.

37 On double-dipped ware, see most recently P. Tagliente 2000b and 2002, 545–47, with relevant bibliography.

38 An analysis of the morphological characteristics of this ceramic class can be found in P. Tagliente 2002, 545–47.

39 De Crescenzo 2002, 193–94.

40 For Ferrara, see Siviero 1972; for Venice, see Gelichi 1992, 65; for Spalato, see Buerger 1973; 1974; 1979; for Brucato, see D'Angelo 1984, 462–63.

41 P. Tagliente 2002, 545–47.

42 P. Tagliente 2000b, 387; 2002, 547.

43 See P. Tagliente 2000b, 386–87.

Glazed pottery with green decoration

A noteworthy find among the Late Medieval artifacts is the base of a bowl with a distinctive light yellow slip of considerable thickness and stylized flower motif in green beneath a colorless glaze. This type of decoration does not appear to have exact parallels in the local Lucanian and Puglian orbits, although slipped pottery for the kitchen bearing floral decoration (the so-called *a rosetta* style) circulates widely beginning in the 16th century.

Amphorae

Transport vessels of the Medieval period are documented only by a handle of an Otranto-type amphora from site 328 and three fragments of unidentifiable amphora body wall recovered at site 300 (Pietra San Giovanni). Little can be gleaned about production or distribution from these few pieces.

The Post-Medieval Period (16th–19th centuries)

There are 110 fragments of Post-Medieval pottery from 29 sites, to which can be added a small group of brick and tile fragments (*Table 35.2*).

Most heavily attested are the coarse wares, with both open forms (bowls and plates, though predominantly large forms like basins) and closed forms (jugs, bottles, jars), followed by enameled wares (white and polychrome). Fine table and cooking wares are represented by very small numbers. Among the fine table wares appear monochrome glazed pottery (in green, brown, and yellow), graffito ware, and slipped ware. The well-refined and hard fabrics, the surface treatments, and the technical qualities indicate a chronological span for these wares between the 17th and the 20th centuries, with a predominance of artifacts with little diagnostic value and wares dating to the beginning of the 19th century.

Catalog of Medieval Ceramics

This catalog includes only fragments with preserved rim or base sufficient for diameter measurement. All dimensions are in cm. Maximum height indicated by "h," maximum diameter by "rim diam." maximum thickness by "th." An asterisk (*) indicates materials from the 1980 ICA pilot survey (sites numbered 1–44), as well as at the intensive survey sites and plots south of the Basento River through the 2001 season (most of the sites numbered 562 and higher). Additional information about the sites can be found in the Gazetteer. Color was recorded with reference to the Munsell Soil Color Charts.

Early Medieval Ceramics (7th–11th centuries)

Coarse ware

1:2

1. MED 388-01 Paiolo

Description: Almond-shaped, emphatically outward-turned rim above which the handle is suspended. Handle is circular in section.

Fabric: Reddish-yellow (surface and section 5YR 6/6), hard; fairly common white inclusions, rare shiny inclusions.

Date: The association of this fragment with a few pieces of coarse ceramics generally dateable to the final centuries of the Early Medieval period (10th–11th c.), as well as with locally produced Late Roman ceramics (painted coarse wares) and African imports (light sigillata ware, cooking ware, amphorae), suggests a date in the 10th–11th c.

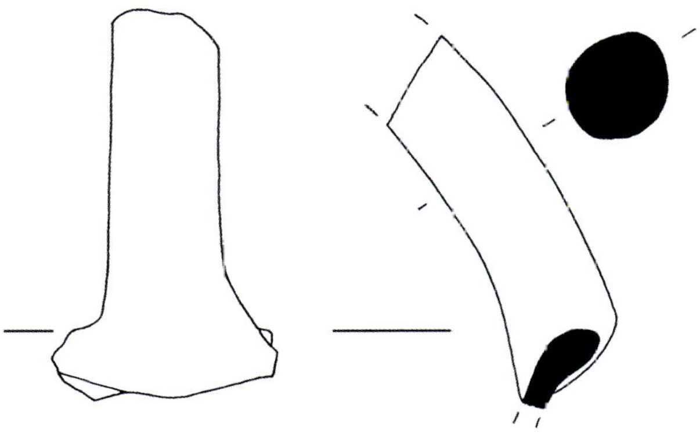

1:2

2. MED 388-02 Jug

Description: Indistinct vertical rim with attachment of ribbon handle.

Fabric: Reddish-yellow section (5YR 6/6) and pale yellow slip (2.5Y 7/3) on the surface, hard, with irregular fractures; rare large and dark inclusions, frequent small, dark and white inclusions.

Date: The form is attested in both coarse and cooking fabrics at numerous sites in central and southern Italy beginning in the 10th–11th c., then spreads widely in the later Medieval period. The association of this fragment with ceramics generally dateable to the final centuries of the Early Medieval period indicates a date in the 10th–11th c.

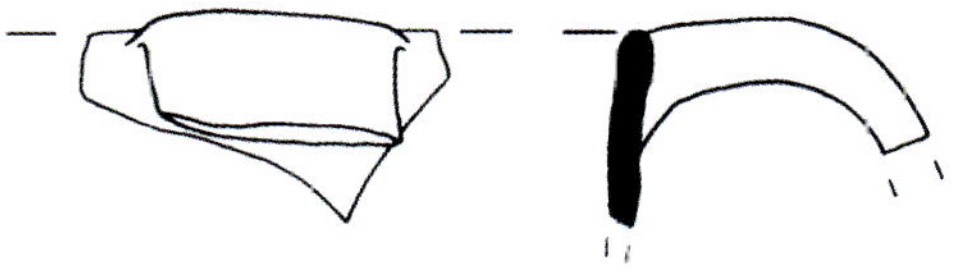

Painted ware

1:2

3. MED 326-01 Unknown closed form

Description: Flattened ribbon handle. The exterior surface is decorated with a broad brown vertical band (7.5YR 5/3–4/3; w 1.8), thinly applied in places.

Dimensions: Handle w 3.8

Fabric: Light brown surface (7.5YR 6/4), hard, refined, with irregular fractures.

Date: 9th–10th c. (M. Tinelli, pers. comm. 2003).

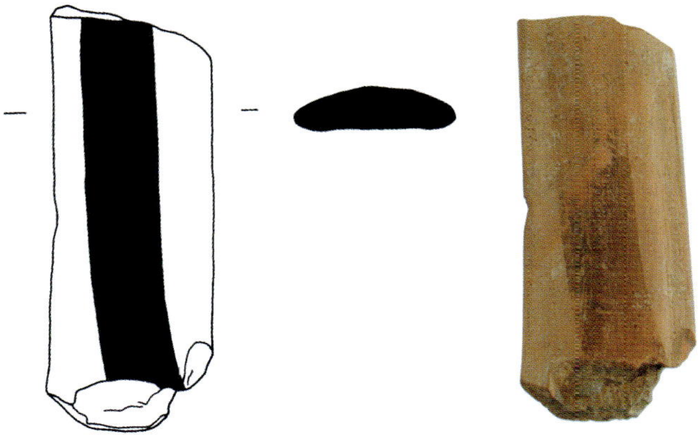

High–Late Medieval Ceramics (12th–15th centuries)

Unglazed cooking ware

1:2

4. MED 381-01 Jar

Description: Rim with triangular exterior profile and rounded edge; a short section of the body wall is likely ovoid.

Fabric: Light gray surface (10YR 7/2) and light red section (2.5YR 6/6), hard; numerous small white and dark inclusions.

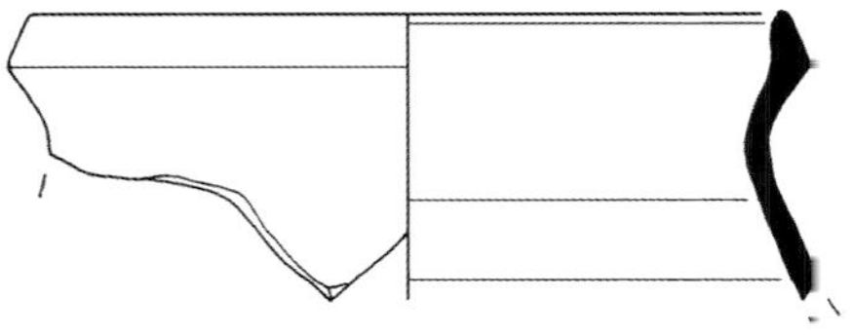

1:2

5. MED 300-01 Jar

Description: Triangular rim with pointed edge.

Fabric: Yellowish-red (5YR 5/6), hard; small white and dark inclusions.

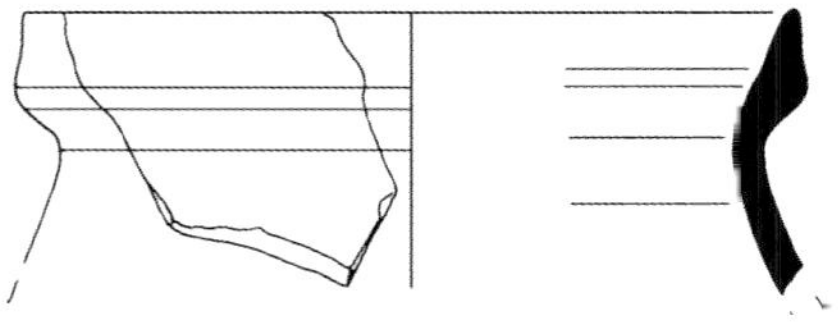

1:3

6. MED 300-02 Jar

Description: Small and fragmentary, with trapezoidal body, slightly convex, unaccentuated rim attached to the short neck, and flat base. The form could be equipped with handles. (The author thanks Giuditta Cavallo for the execution of the illustration.)

Dimensions: h 14.0 rim diam. 8.0 base diam. 6.0.

Fabric: Light yellowish-brown surface (10YR 6/4) and yellowish-red section (5YR 5/6), soft and rough, abundant white and dark inclusions.

Date: This form, produced in both cooking and coarse wares, is widely attested in south Italian contexts of the 12th and 14th c. Close parallels exist at Pietra San Giovanni (Noyé 1976, 706–07, Figs. 5.2 and 6), Anglona (Whitehouse and Whitehouse 1969, 63, Figs. 15.9 and 14), and Policoro (Salvatore 1984, 432, Pl. CXC, Nos. 3–4), and the form is known in several morphological variants also at Monte d'Irsi (Cotton 1971, Figs. 12.18–19, 13.24, and 13.32), Satriano (D. Whitehouse 1970, Figs. 12.7–15), Capaccio (Maetzke 1976, Pls. XXXI.9 and XXXVI.1–6), Naples (Arthur 1986, 551, Figs. 6.28–29), Fiorentino (Laganara Fabiano 1998), Ofanto (Patterson and Whitehouse 1992, 100, Fig. 6.3, Nos. 436–40), Carpignano Salentino (D'Andria 1978b, 162, Pl. LXVIII.5), and Brucato (Maccari Poisson 1984, 293, Pl. 25b).

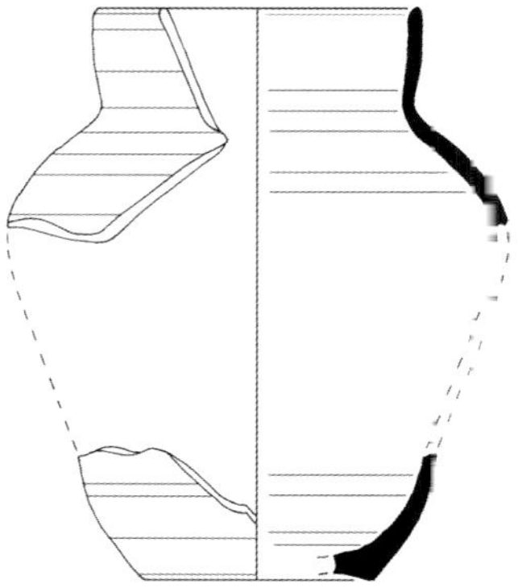

1:2

7. MED 300-03 Jar

Description: Outward-turned rim with rounded edge and groove for the lid. The exterior of the convex wall contains numerous grooves.

Fabric: Reddish-brown (5YR 5/4), fairly hard.

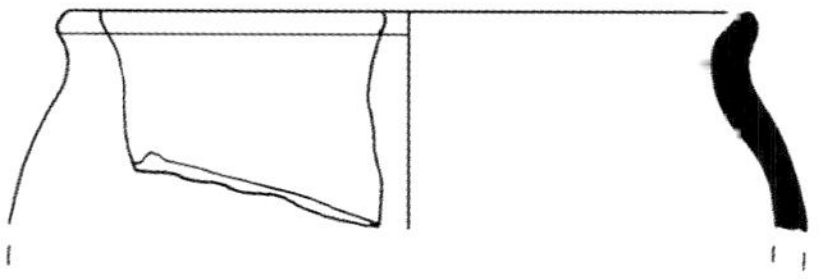

Glazed cooking ware

1:2

8. MED 086-01 Pot

Description: Vertical rim with pointed edge and concave interior profile; ribbon handle attaches just below rim.

Fabric: Red (2.5YR 5/8), hard; numerous small white and dark inclusions.

Date: 15th c.

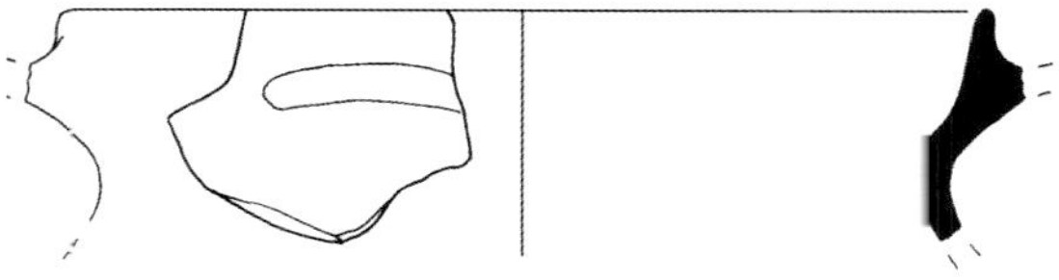

Coarse ware

1:2

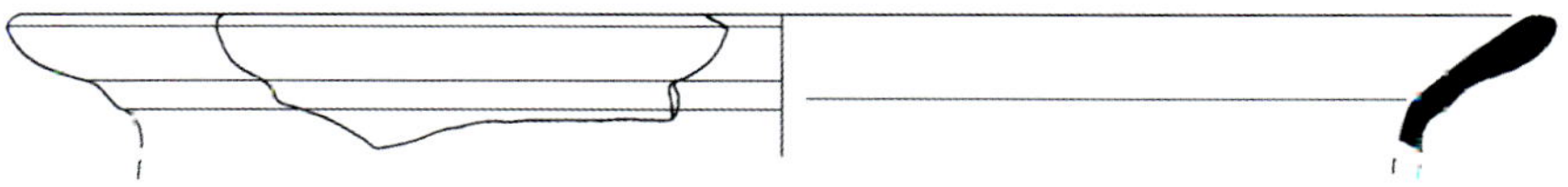

9. MED 381-02 Bowl
Description: Wide-flanged, outward-turned rim probably belonging to a large bowl.
Dimensions: rim diam. 22.4
Fabric: Very pale brown (surface and section 10YR 7/3), hard, with clean fractures; rare white inclusions.

1:2

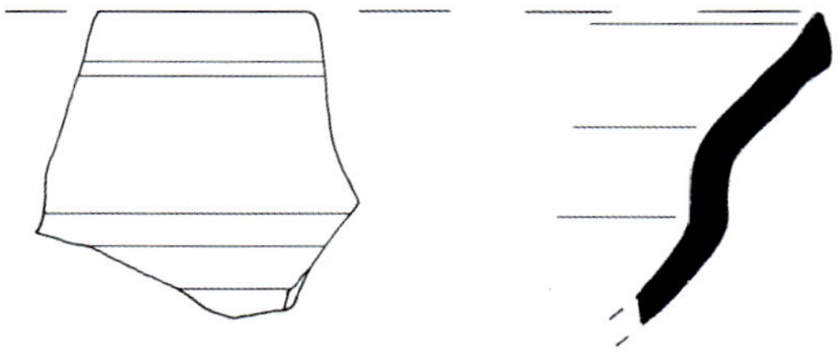

10. MED 300-04 Bowl
Description: Outward-turned rim with pointed edge and carinated wall.
Fabric: Pale yellow (surface 2.5Y 8/2, section 2.5Y 7/4), hard; rare shiny inclusions.

1:2

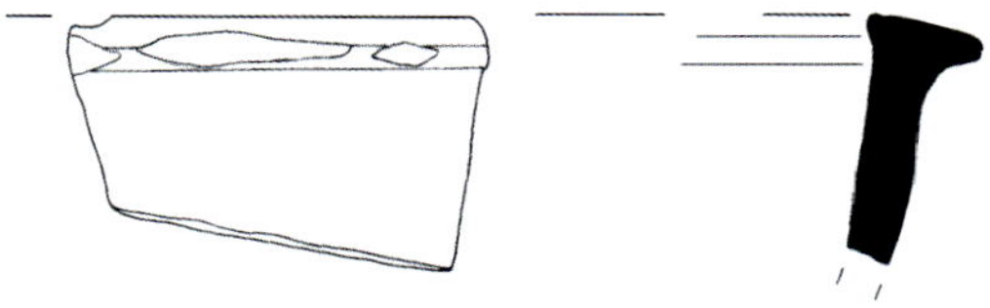

11. MED 300-05 Bowl
Description: Short-flanged rim, slightly inclined towards the exterior, with subtle molding on the inner face to emphasize the interface of the wall and rim.
Fabric: Reddish-yellow (5YR 6/6), hard; very pale brown slip (10YR 8/3) on the surface.
Date: The characteristics of this form suggest two interpretations: (1) coarse domestic ware, thus a simple bowl for the table; or (2) polychrome glazed ware, as indicated by the presence of the slip, thus a vessel subjected to the first firing but left undecorated. The form is dateable to the second half of the 13th and 14th c.

1:2

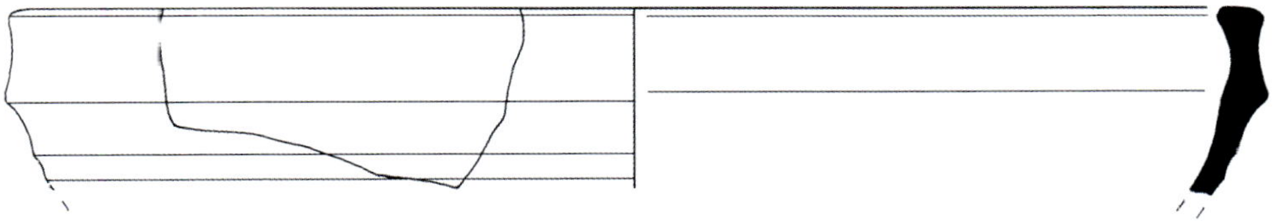

12. MED 300-06 Bowl
Description: Rim in the form of a band with concave exterior profile, emphasized by a prominent carination, and flattened edge.
Fabric: Reddish-yellow surface (7.5YR 7/6) and light brown section (7.5YR 6/4), hard; abundant small, shiny inclusions.

1:2

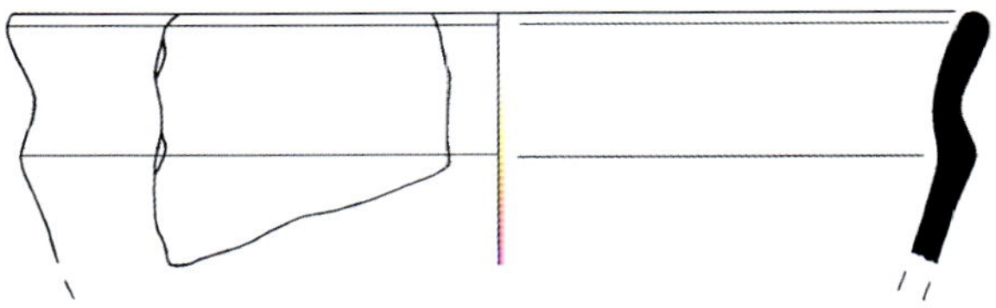

13. MED 300-07 Cup
Description: Slightly outward-turned rim with rounded edge and carinated wall.
Fabric: Pink (7.5YR 7/4), partially refined and hard, with irregular fractures; pale yellow slip (2.5Y 8/2).

1:2

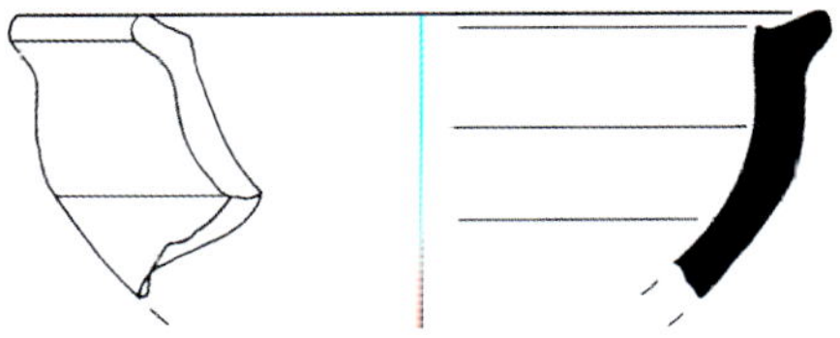

14. MED 300-08 Cup
Description: Short-flanged, inward-turned rim with rounded edge and convex wall.
Dimensions: rim diam. 11.2.
Fabric: Pale yellow exterior surface (2.5Y 8/2) and reddish-yellow section (7.5YR 6/6), hard.

Coarse ware cont.

1:2

15. MED 300-09 Cup

Description: Flanged rim with rounded edge; nearly straight, slightly outward-turned wall has faint grooving at the middle; flat base.

Dimensions: rim diam. 12.6 base diam. 6.1

Fabric: Light brown section (7.5YR 6/4), hard; very pale brown slip (10YR 7/4) on exterior surface.

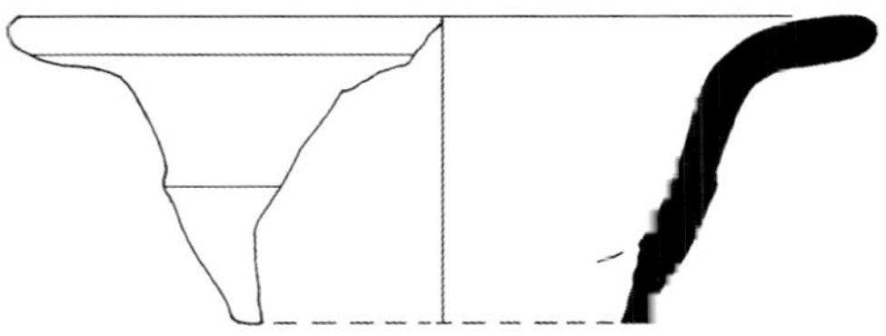

1:2

16. MED 300-10 Basin

Description: Outward-turned rim with flattened edge and grooving that accentuates the attachment to the vertical wall; on the exterior, beneath the rim, there is an incised decoration of wavy lines.

Fabric: Reddish-yellow (surface and section 5YR 6/6), fairly soft; inclusions can be white, shiny, or red (grog).

Date: 14th c.

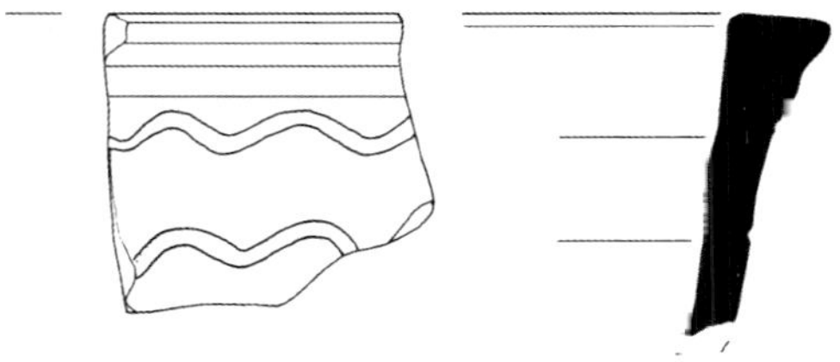

1:2

17. MED 300-11 Basin

Description: Wide-flanged, outward-turned rim; on the exterior, beneath the rim, there is an incised decoration of a wavy line; on the flange there is an incised decoration of straight strokes.

Fabric: Pale yellow exterior surface (5Y 8/2) and very pale brown section (10YR 7/4), refined and hard.

Date: 14th c., with a very close parallel from Anglona (Whitehouse and Whitehouse 1969, 63, Fig. 16.6). Bowls with similar articulation of the rim and wavy line decorative motif have been recovered in contexts of the 13th–14th c. at Otranto. See Patterson and Whitehouse 1992, 100, Fig. 6.4, No. 443.

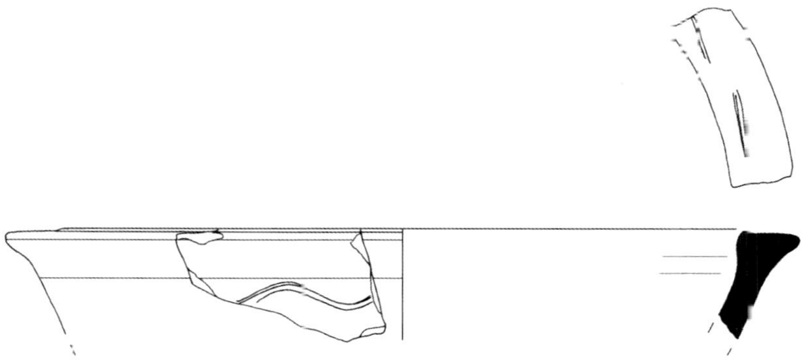

1:2

18. MED 300-12 Basin

Description: Unaccentuated rim with flattened edge; on the exterior, beneath the rim, there is an incised decoration of a wavy line.

Fabric: Pale yellow (surface and section 2.5Y 8/3), hard.

Date: The form is widespread in southern Italy in the 14th c.

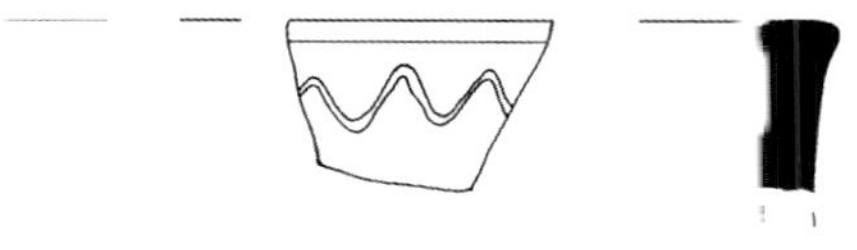

1:2

19. MED 300-13 Basin

Description: Unaccentuated rim with rounded edge; on the exterior, beneath the rim, a strip of clay decorated with finger-impressions has been applied.

Fabric: Pale yellow surface (2.5Y 8/3) and light yellowish-brown section (10YR 6/4), hard.

Date: The applied clay strip decoration with finger-impressions is attested on a closed form found in contexts of the 14th c. at Monte d'Irsi (Cotton 1971, 165, Fig. 12.7).

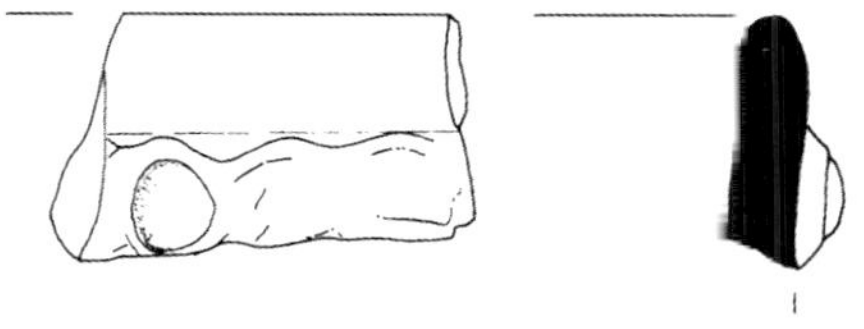

Coarse ware cont.

1:2

20. MED 381-03 Paiolo
Description: Straight, very slightly outward-turned rim to which attaches the overarching vertical rim with circular section.
Fabric: Very pale brown (surface and section 10YR 7/4), refined and very hard, with clean fractures.
Date: The association with coarse and cooking wares, as well as monochrome green and graffito fine wares, dateable between the 13th and 14th c. suggests a date for this form in the central centuries of the Medieval period although exact parallels come from the 10th and 11th c.

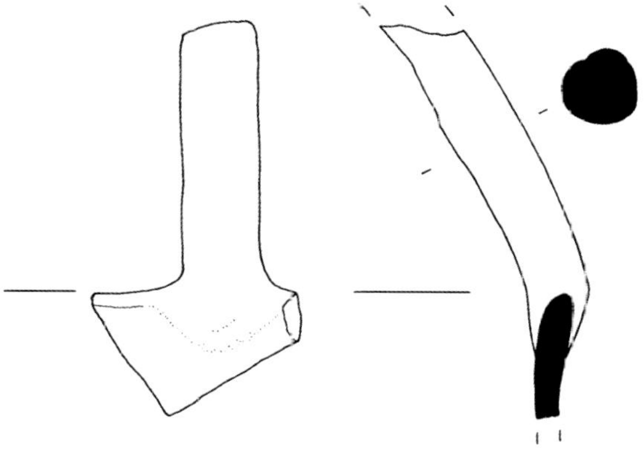

1:2

21. MED 300-14 Unknown closed form
Description: Triangular, unaccentuated rim with pointed edge; straight, slightly outward-turned wall.
Fabric: Pale yellow surface (2.5Y 8/3) and brownish yellow section (10YR 6/6), hard.

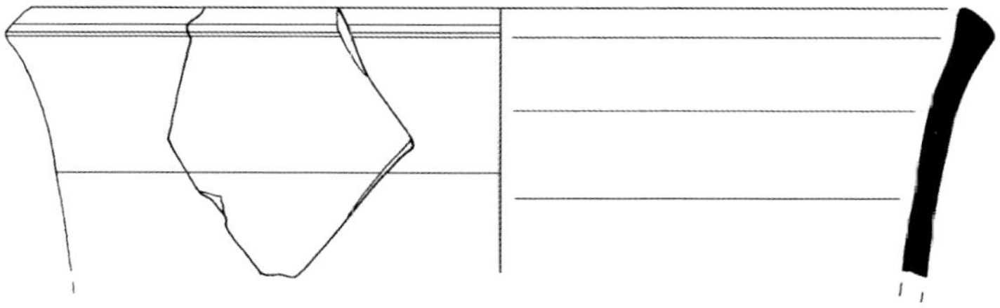

1:2

22. MED 300-15 Bottle (?)
Description: Short-flanged rim, slightly inclined towards the exterior; concave wall.
Fabric: Light brown surface (7.5YR 6/4) and light red section (2.5YR 6/8), hard.
Date: The association with glazed materials dates this form between the end of the 13th and the 14th c.

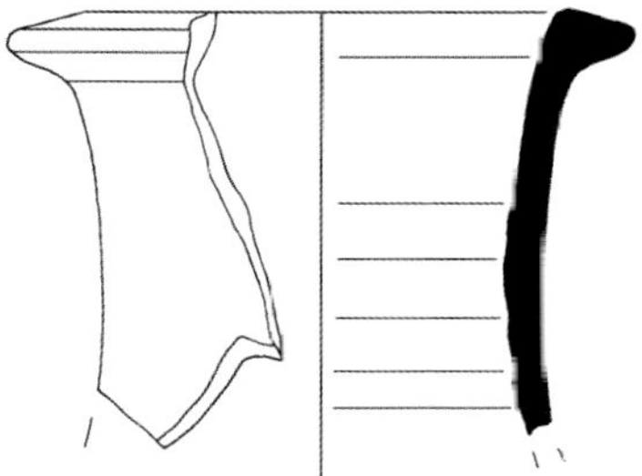

1:2

23. MED 300-16 Bottle (?)
Description: Unaccentuated rim with flattened and outward-turned edge; straight wall, to which attach the handles.
Fabric: Pink section (7.5YR 7/4) with pale yellow slip (2.5Y 8/2) on surface, hard.
Date: End of 13th–14th c. (probable).

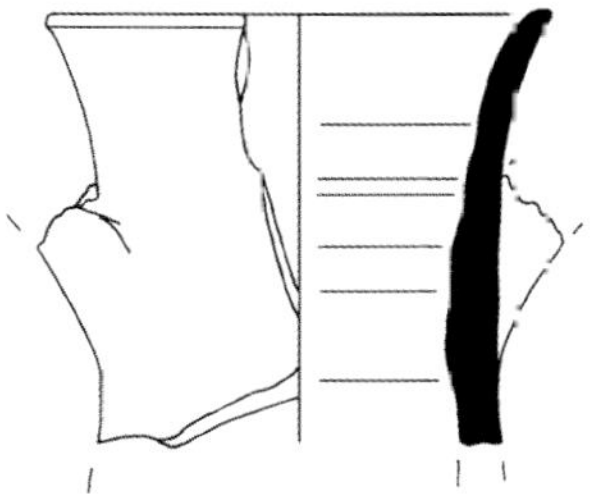

1:2

24. MED 300-17 Jar
Description: Rim in the form of a vertical band, outward-turned with rounded edge; grooved wall. There is a fingerprint of the potter on the interior surface.
Fabric: Pale yellow surface (2.5Y 8/2) and very pale brown section (10YR 8/3), fairly soft.

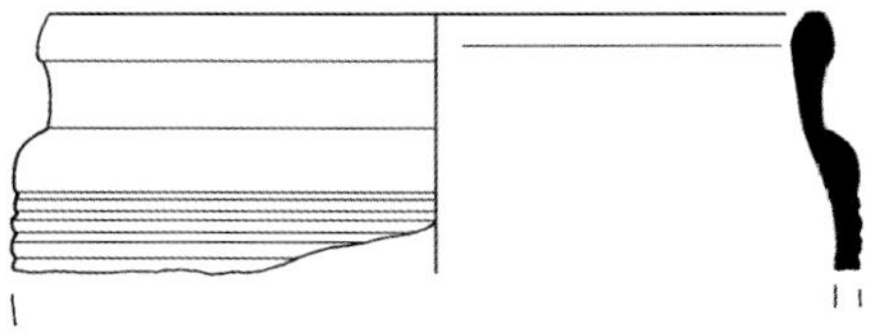

1:2

25. MED 300-18 Jug
Description: Unaccentuated rim with rounded edge with attached slightly overarching ribbon handle.
Fabric: Pink surface (7.5YR 7/4) and red section (2.5YR 5/6), hard.

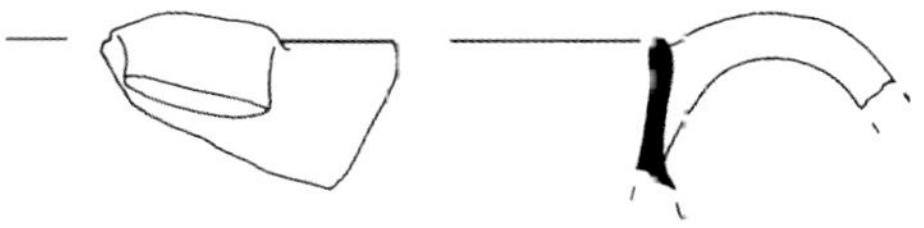

Coarse ware cont.

1:2

26. MED 507-01 Jug

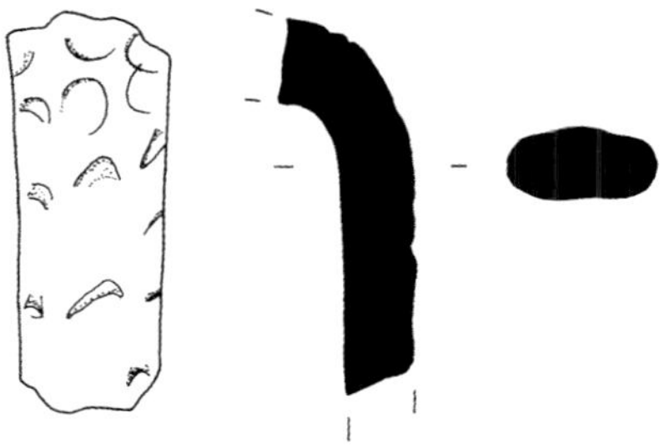

Description: Flattened ribbon handle with decorative motif of small oblique strokes and partial circles impressed into the upper surface.

Fabric: Light yellowish-brown surface (2.5Y 6/4) and yellowish-brown section (10YR 5/4), refined and hard; rare shiny inclusions.

Date: Similar decorations are widely attested on closed forms (jugs, domestic amphorae) and lids in both coarse and glazed wares of the Early Medieval period (Arthur and Patterson 1998, 521). However, the association with later Medieval fragments (coarse, monochrome green glazed, and polychrome glazed wares) within a site entirely lacking Early Medieval artifacts suggests a date between the 13th and 14th c.

1:2

27. MED 300-19 Domestic amphora

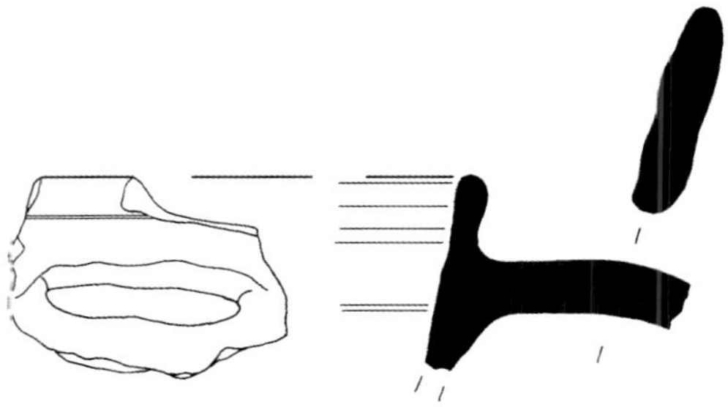

Description: Unaccentuated rim with rounded edge; grooved ribbon handle attached to the neck immediately below the rim.

Fabric: Light yellowish-brown surface (2.5YR 6/4) and pale yellow section (2.5Y 8/2), hard.

Date: This form of container is widespread in later Medieval southern Italy with several variations in shape and size.

1:3

28. MED 300-20 Domestic amphora

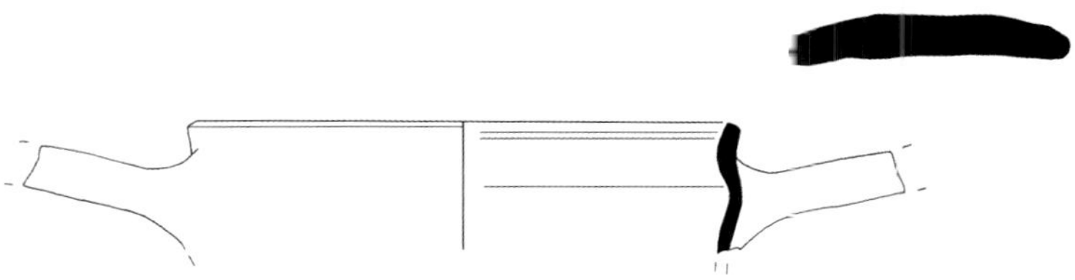

Description: Unaccentuated rim with flattened edge, inclined towards the exterior; wide ribbon handle attached to the neck immediately below the rim.

Dimensions: rim diam. 15.2 handle w 7.6

Fabric: Very pale brown section (10YR 7/3) with pale yellow slip (2.5Y 8/2) on the surface, hard.

Date: Large jars with relatively wide ribbon handles (up to a maximum of 10 cm) are attested in numerous southern Italian sites between the end of the 13th and the 14th c. Pietra San Giovanni (Noyé 1976, 705, Fig. 4.4); Policoro (Salvatore 1984, 431, Pl. CXC, Nos. 1–2); Satriano (D. Whitehouse 1970, 207, Figs. 13, 17a–b); Anglona (Whitehouse and Whitehouse 1969, 63–64, Figs. 16, 18–20); Monte d'Irsi (Cotton 1971, 165, Fig. 11); Brucato (Maccari Poisson 1984, 258–66); Rome (Romei 1986, 526–29, although the examples from the Crypta Balbi display a different articulation of the body).

1:3

29. MED 300-21 Domestic amphora

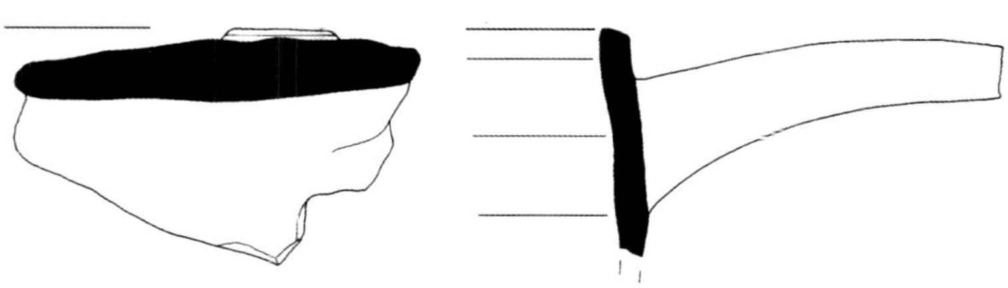

Description: Unaccentuated rim with flattened edge, inclined towards the exterior; wide ribbon handle attached to the neck immediately below the rim.

Dimensions: Handle w 7.9

Fabric: Pale yellow (surface and section 2.5Y 8/2), hard, with clean fractures.

Date: End of the 13th–14th c. (on the diffusion of this form, see **No. 11**).

Coarse ware cont.

1:3

30. MED 300-22 Domestic amphora

Description: Unaccentuated rim with rounded edge and molded exterior profile; ribbon handle attaches to the neck immediately below the rim.

Fabric: Light brownish-grey section (10YR 6/2) with white slip (5Y 8/1) spread over the exterior surface, hard.

Date: See **No. 29**.

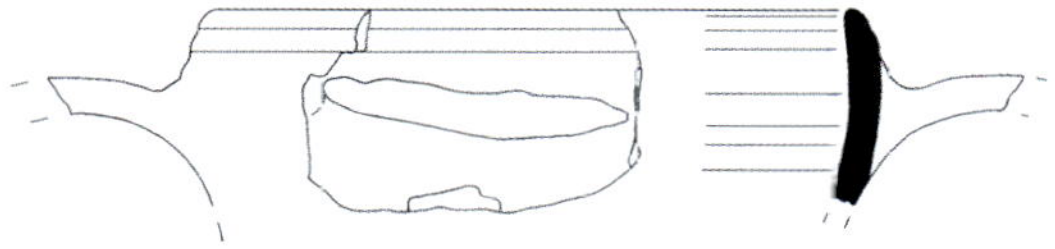

1:3

31. MED 300-23 Domestic amphora

Description: Flattened wide ribbon handle of considerable width (8.0–8.7 cm).

Dimensions: Handle w 8.0–8.7

Fabric: Light yellowish-brown surface (10YR 6/4) and pale yellow section (2.5Y 7/3), hard, with clean fractures.

Date: See **No. 29**.

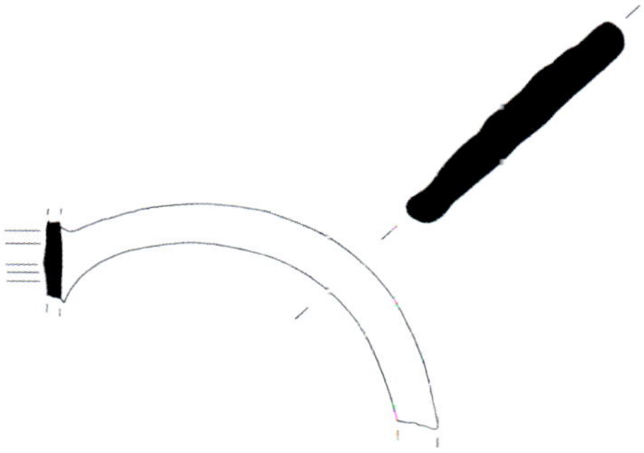

1:2

32. MED 381-04 Lid

Description: Lid in the form of an inverted trapezoid, with outward-leaning wall, flat base, and internal grip.

Dimensions: base diam. 6.8

Fabric: Very pale brown (surface and section 10YR 7/4), hard; rare white inclusions.

Date: This is likely the lid of a domestic amphora (see Romei 1986, 526, 529, Pl. IX.5). The association with imported later Medieval monochrome green and graffito wares suggests a date in the 13th–14th c. This form, widespread in central-southern Italy, took root in Late Roman Mediterranean ceramic production and persisted throughout the Early Medieval period (M. Ricci 1998, 360, Fig. 5.3, 376, Fig. 15.7; Romei 1986, 526). Beginning in the late 12th–13th c., the irregularly glazed vessels were replaced by products in coarse and cooking wares (Manacorda et al. 1986, 527; Pl. VIII.5, 533; Pl. XIII.7–9).

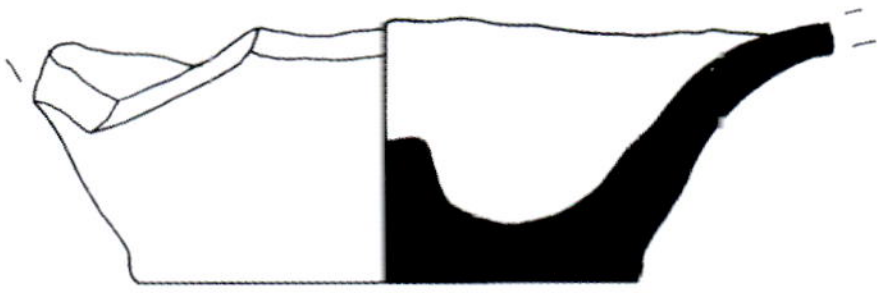

Painted ware

1:2

33. MED 086-02 Unknown closed form

Description: Wall fragment belonging to an unidentifiable closed form (domestic amphora?). On the exterior surface there is a decoration in dark greyish-brown (2.5Y 4/2) consisting probably of a spiral, of which only two curves are preserved.

Fabric: Pale yellow (surface and section 2.5Y 7/3), hard, with clean fractures; numerous shiny inclusions, rare, small white and dark inclusions.

Date: 14th–15th c. Parallels for the decoration have been found in Basilicata at Satriano (D. Whitehouse 1970, 210–11) and in the Salento at Otranto (Patterson and Whitehouse 1992, 116, Fig. 6.8, No. 499); Cutrofiano (contrada Badia: Blattmann D'Amelj 1996, 16, Pl. 3); Quattro Macine (Giuggianello, province of Lecce); and Muro Leccese. (The materials recovered at Lecce [castle of Charles V], Quattro Macine, and Muro Leccese are under study by the team lead by Prof. Paul Arthur of the University of Lecce.)

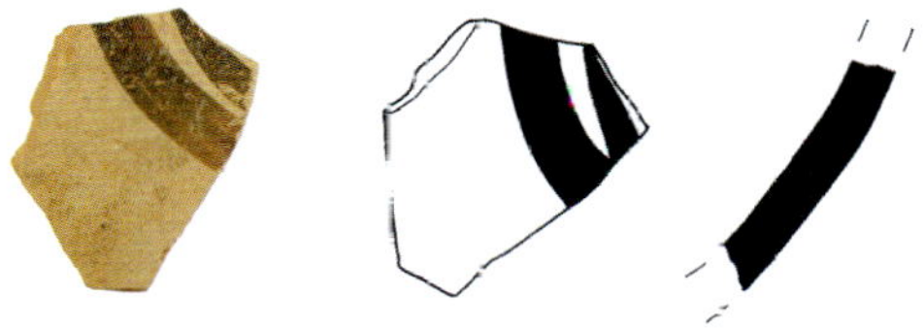

Pottery with monochrome green glaze

34. MED 381-05 Bowl

Description: Short-flanged rim with flattened edge, emphasized on the exterior by an accentuated projection; short section of carinated wall. A thick layer of dark green glaze on the rim and on the interior and exterior surfaces.

Fabric: Pale yellow section (5Y 8/3), refined and hard, with irregular fractures.

Date: The form has exact parallels in contexts at Policoro dated between the 12th and first half of the 14th c. (Salvatore 1984, 433–35, Pl. CXCIII, Nos. 47–48).

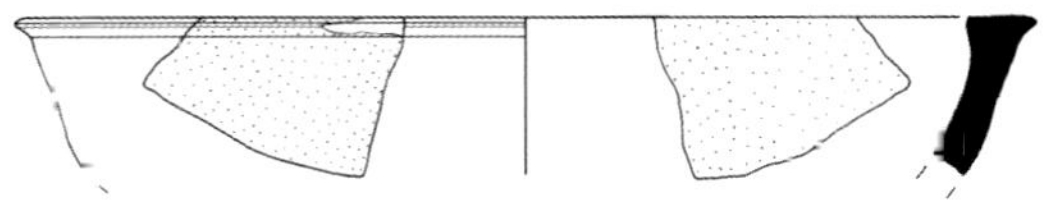

35. MED 086-03 Bowl

Description: Rim with short and flat flange. Green glaze on the interior and exterior surfaces.

Fabric: Light red section (2.5YR 6/8), hard.

Date: The form has very close parallels at Otranto in materials recovered from contexts of the 13th c. (Patterson and Whitehouse 1992, Fig. 6.16, No. 570).

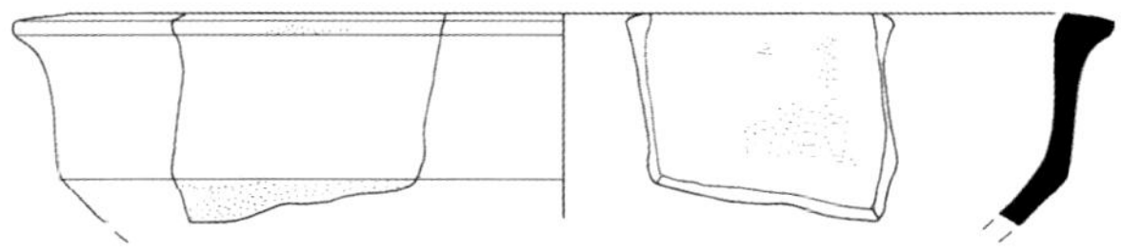

36. MED 300-24 Bowl

Description: Short-flanged rim and carinated wall. Green glaze on the interior surface and the rim.

Fabric: Pale brown section (10YR 6/3), hard, with clean fractures.

Date: The form has exact parallels with materials from Otranto dated to the 13th c. (Patterson and Whitehouse 1992, Fig. 6.16, No. 568).

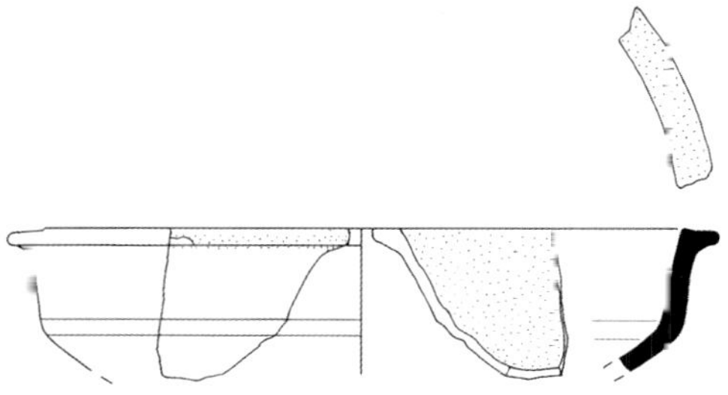

37. MED 300-25 Bowl

Description: Rim with projecting flange and concave upper surface; carinated wall. Green glaze on the interior surface, the rim, and the exterior surface beneath the rim.

Fabric: Light yellowish-brown (10YR 6/4), partially refined; rare white and shiny inclusions.

Date: 13th–14th c.

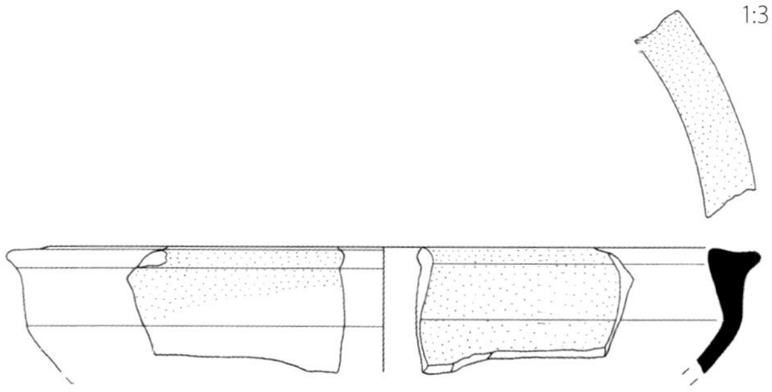

38. MED 300-26 Bowl

Description: Short-flanged rim with slightly concave upper surface; convex wall. Green glaze on the interior surface and the rim.

Fabric: Light gray section (2.5Y 7/2), hard, with clean fractures.

Date: There is a parallel for this form at Otranto (Patterson and Whitehouse 1992, 133, Fig. 6.17, No. 580).

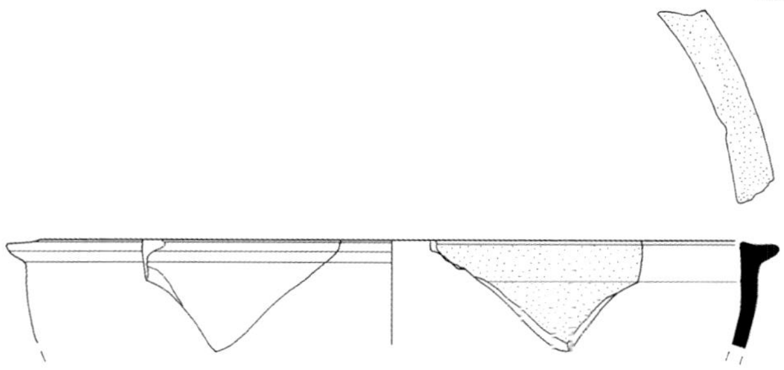

Pottery with monochrome green glaze cont.

1:3

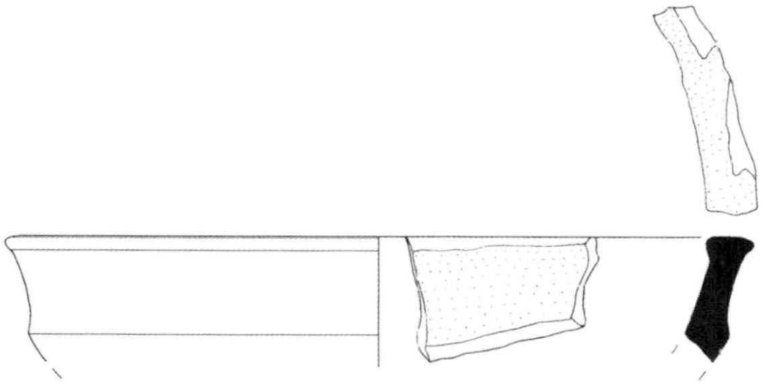

39. MED 300-27 Bowl
Description: Rim in the form of a high band terminating in a flange with flat upper surface; carinated wall. Green glaze on the interior surface and rim.
Fabric: Pale yellow (2.5Y 8/2), refined and hard, with irregular fractures.
Date: The form has very close parallels in materials from Otranto (Patterson and Whitehouse 1992, 133, Fig. 6.17, No. 577) and Monte d'Irsi (Cotton 1971, Fig. 6.12) dated to the 13th–14th c.

1:2

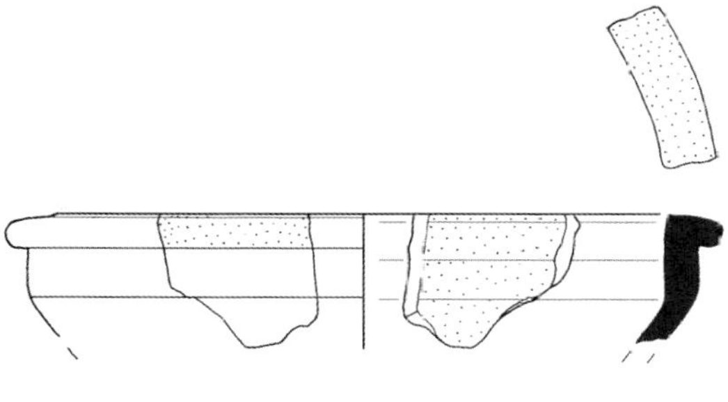

40. MED 300-28 Bowl
Description: Rim with short, projecting flange and right-angle profile; carinated wall. Green glaze on the interior surface and on the rim.
Fabric: Very pale brown section (10YR 7/3), partially refined and hard; rare white inclusions.
Date: The form is very close to examples dated to the 13th–14th c. from Otranto (Patterson and Whitehouse 1992, 132, Fig. 6.16, No. 567) and Torre di Mare (De Crescenzo 2002, 177, No. 26).

1:2

41. MED 300-29 Bowl
Description: Rim with wide, inclined flange and rounded edge. Green glaze on the interior and exterior surfaces.
Fabric: Light brown section (7.5YR 6/4), hard, with clean fractures.
Date: The form has parallels in materials from Policoro (Salvatore 1984, Pl. CXCII, No. 33) dated between the 13th and first half of the 14th c., as well as in materials dated between the 12th and first half of the 13th c. from Anglona (Whitehouse and Whitehouse 1969, Fig. 19.1) and Genoa (Cabona/Gardini/Pizzolo 1986, 461–62, Pl. VI, Nos. 20–21).

1:2

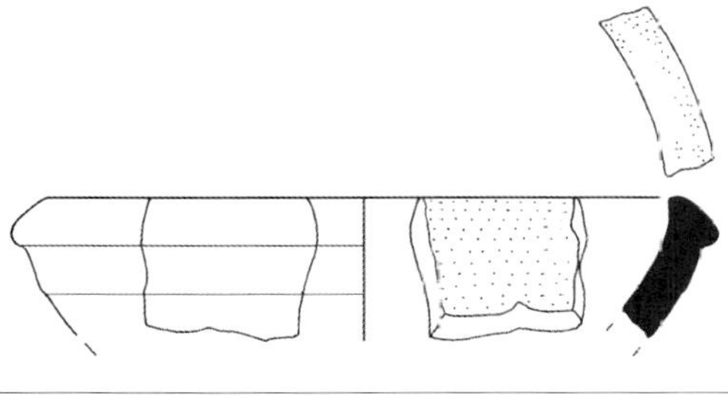

42. MED 300-30 Cup
Description: Rim with flattened edge and accentuated thickening on the exterior surface. Green glaze on the interior surface and the rim.
Fabric: Pink section (7.5YR 7/4), partially refined and fairly soft; red inclusions.

1:2

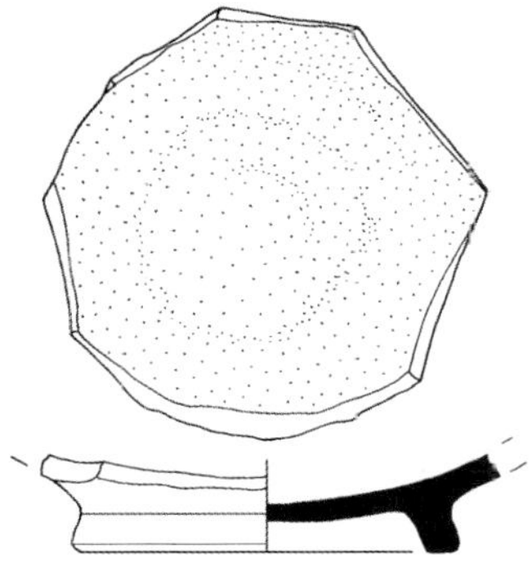

43. MED 300-31 Bowl
Description: Base with ring foot. Green glaze on the interior surface.
Fabric: Pale yellow section (2.5Y 8/2), partially refined and hard, with irregular fractures.

Pottery with monochrome green glaze cont.

1:2

44. MED 090-01 Bottle

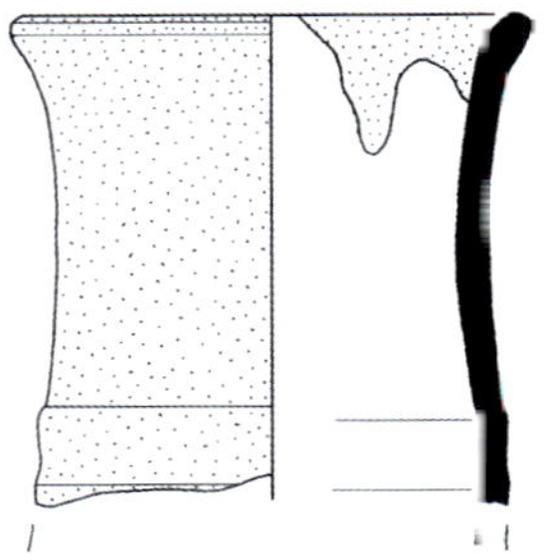

Description: Unaccentuated, outward-turned rim with rounded edge; cylindrical neck with double grooving on the lower part. Dark green glaze distributed uniformly on the exterior surface, dripping below the rim on the interior; golden patina on the exterior surface.
Fabric: Very pale brown section (10YR 7/4), hard; numerous very small white inclusions.
Date: Jugs with globular bodies, cylindrical or outward-flaring necks, and green glazing are attested at Corinth in 13th-c. contexts (*Corinth* XI, 62–63). Products of this type are very often attributed to the Metallic Ware class and considered Byzantine imports. For the definition of Metallic Ware, see Stillwell Mackay 1967, 252. For the proposed Venetian origin of at least a part of these vessels, see Gelichi 1988, 11–13, and 1993, n. 37.

1:2

45. MED 300-32 Bottle?

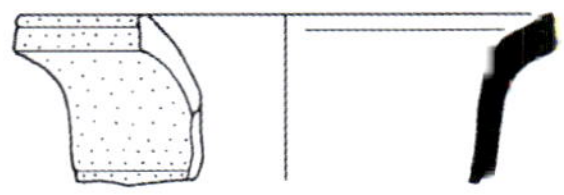

Description: Outward-turned rim with flattened edge. Green glaze on the exterior surface, dripping onto the interior surface.
Fabric: Reddish-yellow section (5YR 6/6), hard, with regular fractures.

Pottery with monochrome brown glaze

1:2

46. MED 086-04 Bowl

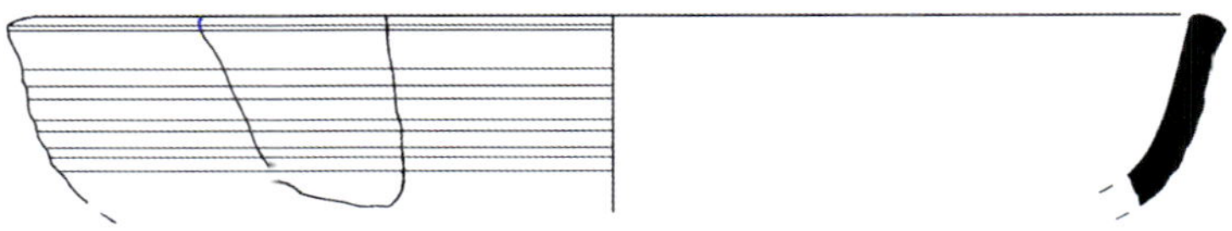

Description: Unaccentuated rim with slightly rounded edge; subtle grooves on the exterior surface beneath the rim. Light brown glaze on the interior and exterior surfaces.
Fabric: Pale yellow section (2.5Y 8/2), hard.
Date: 13th–15th c.

1:2

47. MED 086-05 Bowl

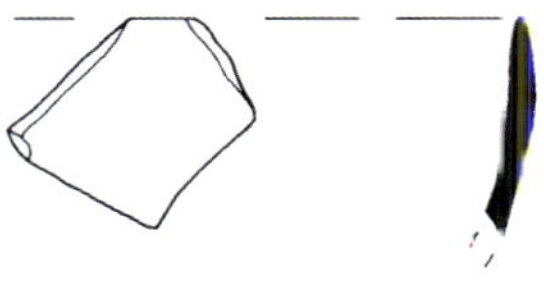

Description: Unaccentuated rim with flattened edge. Brown glaze on the interior and exterior surfaces.
Fabric: Pink section (7.5YR 7/4), soft.
Date: 13th–15th c.

Pottery with green-brown decoration

1:2

48. MED 551-01 Bowl

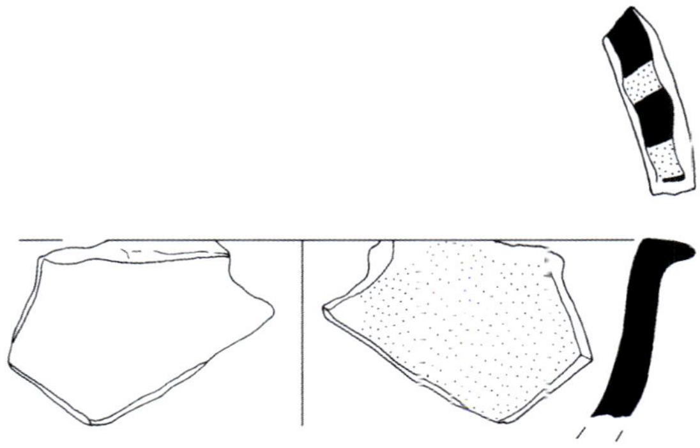

Description: Short-flanged rim, slightly inclined towards the exterior; carinated wall. Thin green glaze on the interior surface and rim; parallel strokes of thick dark green glaze on the flange.
Fabric: Very pale brown section (10YR 7/4), soft; white and shiny inclusions.
Date: The form has a very close parallel at Otranto and probably dates to the 12th c. (Patterson and Whitehouse 1992, 175).

Pottery with green-brown decoration cont.

1:3

49. MED 300-33 Bowl

Description: Molded rim with flattened edge belonging to a carinated bowl. Thick layer of dark green glaze on the exterior and interior surfaces; traces of brown glaze on rim.

Fabric: Pale yellow section (5Y 8/2), hard, with irregular fractures.

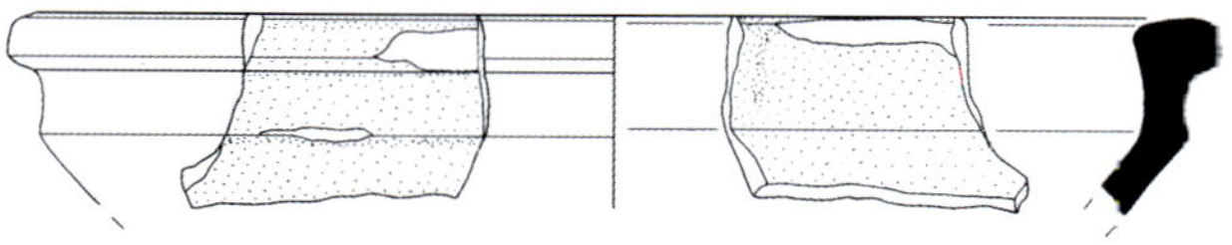

Date: The form probably dates to the 12th c.

50. MED 551-02 Bowl

1:3

Description: Unaccentuated rim with rounded edge. Thin green glaze spread uniformly on the interior and exterior surfaces; drips of thick dark green glaze on the rim.

Fabric: Pale yellow section (2.5Y 7/3), hard, with irregular fractures.

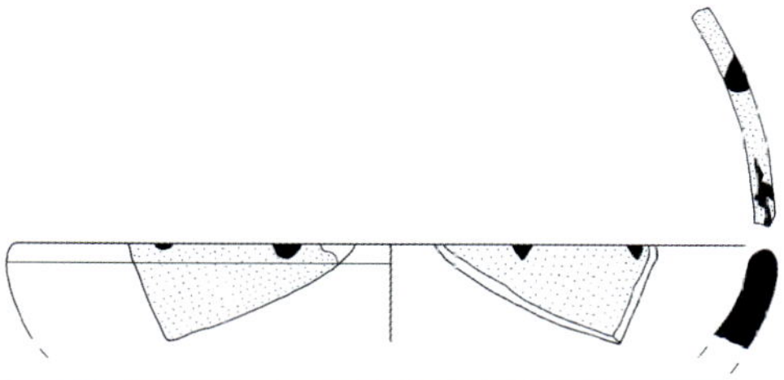

Glazed pottery with brown decoration

1:2

51. MED 140-01 Bowl

Description: Base characterized by ring foot with truncated conic profile and flat resting-surface. The gridiron design at the center of the interior is brown, surrounded by a band of short vertical strokes also in brown. Traces of colorless glaze.

Dimensions: base diam. ca. 8.0

Fabric: Pale yellow (surface and section 2.5Y 8/2), soft; rare red inclusions.

Date: Products bearing the gridiron design at the center of the interior are attested in Puglia (D. Whitehouse 1966b; Salvatore 1984, 436, Pl. CXCV; Dufournier/Flambard/Noyé 1986, 272–75; Patitucci/Uggeri 1995) and Basilicata (Pietra San Giovanni and Monte d'Irsi) in 13th-c. contexts. This example dates to the end of the 13th c. For the discoveries at Pietra San Giovanni in particular, see Noyé 1976, 715, Fig. 11.1–2; for Monte d'Irsi, Cotton 1971, 163, Fig. 9.8.

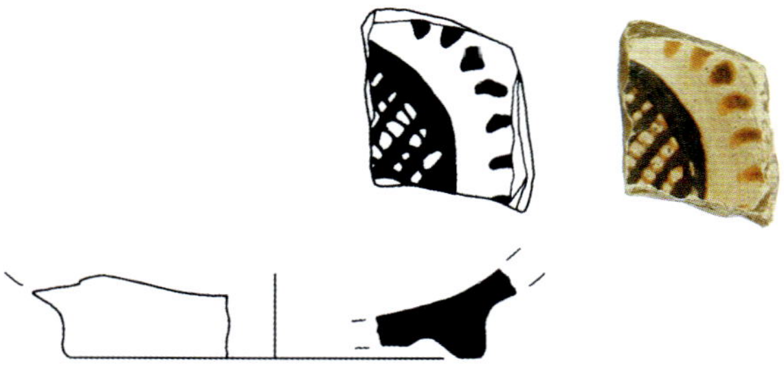

52. MED 300-34 Unknown open form

1:2

Description: Wall fragment bearing motif of two thin concentric lines in brown on interior surface, beneath which runs a wavy line and individual circular elements. Glaze borders on pale yellow.

Fabric: Pale yellow (surface and section 2.5Y 8/2), refined and hard, with irregular fractures.

Date: The decorative motif is attested at Torre di Mare (P. Tagliente, pers. comm. 2003).

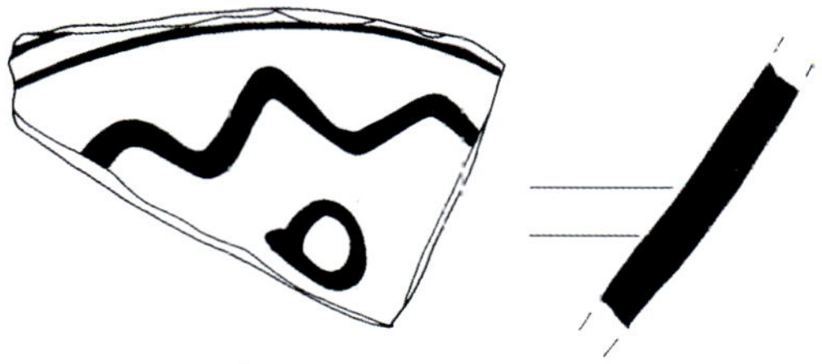

53. MED 313-01 Jug

1:2

Description: Vertical rim with flattened edge and handle attachment on the upper surface; the flattened handle is decorated with parallel strokes in brown oriented perpendicular to the handle.

Fabric: Pale yellow surface (2.5Y 8/3) and light yellowish-brown section (10YR 6/4), hard.

Date: Very close parallels found at Torre di Mare and Pietra San Giovanni belong to the end of the 13th and 14th c. This example dates to the end of the 14th c. (Pietra San Giovanni: Noyé 1976, 717, Fig. 19.4-5. Torre di Mare: P. Tagliente, pers. comm. 2003).

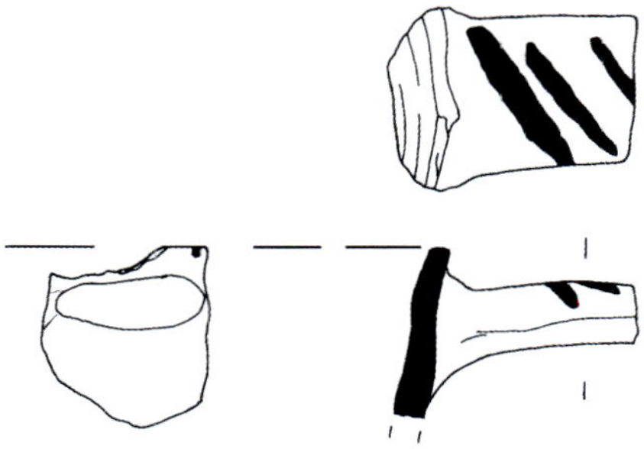

Glazed pottery with painted decoration including red

1:3

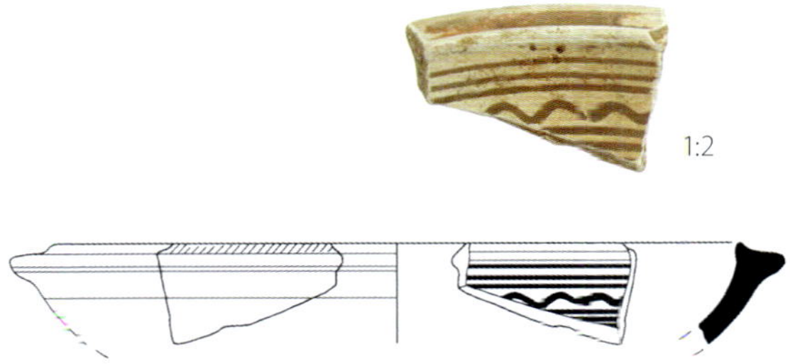

54. MED 140-02 Bowl

Description: Projecting rim characterized by a short flange with concave upper surface. The decoration consists of a red band on the flange and three concentric lines followed by a wavy line and then another three concentric lines in brown on the high part of the interior wall. No trace of glaze.

Dimensions: rim diam. 19.2

Fabric: Pale yellow (surface and section 2.5Y 8/2), soft; rare dark inclusions.

Date: Although the small preserved fragment is decorated only in brown and red, the former presence of green cannot be excluded. An exact parallel for the articulation of the rim has been found at Policoro (Salvatore 1984, Pl. CXCII, No. 39). The decoration is in the Torre di Mare style, so called by Noyé because of its presence in great quantities at that site (Dufournier/Flambard/Noyé 1986, 257). Materials decorated in this style have been recovered at other sites in Basilicata (Anglona, Montescaglioso) and Puglia (Lucera, S. Maria di Bagnolo, Avetrana, Veglie, Galatone, Gallipoli) in late 13th-c. contexts. For Anglona, see Whitehouse and Whitehouse 1969, 66, Fig. 18.6–7; for Montescaglioso, see Castronovi 1996, 221, Pl. II, No. 15; for Lucera, see D. Whitehouse 1988, 299–301, 313, Fig. 58.37–39; for S. Maria di Bagnolo, see Scionti and Tarentini 1990, 167; for Avetrana, Veglie, Galatone, and Gallipoli, see P. Tagliente 2000a, 178, Fig. 7.30.

1:3

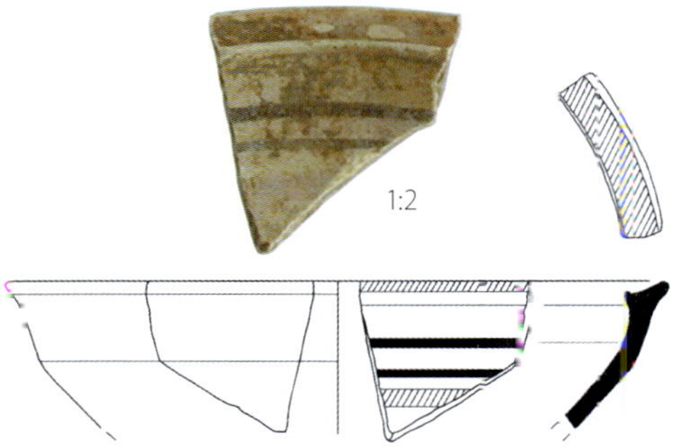

55. MED 300-35 Bowl

Description: Short-flanged rim strongly inclined towards the interior, with rounded edge and raised molding along the inner border. The decoration consists of a red band on the rim and two brown lines followed by a red band on the interior surface of the wall. Transparent/colorless glaze.

Fabric: Pale yellow (surface and section 2.5Y 8/2), hard.

Date: Exact parallels for the decoration and the morphological articulation have been found at Torre di Mare (Dufournier/Flambard/Noyé 1986, Fig. 6.5), Pietra San Giovanni (Noyé 1976, Figs. 16.5, 17.2 and 5), and Otranto (Patterson and Whitehouse 1992, Fig. 6.22, No. 652). For generic comparanda from S. Maria Bagnolo, see Scionti and Tarentini 1990, 167. The form dates between the mid-13th and the 14th c.

1:3

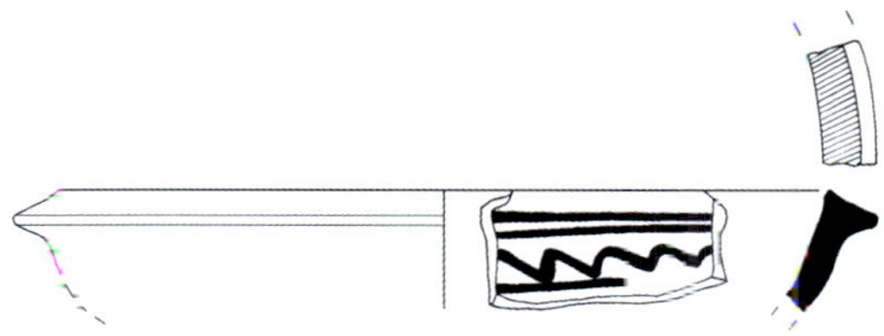

56. MED 034-01 Bowl

Description: Triangular rim with pointed edge; convex wall. The decoration consists of a red band on the rim and a wavy line surrounded by parallel lines in brown on the interior surface. No traces of glaze.

Fabric: Pale yellow (surface and section 2.5Y 8/2), soft.

Date: The decoration in the Torre di Mare style is attested on products found at numerous late 13th-c. sites in Basilicata and Puglia (on the circulation of this type of decoration, see n. 29 in text).

1:3

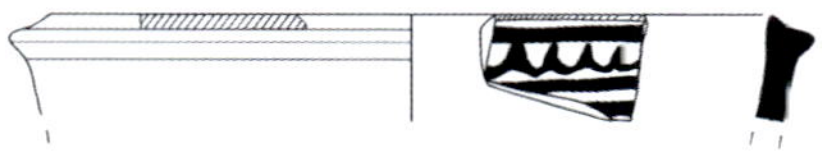

57. MED 033-01 Bowl

Description: Triangular rim with pointed edge and slight concavity on upper surface. The decoration consists of a red band on the rim and a series of concentric lines enclosing a wave design in brown on the interior surface.

Fabric: Pale yellow (surface and section 2.5Y 8/2), soft.

Date: See **No. 56** above.

Glazed pottery with painted decoration including red cont.

1:2

58. MED 300-36 Bowl

Description: Unaccentuated rim with pointed edge. The decoration consists of a red band and brown line on the interior and a brown band running along the rim on the exterior. No traces of glaze.

Fabric: Pale yellow (surface and section 5Y 8/2), fairly soft, with irregular fractures.

Date: Generic parallels for the morphological characteristics exist at Policoro (Salvatore 1984, Pl. CXCII, No. 32), Montella (Ebanista and Fusaro 2000, 126, Fig. 3.4), and Otranto (Patterson and Whitehouse 1992, Fig. 6.24, No. 679). The form dates between the mid-13th and 14th c.

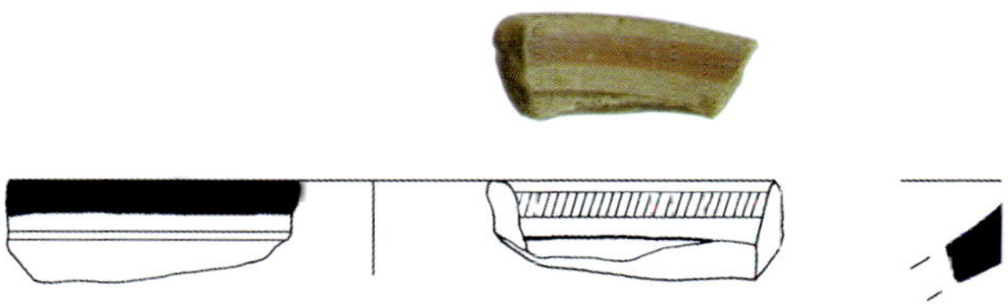

59. MED 300-37 Bowl

1:2

Description: Base characterized by slightly pronounced ring foot. The decoration at the center of the interior consists of a circle enclosing a dense reticulate pattern (gridiron) and surrounded by a red band, only a short part of which is preserved. There are faint traces of colorless glaze. A circular hole at the center of the interior suggests re-use for indeterminate purposes (a *margottiera*, container for rooting cuttings?).

Fabric: Pale yellow section (2.5Y 7/4), hard, with irregular fractures; red inclusions (grog), occasionally large.

Date: The form circulated widely in southern Italy in 12th- and 13th-c. contexts on containers with both lead-based and tin-based slips. Products of Puglian origin have been recovered in numerous parts of Greece, Albania, and Israel as well. Sites in Puglia: Lucera (Salvatore 1982b, 84, Fig. 5), Bari (Salvatore 1980, 155, 168), Brindisi (Patitucci Uggeri 1977; 1984), Mesagne (Patitucci Uggeri 1977, 266; 1979, 252). Sites in Basilicata: Pietra San Giovanni (Noyé 1976, 729; Dufournier/Flambard/Noyé 1986, 272–75), Torre di Mare (Dufournier/Flambard/Noyé 1986, 274; P. Tagliente 1996–1997), Policoro (Salvatore 1984), Anglona (Schlager and Rudiger 1967, 348, Fig. 21), Monte d'Irsi (Cotton 1971, 158, Fig. 9.8); Lagopesole (Fiorillo 2005). For Abbruzzo, see Pannuzi and Staffa 1997, 112; for Calabria (Tropea), see Di Gangi 1997, 168, Fig. 2.20; for Greece, see Dufournier/Flambard/Noyé 1986, 274; for Israel (Atlit), see Riavez 2000, 210–12; for Albania (Spalato, Butrint), see Riavez 2000, 212.

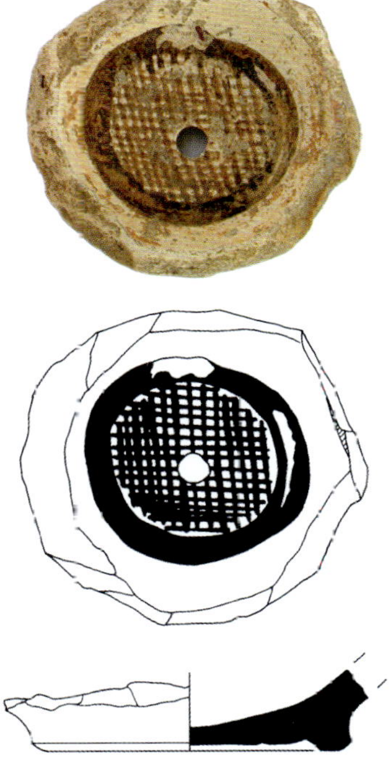

Glazed pottery with painted decoration including brown, yellow, and green

1:2

60. MED 140-03 Bowl

Description: Wall fragment bearing geometric decoration of oblique bands arranged orthogonally, defined in brown, and filled with green and yellow on the interior. There are traces of transparent glaze.

Fabric: Very pale brown section (10YR 7/4), hard, with numerous white inclusions, occasionally large.

Date: The line and band decorative motif has a very close parallel in the decoration of several containers recovered at Cutrofiano (Badia locality); see Blattmann D'Amelj 1996, 20. The fragment is dateable to the 13th–14th c.

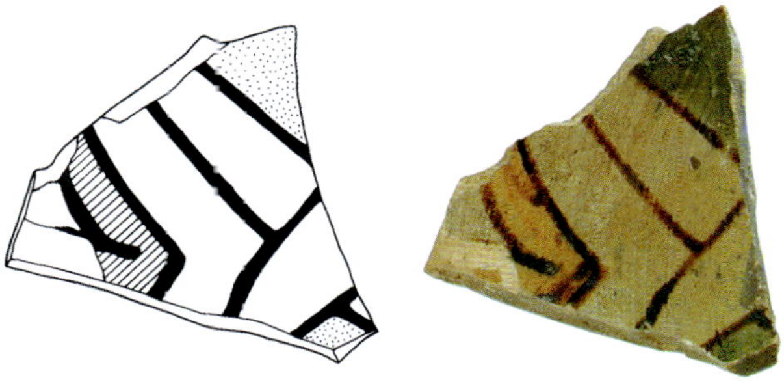

Glazed pottery with painted decoration including brown and green

1:2

61. MED 300-38 Bowl

Description: Base with ring foot. The center of the interior is decorated with a medallion formed by a brown circle divided by three parallel lines into two sectors, each occupied by a wavy line; the medallion is surrounded by green dots. Transparent glaze, verging into pale yellow.

Dimensions: base diam. 6.2

Fabric: Pale yellow surface (2.5Y 8/3) and pink section (7.5YR 7/4), partially refined and hard.

Date: The Taranto-type decoration is widely attested on open forms in the Ionian basin (Scribla, Santa Severina, Anglona, Policoro, Torre di Mare, Montescaglioso, Monte d'Irsi, Satriano, Matera, Ginosa, Taranto, Avetrana, Veglie, Porto Cesareo, Galatone, Gallipoli, Supersano) and the Adriatic (Brindisi, Spalato/Split, Butrint, Malta) in 13th–14th-c. contexts. For Taranto, see Dufournier/Flambard/Noyé 1986, 272–73; for Torre di Mare, see Dufournier/Flambard/Noyé 1986, 272 and P. Tagliente 1996–1997, 68–69; for Scribla, see Flambard and Noyé 1984; for Anglona, see Whitehouse and Whitehouse 1969, 66; for Policoro, see Salvatore 1984, Pl. CCII a–f; for Satriano, see D. Whitehouse 1970, 211, Fig. 15.3; for Brindisi, see Flambard 1983, Pl. 87.1; for Spalato, see Buerger 1974, 253, Fig. 8; for Butrint, see Mazzucato 1985, Figs. 3.42 and 4.60; for Malta, see D'Andria 1969, Pl. 16.3. A distribution map for the Terra d'Otranto appears in P. Tagliente 2001, 464, Fig. 3.

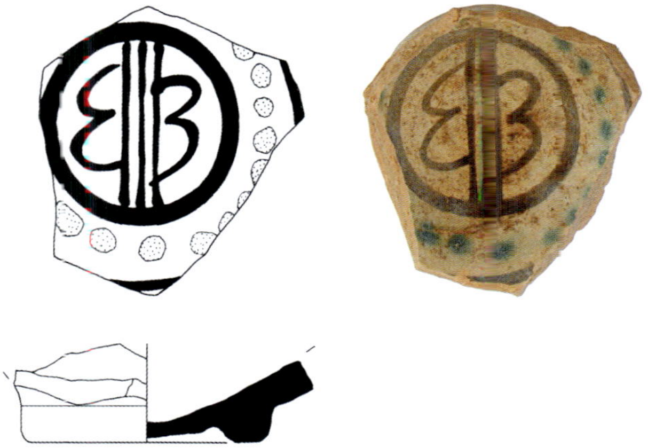

1:2

62. MED 300-39 Bowl

Description: Base characterized by pronounced ring foot. The interior is decorated with two narrow leaves, outlined in brown and filled with green, converging towards the center (there were probably three or four leaves originally). Colorless glaze.

Fabric: Very pale brown section (10YR 7/4), hard.

Date: The stylized leaf design, with variations on the filling color, was widespread principally in the Salento, although there are also attestations in several Lucanian sites. For a synthesis of the diffusion of the narrow leaf design in southern Italy, see Dufournier/Flambard/Noyé 1986, 268, and Patterson and Whitehouse 1992, 154–56. For the finds from Melfi in particular, see Maetzke 1977, Pl. IV, No. 14. For the production centers, see P. Tagliente 2000a. In Basilicata, the stylized leaf design is attested at Policoro (Salvatore 1984, 444, Pl. CXCVI, No. 6c). The decorative scheme dates the form between the second half of the 13th and the mid-14th c.

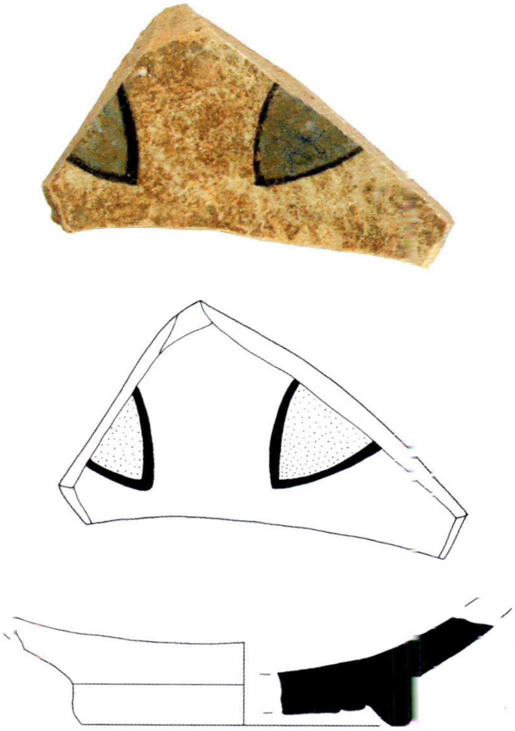

Glazed pottery with painted decoration including brown and green cont.

1:2

63. MED 300-40 Bowl

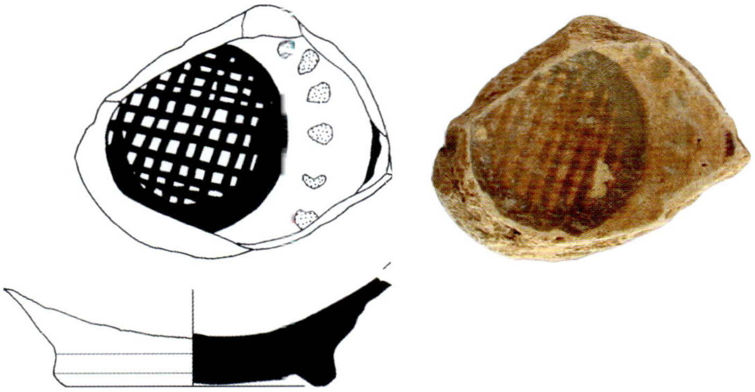

Description: Base characterized by square ring foot. At the center of the interior there is a reticulated design (gridiron) in brown, surrounded by green dots and then a brown line. No traces of glaze.

Fabric: Pale yellow surface (5Y 8/2) and section (2.5Y 7/3), refined and hard, with irregular fractures.

Date: The medallion motif with reticulate pattern surrounded by green dots is attested at numerous sites in Puglia, Basilicata, and the eastern Mediterranean. On the diffusion of this decorative style in southern Italy and the eastern Mediterranean, see Dufournier/Flambard/Noyé 1986, 272–75, with relevant bibliography. There are exact parallels from Torre di Mare (Dufournier/Flambard/Noyé 1986, Fig. 8.3) and S. Maria di Bagnolo (Scionti and Tarentini 1990, 167), and the form dates between the second half of the 13th and beginning of the 14th c.

1:2

64. MED 300-41 Unknown open form

Description: Wall fragment with interior decoration consisting of a circular feature outlined by a brown line (from which branches a linear appendage with angular termination) and filled with short, parallel green strokes. The poor preservation of the fragment does not allow precise identification of the motif, which could be either an animal (e.g., bird) or a geometric design. Colorless glaze. On the exterior the letter "N" has been incised, which is probably a mark of the workshop of origin.

Fabric: Pale yellow (surface and section 2.5Y 8/2), refined and hard, with irregular fractures.

Date: The bird motif is very common on glazed polychrome ceramics and proto-maiolica originating in numerous south Italian sites (decorative motifs bearing animals are examined in Dufournier/Flambard/Noyé 1986, 263–67). The form dates to the mid-13th–14th c.

Glazed pottery with brown, red, and green decoration ("RMR")

1:3

65. MED 300-42 Bowl

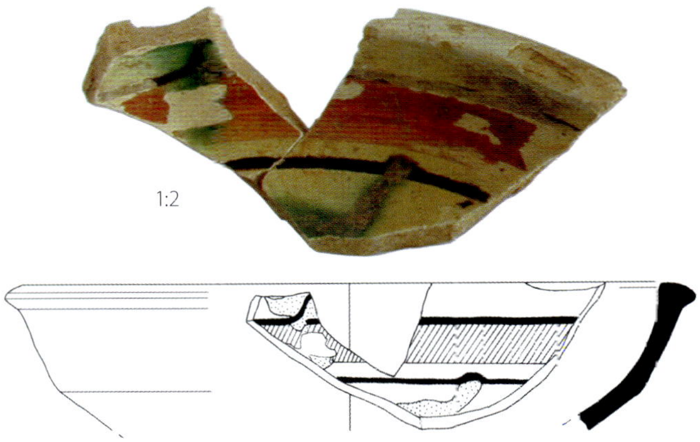

Description: Outward-turned rim with convex edge; carinated wall. The interior surface is decorated with alternating concentric bands in brown, red, and green. The glaze is transparent, verging on pale yellow.

Fabric: Very pale brown section (10YR 7/4), hard, with irregular fractures.

Date: The concentric bands motif has been found at numerous sites in southern Italy. In Basilicata the closest parallels come from Pietra San Giovanni (Noyé 1976, Figs. 16 and 18) and Monte d'Irsi (Cotton 1971, Fig. 9.7, 9.13–14) in mid-13th–14th-century contexts.

Glazed pottery with brown, red, and green decoration ("RMR") cont.

66. MED 300-43 Bowl

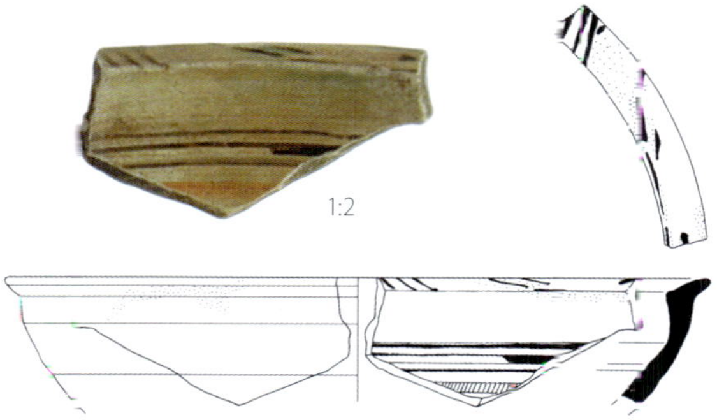

Description: Rim with short, oblique flange and interior molding emphasizing the attachment to the wall. The decoration consists of parallel and oblique strokes in brown on the rim and concentric lines in brown followed by a red band on the interior surface. There are traces of green glaze on the rim. The glaze is transparent, verging on pale yellow.
Fabric: Pale yellow (2.5Y 8/2), refined and hard.
Date: The decoration and morphological articulation have exact parallels at Monte d'Irsi (Patterson and Whitehouse 1992, Fig. 6.22, No. 652) and Otranto (Cotton 1971, Figs. 10.3 and 10.5). Form is dateable to the mid-13th–14th c.

67. MED 300-44 Bowl

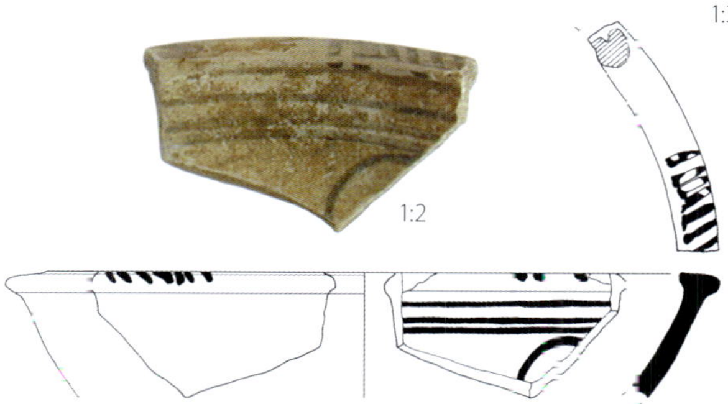

Description: Projecting rim with short, flat flange and interior molding emphasizing the attachment to the wall. The decoration consists of oblique strokes in brown on the rim and three brown lines and a portion of what was probably a vegetal motif (lobe of a leaf?) on the interior surface. No traces of glaze.
Fabric: Pale yellow (2.5Y 8/2), refined and hard.
Date: The decoration has a fairly reasonable parallel at Montella in Campania (Ebanista and Fusaro 2000, 126, Fig. 3.14); materials with similar morphology and decoration have been found at Torre di Mare (P. Tagliente, pers. comm. 2003). The morphological characteristics of the rim and wall have been found in numerous south Italian contexts (see Patterson and Whitehouse 1992 and Dufournier/Flambard/Noyé 1986). The form dates to the mid-13th–14th c.

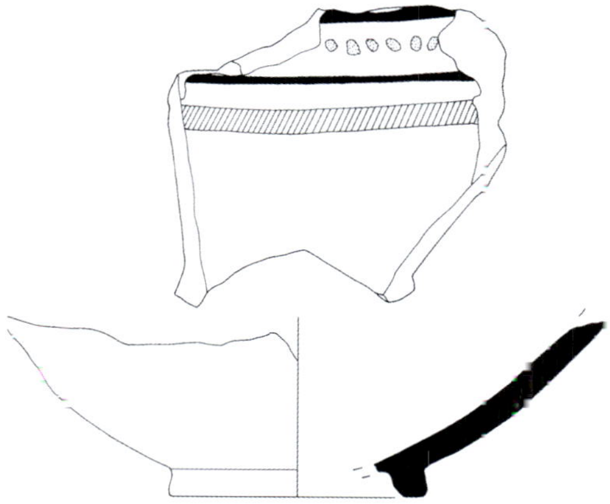

68. MED 300-45 Basin

Description: Base characterized by pronounced ring foot and part of the wall. The interior decoration consists of a geometric motif of (from bottom to top) a red band followed by two thin brown lines enclosing a band of green dots. No traces of glaze.
Fabric: Very pale brown surface (10YR 8/2) and light brown section (7.5YR 6/4), hard; rare white and shiny inclusions.
Date: The geometric design is widely attested in southern Italy in many variants. There is a fairly exact parallel for this form at Torre di Mare, dated between the 13th and the first half of the 14th c. (Dufournier/Flambard/Noyé 1986, 266, Fig. 6.5).

Glazed pottery with green decoration

1:2

69. MED 300-46 Unknown Open Form
Description: Base with ring foot. The decoration consists of a plant motif at the center of the interior surface, probably representing a small branch. Faint traces of transparent/colorless glaze.
Fabric: Reddish-brown (surface and section 5YR 5/4), hard; abundant shiny inclusions.

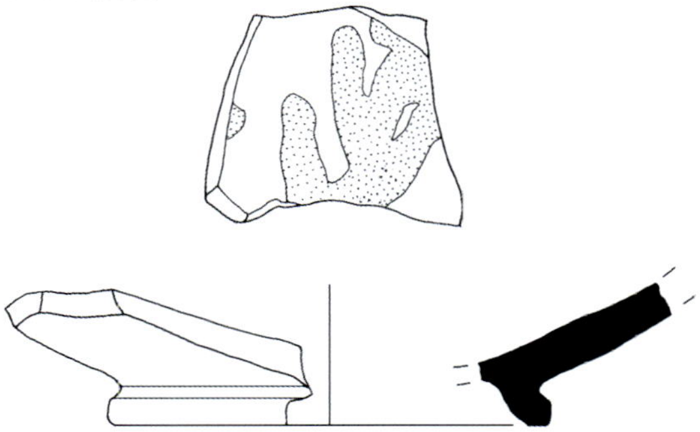

Proto-maiolica

1:2

70. MED 300-47 Bowl
Description: Base with ring foot. The decoration consists of concentric circles formed by pale green-blue dots at the center of the interior.
Fabric: Pale yellow surface (2.5Y 8/2) and very pale brown section (10YR 7/4), hard, with irregular fractures.

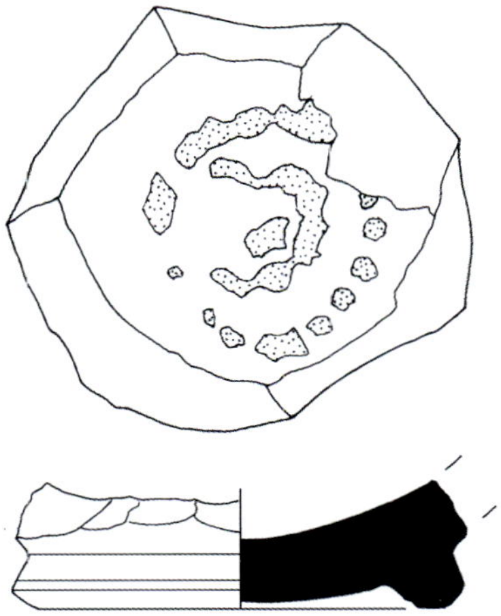

1:2

71. MED 300-48 Bowl
Description: Base with square, outward-turned ring foot. The decoration consists of a blue-green reticulate pattern (gridiron) at the center of the interior.
Dimensions: base diam. 8.8
Fabric: Pale yellow (surface and section 5Y 8/3), refined and hard, with irregular fractures.
Date: The gridiron motif, one of the most prevalent on proto-maiolica products, is widely attested in Puglia and Basilicata, as well as in many parts of the eastern Mediterranean (on the diffusion of the gridiron motif in Puglia, Basilicata, and the eastern Mediterranean, see Salvatore 1984, 444). The decoration in blue, however, is not widely attested, although an example has been documented at Policoro, to which the Metapontine fragment bears a strong resemblance in both morphology and decorative scheme (Salvatore 1984, 444, Pl. CCVI, No. 37d and Pl. CXCIV, No. 81). The fragment can be dated to the first half of the 13th c.

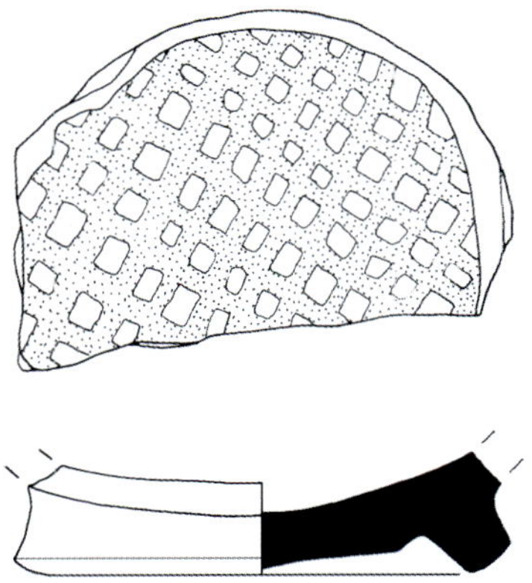

Glazed pottery with graffito decoration

1:3

72. MED 381-06 Bowl

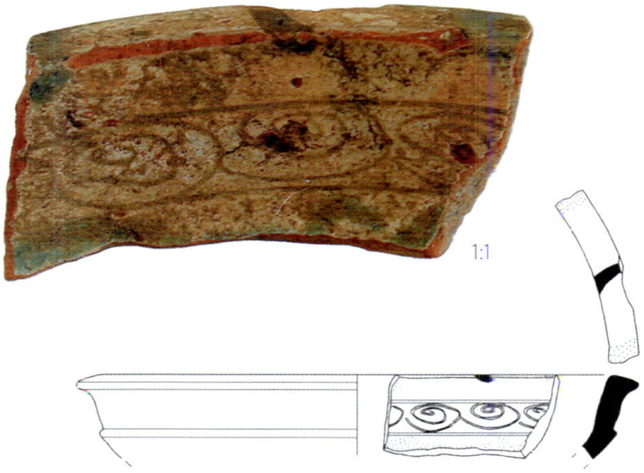

Description: Projecting rim with short flange belonging to a bowl with carinated profile, emphasized by an accentuated molding. The decoration consists of a graffito spiral motif on the interior beneath the rim and painted strokes in green and brown on the flange. Traces of green decoration beneath the band bearing the graffito motif.

Dimensions: rim diam. 20.75

Fabric: Light red (surface and section 2.5YR 6/6); abundant white and shiny inclusions, as well as small dark inclusions. Pale yellow slip (2.5Y 8/2).

Date: The fragment appears to belong to the glazed wares of Adriatic (possibly Venetian) origin, dateable to the end of the 12th or 13th c.

Double-dipped ware

1:2

73. MED 086-06 Bowl

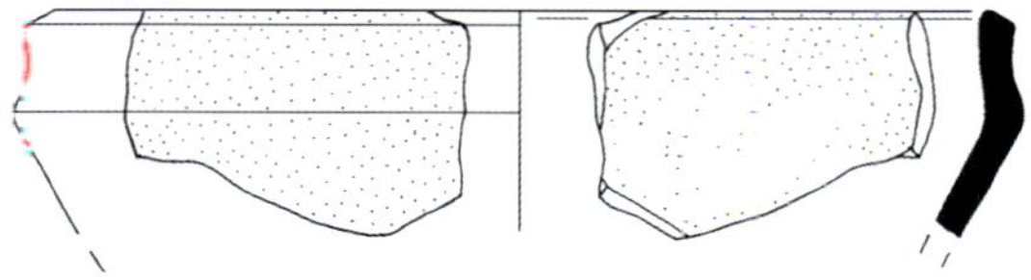

Description: Flanged rim with concave exterior profile and pointed edge. The decoration consists of green glaze on the interior and exterior surfaces.

Fabric: Pale yellow section (2.5Y 7/3), soft.

1:2

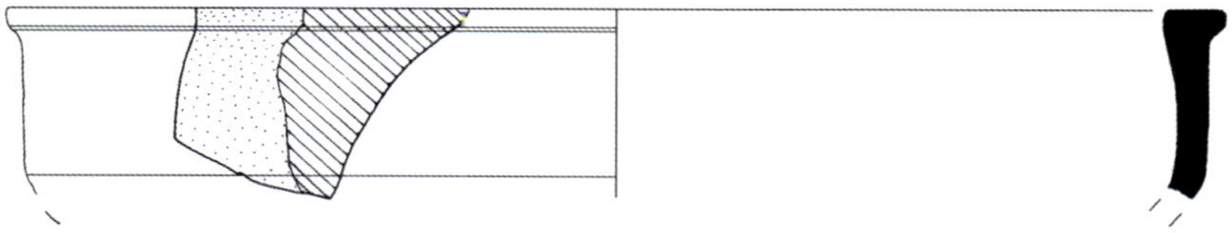

74. MED 140-04 Bowl

Description: Rim with short, flat flange belonging to a carinated bowl. The decoration consists of green and brown glaze on both the interior and exterior surfaces.

Fabric: Light brown section (7.5YR 6/4), soft; small white inclusions.

Date: 15th c.

1:2

75. MED 086-07 Bowl

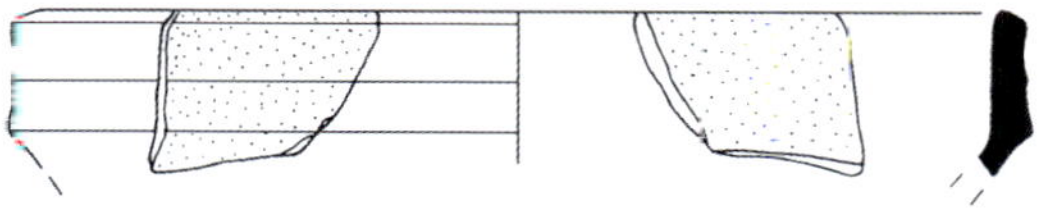

Description: Rim with vertical flange and two shallow grooves on the exterior surface. The decoration consists of green glaze on the exterior surface and faint traces of glaze on the interior surface.

Fabric: Very pale brown section (10YR 7/4), hard; rare small, white inclusions.

Date: Despite the small size of the artifact and the presence of only green glaze, the morphological characteristics of the rim place the artifact among the double-dipped wares and date it to the 15th c.

76. MED 086-08 Bowl

Description: Flanged rim with concave exterior profile and rounded edge. The decoration consists of green glaze on the interior and exterior surfaces.

Fabric: Light gray section (10YR 7/2), hard.

Date: The morphological characteristics of the rim place the artifact among the double-dipped wares and date it to the 15th c.

Double-dipped ware cont.

77. MED 762-01 Bowl
Description: Flanged rim with concave exterior profile, pointed edge, and accentuated carination emphasizing the attachment of the wall. The decoration consists of thin green glaze on the interior surface and rim, with faint traces on the exterior surface as well.
Fabric: Very pale brown (10YR 7/4), hard.

Glazed pottery with green decoration

1:2

78. MED 300-49 Bowl
Description: Base with ring foot. At the center of the interior there is a vegetal motif in green, probably a four-petaled flower. The glaze is colorless and fairly thick.
Fabric: Pale brown section (10YR 6/3), refined and fairly soft. A pale yellow slip (2.5Y 8/2) covers the interior and exterior surfaces.
Date: Second half of the 15th c. (possibly).

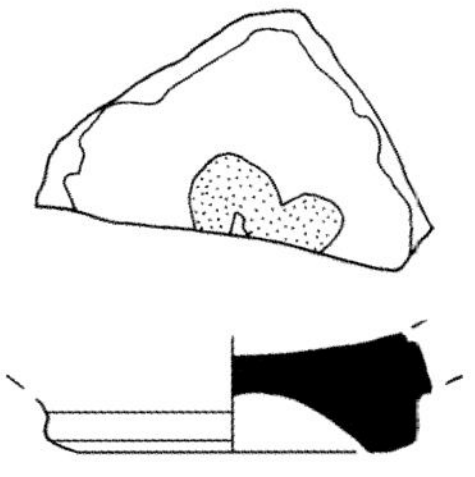

1:2

Post-Medieval Ceramics (16th–19th century)

Coarse ware

79. PMED 551-03 Unknown closed form
Description: Wall fragment bearing circular appliqué decorated with a floral motif.
Fabric: Light yellowish-brown (10YR 6/4), hard, with clean fractures. Pale yellow slip (2.5Y 8/2).
Date: There is an exact parallel for this fragment from the castle of Santa Severina in Calabria, dating it to the 18th c. (Morrone 1998, 126).

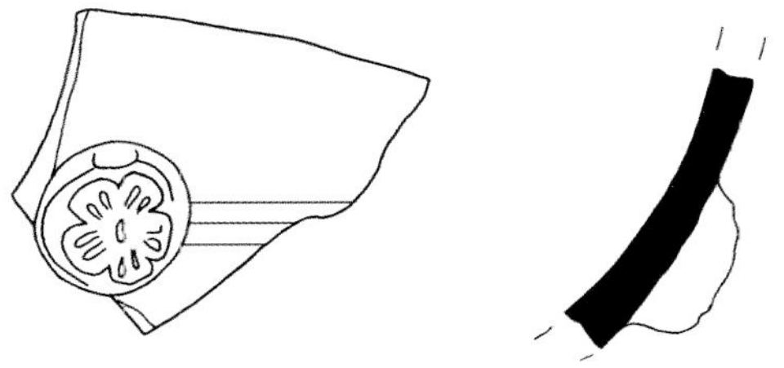

Pottery with monochrome brown glaze

1:2

80. PMED 300-50 Plate
Description: Flanged rim bearing incised "S" motif. The decoration consists of a thick brown glaze on the upper surface of the rim.
Fabric: Pale yellow (5Y 7/3), hard.
Date: Late 15th–16th c.

Painted graffito ware

1:3

81. PMED 507-02 Bowl
Description: Flanged rim with short section of wall. The decoration consists of yellow and green glaze under which, on the interior surface beneath the rim, appears an incised wave motif.
Fabric: Light brown (7.5YR 6/4), hard, with irregular fractures. Pale yellow slip (2.5Y 8/3).
Date: 16th c. (probably).

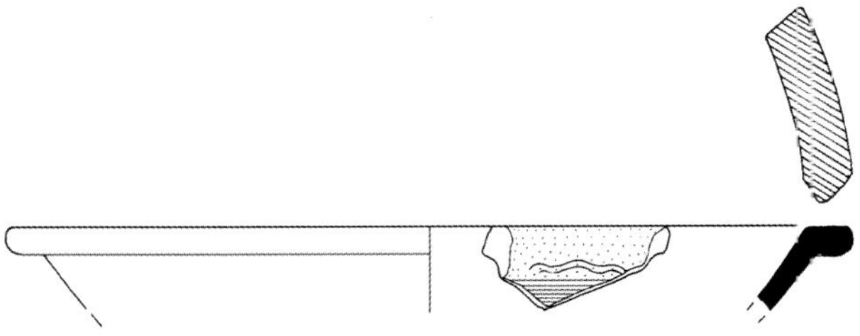

IX. Reference Materials

References

Periodicals/Series/Compendia

AA	*Archäologischer Anzeiger*
AAGSNC	*Atti dell'Accademia Gioenia di scienze naturali di Catania*
AAPG	*Bulletin of the American Association of Petroleum Geologists*
AFLFB	*Annali della Facoltà di Lettere e Filosofia, Università degli Studi di Bari*
AION	*Annali dell'Istituto Universitario Orientale di Napoli, Sezione*
ArchStAnt	*Archeologia e Storia Antica*
AJA	*American Journal of Archaeology*
AJAH	*American Journal of Ancient History*
AJPA	*American Journal of Physical Anthropology*
AM	*Mitteilungen des Deutschen Archäologischen Instituts, Athenische Abteilung*
AmerAnt	*American Antiquity*
AmerGeol	*American Geology*
AnnAStorAnt	*Annali. Sezione di archeologia e storia antica. Istituto universitario orientale di Napoli. Di partimento di studi del mondo classico e del Mediterraneo antico*
ArchCl	*Archeologia classica*
ArchMed	*Archeologia medievale*
ASAA	*Annuario della Scuola Archeologica di Atene*
ASNSP	*Annali della Scuola Normale Superiore di Pisa, Classe di Lettere e Filosofia*
ASP	*Archivio Storico Pugliese*
ASTSN	*Atti della Società Toscana di Scienze Naturali, Mem. Ser. A*
AttiMGrecia	*Atti e memorie della Società Magna Grecia*
Atti Taranto	*Atti del Convegno di Studi sulla Magna Grecia (Taranto)*
AWE	*Ancient West & East*
BABesch	*Bulletin Antieke Beschaving / Annual Papers in Classical Archaeology*
BAGF	*Bulletin de l'Association de géographes français*
BAR-IS	*British Archaeological Reports – International Series*
BAR-SS	*British Archaeological Reports – Supplementary Series*
BBasil	*Bollettino storico della Basilicata*
BdA	*Bollettino d'Arte*
BdArch	*Bollettino di archeologia*
BGSA	*Bulletin of the Geological Society of America*
BICS	*Bulletin of the Institute of Classical Studies, University of London*
BOTA	*Bollettino di Oceanologia teorica ed applicata*
BPI	*Bulletino di Paletnologia Italiana*
BSA	*Annual of the British School at Athens*
BSGLI	*Bollettino della Società Geologica Italiana*
BSGRI	*Bollettino della Società Geografica Italiana*
BStM	*Bollettino dell'Associazione Internazionale di Studi Mediterranei*
BWHO	*Bulletin of the World Health Organization*
COGI	*Canadian Oil and Gas Industry*
CR	*Classical Review*
CRAI	*Comptes rendus des séances de l'Académie des inscriptions et belles-lettres*
CSCA	*University of California Studies in Classical Antiquity*
CVA	*Corpus Vasorum Antiquorum*
DGS	*Dusseldorfer geographische Schriften*
DialArch	*Dialoghi di archeologia*
EJA	*European Journal of Archaeology*
ESR	*Earth Science Review*
Exploration	*Research and Exploration, National Geographic Society*
GAIG	*Geologia applicata e idrogeologia*
GeogAnn	*Geografiska Annaler*
GeogJ	*Geographical Journal*
GeogM	*Geography Magazine*
GeoRom	*Geologica romana*
GFDQ	*Geografia fisica e dinamica quaternaria*
GGeol	*Giornale di geologia*
GIF	*Giornale italiano di filologia*
GSIB	*Geological Survey of Israel Bulletin*
HSR	*Histoire et societés rurales*
JA	*Journal of Anatomy*
JAS	*Journal of Archaeological Science*
JFA	*Journal of Field Archaeology*
JMA	*Journal of Mediterranean Archaeology*
JP	*Journal of Periodontology*
JSP	*Journal of Sedimentary Petrology*
JWP	*Journal of World Prehistory*
LIMC	*Lexicon Iconographicum Mythologiae Classicae*
LMGJ	*Liverpool and Manchester Geological Journal*
MAAR	*Memoirs of the American Academy in Rome*
MAP	*Memorie dell'Accademia Patavina*
MarGeol	*Marine Geology*

MedArch	*Mediterranean Archaeology*
MedievArch	*Medieval Archaeology*
Méditerranée	*Méditerranée. Revue géographique des pays méditerranées*
MEFRA	*Mélanges de l'École française de Rome, Antiquité*
MEFRM	*Mélanges de l'École française de Rome, Moyen âge*
MGrecia	*Magna Grecia. Rassegna di archeologia, storia, arte, attualità*
MHR	*Mediterranean Historical Review*
ModGeol	*Modern Geology*
MSGLI	*Memorie della Società Geologica Italiana*
NGCG	*Northeast Gulf Coast Geology*
NSc	*Notizie degli scavi di antichità*
NUB	*Nebraska University Bulletin*
NumAntCl	*Numismatica e Antichità Classiche*
OJA	*Oxford Journal of Archaeology*
PAPS	*Proceedings of the American Philosophical Society*
PBSR	*Papers of the British School at Rome*
PCE	*Physics and Chemistry of the Earth*
PE	*Photogrammetric Engineering*
Polis	*Polis. Studi interdisciplinari sul mondo antico*
PotTech	*Newsletter of the Department of Pottery Technology, Leiden University*
PP	*La Parola del passato*
PPP	*Palaeogeography, Palaeoclimatology, Palaeoecology*
PPS	*Proceedings of the Prehistoric Society*
QR	*Quaternary Research*
QSAP	*Quaderni della Soprintendenza Archeologica del Piemonte*
RA	*Revue archéologique*
RasArch	*Rassegna di archeologia*
RassAPiomb	*Rassegna di archeologia, associazione archeo logica piombinese*
REG	*Revue des études grecques*
RendAccLinc	*Rendiconti dell'Accademia Nazionale dei Lincei, Classe di Scienze Morali, Storiche e Filologiche*
RendNap	*Rendiconti dell'Accademia di Archeologia, Lettere e Belle Arti (Naples)*
RGI	*Rivista geografica italiana*
RIN	*Rivista italiana di numismatica e scienze affini*
RivFil	*Rivista di filologia e di istruzione classica*
RM	*Mitteilungen des Deutschen Archäologischen Instituts, Römische Abteilung*
RS	*Ricerche e studi*
RSB	*Ricerche e Studi, Museo Provinciale Francesco Ribezzo, Brindisi*
RSP	*Rivista di Scienze Preistoriche*
ScAnt	*Scienze dell'antichità: Storia, archeologia, antropologia*
SciAm	*Scientific American*
SicArch	*Sicilia archeologica*
StAnt	*Studi di antichità*
StBiz	*Rivista di studi bizantini e neoellenici*
StStor	*Studi storici*
TGM	*The Geographical Magazine*
WorldArch	*World Archaeology*
YPA	*Yearbook of Physical Anthropology*
ZG	*Zeitschrift für Geomorphologie*
ZPE	*Zeitschrift für Papyrologie und Epigraphik*

Short Title Abbreviations

L'Africa romana
Convegno di studio su "L'Africa romana." 2002. *L'Africa romana*. Atti del XVI convegno di studio, Sassari, 7-10 dicembre 2000. Rome.

Agora IV
Howland, R.H. 1958. *The Athenian Agora* IV. *Greek Lamps and Their Survival*. Princeton.

Agora V
Robinson, H.S. 1959. *The Athenian Agora* V. *Pottery of the Roman Period: Chronology*. Princeton.

Agora XII
Sparkes, B.A., and L. Talcott. 1970. *The Athenian Agora XII. Black and Plain Pottery of the 6th, 5th, and 4th Centuries* BC. 2 vols. Princeton.

Agora XXIX
Rotroff, S.I. 1997. *The Athenian Agora* XXIX. *Hellenistic Pottery. Athenian and Imported Wheelmade Table Ware and Related Material*. Princeton.

Agora XXXIII
Rotroff, S.I. 2006. *The Athenian Agora* XXXIII. *Hellenistic Pottery. The Plain Wares*. Princeton.

Agricoltura e scambi
Carlsen, J., and E. Lo Cascio, eds. 2009. *Agricoltura e scambi nell'Italia tardo-repubblicana*. Bari.

Agriculture in Ancient Greece
Wells, B., ed. 1992. *Agriculture in Ancient Greece*. (Proceedings of the Seventh International Symposium at the Swedish Institute at Athens, 16-17 May 1990. *Acta Instituti Atheniensis Regni Sueciae* Ser. 4, No. 42.) Stockholm.

Ancient Territories 1999
Carter, J.C., ed. 1999. *Ancient Territories. Metaponto and Chersonesos. 1999 Annual Report, Institute of Classical Archaeology*. Austin.

Ancient Territories 2000
Carter, J.C., ed. 2000. *The Study of Ancient Territories. Chersonesos and Metaponto. 2000 Annual Report, Institute of Classical Archaeology*. Austin.

Ancient Territories 2001
Carter, J.C., ed. 2001. *The Study of Ancient Territories. Chersonesos and Metaponto. 2001 Annual Report, Institute of Classical Archaeology*. Austin.

Ancient Territories 2003
Carter, J.C., ed. 2003. *The Study of Ancient Territories. Chersonesos and Metaponto. 2003 Field Report, Institute of Classical Archaeology*. Austin.

Ancient Territories 2004
Carter, J.C., ed. 2004. *The Study of Ancient Territories. Chersonesos and Metaponto. 2004 Annual Report, Institute of Classical Archaeology.* Austin.
Ancient Territories 2006-2007
Carter, J.C., ed. 2008. *The Study of Ancient Territories. Chersonesos and Southern Italy. 2006-2007 Annual Report, Institute of Classical Archaeology.* Austin.
Archaeological Survey
Keller, D.R., and D.W. Rupp, eds. 1983. *Archaeological Survey in the Mediterranean Area.* (BAR-IS 155.) Oxford.
Archeologia dell'acqua
Soprintendenza Archeologica della Basilicata. 1999. *Archeologia dell'acqua in Basilicata.* Lavello.
Archeologia di un sapere
Settis, S., and M.C. Parra, eds. 2005. *Magna Graecia. Archeologia di un sapere.* Milan.
ArcMap
Harlow, M. 2004. *ArcGIS* 9: *Using ArcMap.* Redlands, CA.
Archeoviva
AA. VV. 2005. *Archeoviva. Siti archeologici di Acquaviva delle Fonti.* Modugno.
Arpi
Mazzei, M., ed. 1995. *Arpi. L'ipogeo della Medusa e la necropoli.* Bari.
Arte e artigianato
Lippolis, E., ed. 1996. *I Greci in Occidente. Arte e artigianato in Magna Grecia.* Naples.
ARV2
Beazley, J.D. 1963. *Attic Red-figure Vase Painters.* 2nd ed. Oxford.
Athenian Potters and Painters
Oakley, J.H., W.D.E. Coulson, and O. Palagia, eds. 1997. *Athenian Potters and Painters: The Conference Proceedings.* Oxford.
Atlante
Istituto della Enciclopedia Italiana. 1985. *Atlante delle Forme Ceramiche* II. (Supplemento all'*Enciclopedia dell'Arte Antica.*) Rome.
Attività archeologica
Lattanzi, E., ed. 1980. *Attività archeologica in Basilicata 1964-1977. Scritti in onore di Dinu Adamesteanu.* Matera.
Bari
Andreassi, G., and F. Radina, eds. 1988. *Archeologia di una città. Bari dalle origini al X secolo.* Bari.
Basento-Cavone
Carter, J.C., ed. Forthcoming. *The Chora of Metaponto: Archaeological Field Survey II – Basento to Cavone.* Austin.
Basilicata/espansionismo romano
Salvatore, M., ed. 1990. *Basilicata. L'espansionismo romano nel sud-est d'Italia: il quadro archeologico.* (Atti del convegno – Venosa, 23-25 aprile 1987.) Venosa.
Bitonto
Riccardi, A., ed. 2003. *Gli antichi Peucezi a Bitonto. Documenti ed immagini dalla necropoli di via Traiana.* Bari.
Brucato
Pesez, J.M., ed. 1984. *Brucato. Histoire et archéologie d'un habitat médiéval en Sicile.* 2 vols. Rome.
Carta Geologica
2004. *Carta Geologica d'Italia alla scala 1:100,000.* Istituto Superiore per la Protezione e la Ricerca Ambientale.
Carta pedologica
Rosa, R., et al. 2006. *I suoli della Basilicata. Carta pedologica della Regione Basilicata in scala 1:250.000.* Potenza.
Cavallino
Pancrazzi, O., ed. 1979. *Cavallino.* Galatina.
Ceglie Peuceta I
Miroslav Marin, M., R. Moreno Cassano, A. Fornaio, and M. Chelotti. 1982. *Ceglie Peuceta* I. Bari.
Ceramica in Italia
Saguì, L., ed. 1998. *La ceramica in Italia: VI-VII secolo.* (Atti del Convegno in onore di John W. Hayes, Roma, 11-13 maggio 1995.) 2 vols. Florence.
Ceramica invetriata
Patitucci Uggeri, S., ed. 2000. *La ceramica invetriata tardomedievale dell'Italia centro-meridionale.* (*Quaderni di Archeologia Medievale* III.) Florence.
Ceramica medievale
Università di Siena and Museo internazionale delle ceramiche. 1986. *Atti del III Congresso Internazionale sulla ceramica medievale nel Mediterraneo occidentale* (Siena-Faenza 1984). Florence.
Ceramica romana e archeometria
Olcese, G., ed. 1994. *Ceramica romana e archeometria: lo stato degli studi.* (Atti delle giornate internazionali di studio, Castello di Montegufoni, Firenze, 26-27 aprile 1993.) Florence.
Ceramiche fini
Frère, D., ed. 2007. *Ceramiche fini a decoro subgeometrico del VI secolo a.C. in Etruria meridionale e in Campania.* (*Collection de l'École française de Rome* 389.) Rome.
Céramique apulienne
Denoyelle, M., E. Lippolis, M. Mazzei, and C. Pouzadoux, eds. 2005. *La céramique apulienne: bilan et perspectives.* (Actes de la table ronde, Naples, 30 novembre–2 décembre 2000.) Naples.
Céramiques
Centre Jean Bérard and Institut Français de Naples. 1978. *Les céramiques de la Grèce de l'Est et leur diffusion en Occident.* (Centre Jean Bérard/Institut Français de Naples, 6-9 juillet 1976. *Colloques internationaux du Centre National de la Recherche Scientifique* 569.) Paris/Naples.
Céramiques communes
Bats, M., ed. 1996. *Les céramiques communes de Campanie et de Narbonnaise (Ier s. av. J.-C. – IIe s. ap. J.-C.). La vaisselle de cuisine et de table.* (Actes des journées d'étude organisées par le Centre Jean Bérard et la Soprintendenza

Archeologica per le Province di Napoli e Caserta, Naples, 27-28 mai 1994.) Naples.
Céramiques hellénistiques et romaines I
Centre de Recherches d'Histoire. 1980. *Céramiques hellénistiques et romaines* I. (*Annales Littéraires de l'Université de Besançon* 242; *Centre de Recherches d'Histoire Ancienne* 36.) Paris.
Céramiques hellénistiques et romaines II
Lévêque, P., and J.-P. Morel, eds. 1987. *Céramiques hellénistiques et romaines* II. (*Annales Littéraires de l'Université de Besançon* 331; *Centre de Recherches d'Histoire Ancienne* 70.) Paris.
Céramiques hellénistiques et romaines III
Lévêque, P., J.-P. Morel, and É. Geny, eds. 2001. *Céramiques hellénistiques et romaines* III. Paris.
Chora und Polis
Kolb, F., ed. 2005. *Chora und Polis.* (*Schriften des Historischen Kollegs Kolloquien* 54.) Munich.
Collezione Banca Intesa
Sena Chiesa, G., and F. Slavazzi, eds. 2006. *Ceramiche attiche e magnogreche. Collezione Banca Intesa. Catalogo Ragionato.* 3 vols. Milan.
Collezione Chini
1995. *Ceramica sovraddipinta, ori, bronzi, monete della Collezione Chini nel Museo civico di Bassano del Grappa.* (*Collezioni e Musei archeologici del Veneto* 39.) Rome.
Collezione Lagioia
Sena Chiesa, G., ed. 2004. *La Collezione Lagioia. Una raccolta storica dalla Magna Grecia al Museo Archeologico di Milano.* Milan.
Conspectus
Ettlinger, E., B. Hedinger, B. Hoffmann, P.M. Kenrick, G. Pucci, K. Roth-Rubi, G. Schneider, S. von Schnurbein, C.M. Wells, and S. Zabehlicky-Scheffenegger. 1990. *Conspectus Formarum Terrae Sigillatae Italico Modo Confectae.* Bonn.
Corinth IV.2
Broneer, O. 1930. *Corinth* IV.2: *Terracotta Lamps*. Cambridge, Mass.
Corinth VII.2
Amyx, D.A. and P. Lawrence. 1975. *Corinth* VII.2: *Archaic Corinthian Pottery and the Anaploga Well.* Princeton.
Corinth VII.3
Edwards, G.R. 1975. *Corinth* VII.3: *Corinthian Hellenistic Pottery.* Princeton.
Corinth VII.5
Risser, M.K. 2001. *Corinth* VII.5. *Corinthian Conventionalizing Pottery.* Princeton.
Corinth XI
Morgan, C.H. 1942. *Corinth* XI. *The Byzantine Pottery.* Cambridge, Mass.
Corinth XII
Davidson, G.R. 1952. *Corinth* XII. *The Minor Objects.* Princeton.
Corinth XIII
Blegen, C.W., R.S. Young, and H. Palmer. 1964. *Corinth* XIII. *The North Cemetery.* Princeton.
Corinth XV.2
Stillwell, A.N. 1952. *Corinth* XV.2. *The Potters' Quarter. The Terracottas.* Princeton.
Corinth XVIII.1
Pemberton, E.G. 1989. *Corinth* XVIII.1. *The Sanctuary of Demeter And Kore: The Greek Pottery.* Princeton.
The Coroplast's Art
Uhlenbrock, J., ed. 1990. *The Coroplast's Art: Greek Terracottas of the Hellenistic World.* New Rochelle, NY.
Cozzo Presepe
du Plat Taylor, J., E. Macnamara, J. Ward-Perkins, et al. 1983. The Excavations at Cozzo Presepe (1969-72). In *Metaponto* II, 191–406.
Cuma
Cuozzo, M., B. d'Agostino, and L. Del Verme, eds. 2006. *Cuma: le fortificazioni 2. I materiali dai terrapieni arcaici.* Naples.
Da Leukania a Lucania
Torelli, M., and L. de Lachenal, eds. 1993. *Da Leukania a Lucania. La Lucania centro-orientale fra Pirro e i Giulio-Claudii.* Venosa.
Dea di Sibari I.1
van der Wielen-van Ommeren, F., L. de Lachenal, Nina Mekacher, Jette Christiansen, and L. Raselli-Nydegger. 2006. *La dea di Sibari e il santuario ritrovato. Studi sui rinvenimenti dal Timpone Motta di Francavilla Marittima* I.1. *Ceramiche di importazione, di produzione coloniale e indigena. (Tomo I.)* (BdA, Volume Speciale.) Rome.
Dea di Sibari I.2
van der Wielen-van Ommeren, F., L. de Lachenal, D. Tsiafakis, Jette Christiansen, A.W. Johnston, J.K. Papadopoulos, M. Kleibrink, L. Raselli-Nydegger, and J. Heiden. 2008. *La dea di Sibari e il santuario ritrovato. Studi sui rinvenimenti dal Timpone Motta di Francavilla Marittima* I.2. *Ceramiche di importazione, di produzione coloniale e indigena.* (*Tomo* 2.) (BdA, Volume Speciale.) Rome.
Dea lucana
Nava, M.L., and M. Osanna, eds. 2001. *Rituali per un dea lucana. Il santuario di Torre di Satriano.* Potenza.
Die Ägäis und das westliche Mittelmeer
Krinzinger, F., ed. 2000. *Die Ägäis und das westliche Mittelmeer. Beziehungen und Wechselwirkungen 8. bis 5. Jh. v. Chr.* (Akten des Symposions, Vien, 24 bis 27 März 1999. *Österreichische Akademie der Wissenschaften, Archäologische Forschungen* B.4.) Vienna.
Emergenze e problemi
Regione Puglia Centro regionale servizi educativi e culturali. 1990. *Emergenze e problemi archeologici. Manduria-Taranto-Heraclea.* Manduria.

Eraclea lucana
Pianu, G. 2002. *L'agorà di eraclea lucana*. Rome.
ESIV
Trendall, A.D. 1974. *Early South Italian Vase-painting*. Mainz.
Forentum I
Giorgi, M., S. Martinelli, M. Osanna, and A. Russo. 1988. *Forentum* I. *Le necropoli di Lavello*. Venosa.
Forentum II
Bottini, A., and M.P. Fresa. 1991. *Forentum* II. *L'acropoli in età classica*. Venosa.
Fouilles de Delphes III.2
Colin, M.G. 1909–1913. *Fouilles de Delphes* III.2. *Inscriptions du trésor des Athéniens*. Paris.
Fouilles de Delphes III.4
Pouilloux, J. 1976. *Fouilles de Delphes* III.4. *Les inscriptions de la terrasse du temple et de la région nord du sanctuaire, nos. 351 à 516*. Paris.
Fratte
Greco, G., and A. Pontrandolfo, eds. 1990. *Fratte. Un insediamento etrusco-campano*. Modena.
Geology of Italy
Squyres, C.H., ed. 1975. *Geology of Italy*. 2 vols. Tripoli.
Giornate di studio
Cotecchia, V., ed. 1979. *Giornate di studio sui movimenti franosi e dinamica dei versanti*. (Atti del convegno internazionale, Salice Terme [Pavia], 3-4-5 maggio 1979. GAIG 14[III].) Bari.
Gli ori
De Juliis, E.M., A. Alessio, A. D'Amicis, A. Dell'Aglio, D. Graepler, E. Lippolis, L. Masiello, and T. Schojer, eds. 1986. *Gli ori di Taranto in età ellenistica*. Taranto. (Also published in French as *Les ors hellénistiques de Tarente*, Milan 1986.)
Gravina
Small, A.M., ed. 1992. *An Iron Age and Roman Republican Settlement on Botromagno, Gravina di Puglia. Excavations of 1965–1974*. (Archaeological Monographs of the British School at Rome 5.) 2 vols. London.
Gravina III.2
du Plat Taylor, J., P.G. Dorrell, and A.M. Small. 1977. Gravina-di-Puglia III: Houses and a Cemetery of the Iron Age and Classical Periods. Part Two. *PBSR* 45: 69–137.
Gravina in Puglia
Cooperative Petra Magna, ed. s.d. [1990]. *Gravina in Puglia. Alla ricerca del passato*. Gravina.
Gravisca 4
Boldrini, S. 1994. Gravisca. *Scavi nel santuario greco* 4. *Le ceramiche ioniche*. Bari.
Gravisca 9
Valentini, V. 1993. Gravisca. *Scavi nel santuario greco 9. Le ceramiche a vernice nera*. Bari.
Gravisca 12.I
Gori, B., and T. Pierini. 2002. *Gravisca. Scavi nel santuario greco 12. La ceramica comune I. Ceramica comune di impasto*. Bari.
Greci, Enotri e Lucani
Bianco, S., A. Bottini, A. Pontrandolfo, A. Russo Tagliente, and E. Setari, eds. 1996. *I Greci in Occidente. Greci, Enotri e Lucani nella Basilicata meridionale*. Naples.
Greci in Occidente
Pugliese Carratelli, G., ed. 1996. *I Greci in Occidente*. Milan.
Greci sul Basento
Orlandini, P., ed. 1986. *I Greci sul Basento. Mostra degli scavi archeologici all'Incoronata di Metaponto 1971-1984*. Como.
Greek Colonists and Native Populations
Descœudres, J.-P., ed. 1990. *Greek Colonists and Native Populations*. (Proceedings of the First Australian Congress of Classical Archaeology, held in honour of Emeritus Professor A.D. Trendall, Sydney 9-14 July 1989.) Canberra/Oxford.
Greek Countryside
Jameson, M.H., C.N. Runnels, and T.H. van Andel. 1994. *A Greek Countryside: The Southern Argolid from Prehistory to the Present Day*. Stanford.
Herakleiastudien
Neutsch, B., ed. 1967. *Archäologische Forschungen in Lukanien* II. *Herakleiastudien*. (RM Suppl. 11) Heidelberg.
Human Landscape
Malone, C., and S. Stoddart, eds. 1985. *Papers in Italian Archaeology* IV, Part i: *The Human Landscape*. (BAR-IS 243.) Oxford.
IGCH
Thompson, M., O. Mørkholm, and C. Kraay. 1973. *An Inventory of Greek Coin Hoards*. New York.
I Greci 4.II
Settis, S., ed. 2002. *I Greci. Storia, cultura, arte, società* 4, Atlante II. Turin.
Il vino di Dioniso
Nava, M.L., M. Nota Santi, and M. Piranomonte, eds. 1999. *Il vino di Dioniso. Dei e uomini a banchetto in Basilicata*. (Catalogo della mostra, Roma.) Rome.
Immagine e mito
Nava, M.L., and M. Osanna, eds. 2002. *Immagine e mito nella Basilicata antica*. Potenza.
Incoronata 1
1991. *Ricerche archeologiche all'Incoronata di Metaponto* 1. *Le fosse di scarico del saggio P. Materiali e problematiche*. Milan.
Incoronata 2
1992. *Ricerche archeologiche all'Incoronata di Metaponto 2. Dal villaggio indigeno all'emporio greco. Le strutture e i materiali del saggio T.* Milan.
Incoronata 3
Orlandini, P. and M. Castaldi, eds. 1995. *Ricerche archeo-*

logiche all'Incoronata di Metaponto 3. *L'oikos greco del saggio S. Lo scavo e i reperti*. Milan.

Incoronata 4
2000. *Ricerche archeologiche all'Incoronata di Metaponto* 4. *L'oikos greco del grande perirrhanterion nel contesto del saggio G.* Milan.

Incoronata 5
Orlandini, P. and M. Castaldi, eds. 1997. *Ricerche archeologiche all'Incoronata di Metaponto* 5. *L'oikos greco del saggio H. Lo scavo e i reperti*. Milan.

Incoronata 6
2003. *Ricerche archeologiche all'Incoronata di Metaponto* 6. *L'oikos greco del saggio E. Lo scavo e i reperti*. Milan.

Kaulonia I
Trèziny, H. 1989. Kaulonia I. *Sondages sur la fortification nord (1982-1985). (Cahiers du Centre Jean Bérard XIII.)* Naples.

Keos
Cherry, J.F., J.L. Davis, and E. Mantzourani, eds. 1991. *Landscape Archaeology as Long-Term History: Northern Keos in the Cycladic Islands from Earliest Settlement until Modern Times*. Los Angeles.

Kerameikos VII.2
Kunze Götte, E., K. Tancke, and K. Vierneisel. 1999. *Kerameikos* VII.2. *Die Nekropole von der Mitte des 6. bis zum Ende des 5. Jahrhunderts*. Munich.

Κοινά
Castoldi, M., ed. 1999. Κοινά. *Miscellanea di studi archeologici in onore di Piero Orlandini*. Milan.

Laconia I-II
Cavanagh, W., J. Crouwel, R.W.V. Catling, and G. Shipley, eds. 2002/1996. *The Laconia Survey: Continuity and Change in a Greek Rural Landscape*. (*Annual of the British School at Athens* Suppl. 26-27.) 2 vols. London.

Lagonegrese
Greco, G., ed. 1982. *L'evidenza archeologica nel Lagonegrese*. (Catalogo della mostra documentaria, Rivello 1981.) Matera.

Laos I
Greco, E., S. Luppino, and A. Schnapp, eds. 1989. *Laos I. Scavi a Marcellina* 1973-1985. Taranto.

Laos II
Greco, E., and P.G. Guzzo, eds. 1992. *Laos* II. *La tomba a camera di Marcellina*. Taranto.

La protomaiolica
Patitucci Uggeri, S., ed. 1997. *La protomaiolica. Bilancio e aggiornamenti*. (Atti del Convegno di Studi, Roma 23 novembre 1995. *Quaderni di Archeologia Medievale* II.) Florence.

LCS
Trendall, A.D. 1967. *The Red-figured Vases of Lucania, Campania and Sicily*. 2 vols. Oxford.

LCS Suppl. 2
Trendall, A.D. 1973. *The Red-figured Vases of Lucania, Campania and Sicily*, Suppl. 2. (BICS Suppl. 31.) Oxford.

LCS Suppl. 3
Trendall, A.D. 1983. *The Red-figured Vases of Lucania, Campania and Sicily*, Suppl. 3. (BICS Suppl. 41.) Oxford.

Leuca
Van Compernolle, René, ed. 1978. *Leuca*. Galatina.

Locri 1983
1983. *Locri Epizefiri. Ricerche archeologiche su un abitato della Magna Grecia. Mostra documentaria*. Locri.

Locri II
Barra Bagnasco, M., ed. 1989. *Locri Epizefiri* II. *Gli isolati I2 e I3 dell'area di Centocamere*. Florence.

Locri III
Barra Bagnasco, M., ed. 1989. *Locri Epizefiri III. Cultura materiale e vita quotidiana*. Florence.

Locri IV
Barra Bagnasco, M., ed. 1992. *Locri Epizefiri* IV. *Lo scavo di Marasà sud. Il sacello tardo arcaico e la "casa dei leoni."* Florence.

Lo sguardo di Icaro
Guaitoli, M., ed. 2003. *Lo sguardo di Icaro. Le collezioni dell'Aerofototeca Nazionale per la conoscenza del territorio*. Rome.

Magna Grecia II
Pugliese Carratelli, G., ed. 1987. *Magna Grecia* II. *Lo sviluppo politico, sociale ed economico*. Milan.

Magna Grecia e Sicilia
Barra Bagnasco, M., E. De Miro, and A. Pinzone, eds. 1999. *Magna Grecia e Sicilia. Stato degli studi e prospettive di ricerca*. (Atti dell'incontro di studi, Messina, 2-4 dicembre 1996.) Messina.

Man's Role
Bottema, S., G. Entjes-Nieborg, and W. van Zeist, eds. 1990. *Man's Role in Shaping the Eastern Mediterranean Landscape*. Rotterdam.

Mediterranean Quaternary River Environments
Lewin, J., M.G. Macklin, and J.C. Woodward, eds. 1995. *Mediterranean Quaternary River Environments*. Rotterdam.

Mégara Hyblaea 2
Vallet, G., and F. Villard. 1964. *Mégara Hyblaea* 2. *La céramique archaïque*. (*Mélanges d'archéologie et d'histoire* Suppl. 1, École française de Rome.) Rome.

Meligunìs-Lipára II
Bernabò Brea, L., and M. Cavalier. 1965. *Meligunìs-Lipára* II. *La necropoli greca e romana nella contrada Diana*. Palermo.

Meligunìs-Lipára V
Bernabò Brea, L. and M. Cavalier. 1991. *Meligunìs-Lipára* V. *Scavi nella necropolis greca di Lipara*. Rome.

Meligunìs-Lipára X
Bernabò Brea, L. and M. Cavalier. 2000. *Meligunìs-*

Lipára X. *Scoperte e scavi archeologici nell'area urbana e suburbana di Lipari*. Rome.

Meligunìs-Lipára XI
Bernabò Brea, L., M. Cavalier, and F. Villard. 2001. *Meligunìs-Lipára* XI. *Gli scavi nella necropoli greca e romana di Lipari nell'area del terreno vescovile*. 2 vols. Palermo.

Metaponto I
Adamesteanu, D., D. Mertens, and F. D'Andria. 1980. *Metaponto* I. NSc Ser. 8, Vol. 29 (1975) Suppl. Rome.

Metaponto II
Adamesteanu, D., B. Chiartano, J.C. Carter, and E. Macnamara. 1983. *Metaponto* II. NSc Ser. 8, Vol. 31 (1977) Suppl. Rome.

Metaponto/Archeologia
De Siena, A., ed. 2001. *Metaponto. Archeologia di una colonia greca*. Taranto.

Minturnae
Kirsopp Lake, A. 1934–1935. *Campana suppellex*. The Pottery Deposit at Minturnae. *BStM* 5: 97–114.

Modalità insediative
Lo Cascio, E., and A. Storchi Marino, eds. 2001. *Modalità insediative e strutture agrarie nell'Italia meridionale in età romana*. Bari.

Monasteri italogreci e benedittini II
Bubbico, L., F. Caputo, and A. Maurano, eds. 1996. *Monasteri italogreci e benedittini in Basilicata* II. *Le architetture*. Matera.

Monte Irsi
Small, A., ed. 1977. *Monte Irsi, Southern Italy. The Canadian Excavations in the Iron Age and Roman Sites, 1971-1972*. (*BAR-SS* 20.) Oxford.

Montemurro
Russo, A., ed. 2006. *Con il fuso e la conocchia. La fattoria lucana di Montemurro e l'edilizia domestica nel IV secolo a.C.* Lavello.

Monte Sannace
Ciancio, A., E.M. de Juliis, A. Riccardi, and F. Rossi, eds. 1989. *Monte Sannace. Gli scavi dell'acropoli (1978-1983)*. Galatina.

Museo Policoro
Bianco, S., and M. Tagliente, eds. 1985. *Il Museo Nazionale della Siritide di Policoro. Archeologia della Basilicata meridionale*. Rome/Bari.

Museo di Taranto
Alessio, A. et al, eds. 1988. *Museo di Taranto. Cento anni di archeologia*. Taranto.

Museo Taranto I.2
Museo Nazionale di Taranto. 1990. *Catalogo del Museo Nazionale Archeologico di Taranto* I.2. *Il progetto del Museo*. Taranto.

Museo Taranto I.3
Museo Nazionale di Taranto. 1997. *Catalogo del Museo Nazionale Archeologico di Taranto* I.3. *Atleti e guerrieri. Tradizioni aristocratiche a Taranto tra VI e V sec. a.C.* Taranto.

Museo Taranto II.1
Dell'Aglio, A., and E. Lippolis. 1992. *Catalogo del Museo Nazionale Archeologico di Taranto* II.1. *Ginosa e Laterza. La documentazione archeologica dal VII al III sec. a.C. Scavi 1900-1980*. Taranto.

Museo Taranto II.2
De Juliis, E.M., ed. 2006. *Catalogo del Museo Nazionale Archeologico di Taranto* II.2. *Rutigliano I. La necropoli di Contrada Purgatorio. Scavo 1978*. Taranto.

Museo Taranto III.1
Lippolis, E., ed. 1994. *Catalogo del Museo Nazionale Archeologico di Taranto* III.1. *Taranto. La necropoli: aspetti e problemi della documentazione archeologica tra VII e I secolo a.C.* Taranto.

Museo Taranto III.2
Colivicchi, F. 2001. *Catalogo del Museo Nazionale Archeologico di Taranto* III.2. *Alabastra tardo-ellenistici e romani dala necropoli di Taranto. Materiali e contesti*. Taranto.

Museo Venosa
Salvatore, M., ed. 1991. *Il Museo Archeologico Nazionale di Venosa*. Matera.

Necropoleis
Carter, J.C., ed. 1998. *The Chora of Metaponto: The Necropoleis*. 2 vols. Austin.

Nobili e guerrieri
Nava, M.L., and V. Cracolici, eds. 2004. *Nobili e guerrieri tra Bradano e Ofanto*. Lavello.

Olynthus XIII
Robinson, D.M. 1950. *Excavations at Olynthus XIII. Vases Found in 1934 and 1938*. London.

Oppido Mamertina
Costamagna, L., and P. Visonà, eds. 1999. *Oppido Mamertina. Ricerche archeologiche nel territorio e in contrada Mella*. Rome.

Ordona IV
Mertens, J., ed. 1974. *Ordona* IV. *Rapports et études*. Brussels/Rome

Ordona V
Mertens, J., ed. 1976. *Ordona* V. *Rapports et études*. Brussels/Rome

Ordona VI
Mertens, J., ed. 1979. *Ordona* VI. *Rapports et études*. Brussels/Rome

Ordona VII.2
Iker, R. 1986. *Ordona* VII.2. *Les tombes dauniennes. Les tombes du IVe et du debut du IIIe siècles avant notre ère*. Brussels/Rome

Ordona VIII
Mertens, J., ed. 1988. *Ordona* VIII. *Rapports et études*. Brussels/Rome

Ordona IX

Mertens, J., ed. 1997. *Ordona* IX. *Rapports et études*. Brussels/Rome.

Ordona X
Volpe, G., ed. 2000. *Ordona* X. *Ricerche archeologiche a Herdonia (1993-1998)*. Bari.

Oria
Maruggi, G.A., ed. 2001. *Oria e l'archeologia. Percorsi di una ricerca*. Oria.

Otranto I
Michaelides, D., and D. Wilkinson, eds. 1992. *Excavations at Otranto* I. *The Excavation*. Galatina.

Otranto II
D'Andria, F., and D. Whitehouse, eds. 1992. *Excavations at Otranto* II. *The Finds*. Galatina.

Paesaggi e insediamenti rurali
Volpe, G., and M. Turchiano, eds. 2006. *Paesaggi e insediamenti rurali in Italia meridionale fra tardoantico e altomedioevo*. (Atti del Primo Seminario sul Tardoantico e l'Altomedioevo in Italia meridionale, Foggia, 12-14 febbraio 2004.) Bari.

Panionion und Melie
Kleiner, G., P. Hommel, and W. Müller-Wiener. 1967. *Panionion und Melie*. (*Jahrbuch des Deutschen Archäologischen Instituts 23. Ergänzungsheft*.) Berlin.

Perachora II
Dunbabin, T.J., ed. 1962. *Perachora* II. *The Sanctuaries of Hera Akraia and Limenia: The Pottery, Ivories, Scarabs, and Other Objects from the Votive Deposit of Hera Limenia*. Oxford.

Phlyax Vases
Trendall, A.D. 1967. *Phlyax Vases*. 2nd ed. (*BICS* Suppl. 19.) Oxford.

Pithekoussai I
Buchner, G., and D. Ridgway, eds. 1993. *Pithekoussai* I. *La necropoli: Tombe 1-723, scavate dal 1952 al 1961*. (Accademia Nazionale dei Lincei, *Monumenti Antichi, Serie monografica* 4.) Rome.

Ploughsoil Assemblages
Francovich, R., H. Patterson, and G. Barker, eds. 2000. *Extracting Meaning from Ploughsoil Assemblages*. (*The Archaeology of Mediterranean Landscapes* 5.) Oxford.

Ploughzone Archaeology
Schofield, A.J., ed. 1991. *Interpreting Artefact Scatters: Contributions to Ploughzone Archaeology*. (*Oxbow Monographs* 4.) Oxford.

Pomarico Vecchio I
Barra Bagnasco, M., ed. 1997. *Pomarico Vecchio* I. *Abitato, mura, necropoli, materiali*. 2 vols. Galatina.

Poseidonia-Paestum II
Greco, E., and D. Theodorescu, eds. 1983. *Poseidonia-Paestum* II. *L'agorà*. Rome.

Poseidonia-Paestum III
Greco, E., and D. Theodorescu, eds. 1987. *Poseidonia-Paestum* III. *Forum Nord*. Rome.

Principi imperatori vescovi
Cassano, R., ed. 1992. *Principi imperatori vescovi. Duemila anni di storia a Canosa*. Venice.

Problèmes de la terre
Finley, M.I., ed. 1973. *Problèmes de la terre en Grèce ancienne*. Paris/La Haye.

Quellheiligtum der Demeter
Otto, B., and M. Müller-Dürr, eds. 1996. *Herakleia in Lukanien und das Quellheiligtum der Demeter*. Innsbruck.

QuickBird
DigitalGlobe, Inc. 2006. *QuickBird Imagery Products: Product Guide*. (Rev. 4.7.2.) Longmont, CO.

Ravitaillement en blé
Centre Jean Bérard and Centre national de la recherche scientifique. 1994. *Le ravitaillement en blé de Rome et des centres urbains des débuts de la République jusqu'au haut empire*. (Actes du colloque international, Naples 1991. *Collection Centre Jean Bérard* 11. *Collection de l'École française de Rome* 196.) Naples/Rome.

Ricerche sulla casa
D'Andria, F., and K. Mannino, eds. 1996. *Ricerche sulla casa in Magna Grecia e in Sicilia*. (Atti del Colloquio – Lecce, 23-24 giugno 1992. Università di Lecce, Scuola di Specializzazione in Archeologia Classica e Medioevale, *Archeologia e Storia* 5.) Galatina.

Roccagloriosa I
Gualtieri, M., and H. Fracchia, eds. 1990. *Roccagloriosa* I. *L'abitato: scavo e ricognizione topografica (1976-1986)*. Naples.

Roccagloriosa II
Gualtieri, M., and H. Fracchia, eds. 2002. *Roccagloriosa* II. *L'oppidum lucano e il territorio*. Naples.

RVAp
Trendall, A.D., and A. Cambitoglou. 1978. *The Red-figured Vases of Apulia*. 2 vols. London.

RVAp Suppl. 1
Trendall, A.D., and A. Cambitoglou. 1983. *The Red-figured Vases of Apulia*, Suppl. 1. (*BICS* Suppl. 42.) London.

San Giovanni di Ruoti I
Small, A.M., and R.J. Buck, eds. 1994. *The Excavations of San Giovanni di Ruoti. The Villas and Their Environment*. Toronto.

San Lorenzo Maggiore
Fontana, M.V., and G. Ventrone Vassallo, eds. 1984. *La ceramica medievale di San Lorenzo Maggiore in Napoli*. (Atti del Convegno di Napoli, 25-27 giugno 1980.) Naples.

Satriano
Greco, E., ed. 1988. *Satriano 1987-88. Un biennio di ricerche archeologiche (mostra documentaria)*. Naples.

Sibari I
Accademia nazionale dei Lincei. 1969. *Sibari I. Saggi di scavo al Parco del Cavallo (1969)*. *NSc* Ser. 8, Vol. 23 Suppl. Rome.

Sibari II

Accademia nazionale dei Lincei. 1970. *Sibari* II. *Sibari: scavi al Parco del Cavallo (1960-62; 1969-70) e agli Stombi (1969-70). NSc* Ser. 8, Vol. 24 Suppl. 3. Rome.

Sibari III
Accademia nazionale dei Lincei. 1972. *Sibari* III. *Rapporto preliminare della campagna di scavo: Stombi, Casa Bianca, Parco del Cavallo, San Mauro (1971). NSc* Ser. 8, Vol. 26 Suppl. Rome.

Sibari IV
Accademia nazionale dei Lincei. 1974. *Sibari* IV. *Relazione preliminare della campagna di scavo: Stombi, Parco del Cavallo, Prolungamento Strada, Casa Bianca (1972). NSc* Ser. 8, Vol. 28 Suppl. Rome.

Sibari V
Accademia nazionale dei Lincei. 1974. *Sibari. 5, Relazione preliminare delle campagne di scavo 1973 (Parco del Cavallo; Casa Bianca) e 1974 (Stombi; Incrocio; Parco del Cavallo; Prolungamento Strada; Casa Bianca). Notizie degli scavi di antichità*, 8th series, 28, Suppl. no. 3. Rome.

Side-by-side Survey
Alcock, S.E., and J.F. Cherry, eds. 2004. *Side-by-side Survey: Comparative Regional Studies in the Mediterranean World.* Oxford.

Siris-Polieion
De Siena, A., and M. Tagliente, eds. 1986. *Siris-Polieion. Fonti letterarie e nuova documentazione archeologica.* (Incontro di studi, Policoro 8-10 giugno 1984.) Galatina.

Siritide e Metapontino
1998. *Siritide e Metapontino. Storie di due territori coloniali.* (Atti dell'incontro di studio, Policoro 1991. *Cahiers du Centre Jean Bérard* 20.) Naples.

Società romana I
Giardina, A., and A. Schiavone, eds. 1981. *Società romana e produzione schiavistica* I. *L'Italia: insediamenti e forme economiche.* Rome/Bari.

Società romana II
Giardina, A., and A. Schiavone, eds. 1981. *Società romana e produzione schiavistica* II. *Merci, mercati e scambi nel Mediterraneo.* Rome/Bari.

Spazio del rito
Nava, M.L., and M. Osanna, eds. 2005. *Lo spazio del rito.* Santuari e culti in Italia meridionale tra indigeni e greci. (Atti delle giornate di studio, Matera, 28 e 29 giugno 2002. SIRIS Suppl. I.) Bari.

Storia della Basilicata 1
Adamesteanu, D., ed. 1999. *Storia della Basilicata* 1. *L'Antichità.* Bari.

Storia della Basilicata 2
Fonseca, C.D., ed. 2006. *Storia della Basilicata* 2. *Il Medioevo.* Bari.

Storia della Magna Grecia
Ciaceri, E. 1924–1940. *Storia della Magna Grecia.* 3 vols. Naples.

Storia dell'alto Medioevo
Francovich, R., and G. Noyé, eds. 1994. *La storia dell'alto Medioevo italiano (VI-X secolo) alla luce dell'archeologia.* Florence.

Structural Model of Italy
Ogniben, L., M. Parotto, and A. Praturlon, eds. 1975. *Structural Model of Italy.* (*Quaderni de "La ricerca scientifica"* 90.) Rome.

Structures rurales
Doukellis, P.N., and L.G. Mendoni, eds. 1994. *Structures Rurales et Sociétés Antiques.* (Actes du colloque de Corfou, 14-16 mai 1992. *Annales Littéraires de l'Université de Besançon* 508. *Centre de Recherches d'Histoire Ancienne* 126.) Paris.

Studia Ietina IV
Caflisch, R.B. 1991. *Studia Ietina IV. Die Firniskeramik vom Monte Iato. Funde 1971-1982.* Zurich.

Studi in Onore di Dinu Adamesteanu
1983. *Studi in Onore di Dinu Adamesteanu.* Galatina.

Sur les pas des Grecs
Arcelin, P., M. Bats, D. Garcia, G. Marchand, and M. Schwaller, eds. 1995. *Sur les pas des Grecs en Occident. Hommages à André Nickels.* (*Études Massaliètes* 4.) Paris.

Terra degli Enotri
La Torre, G.F., and A. Colicelli, eds. 1999. *Nella terra degli Enotri.* (Atti del Convegno di studi, Tortora 18-19 aprile 1998.) Paestum.

Territori tarantini
Istituto per la Storia e l'Archeologia della Magna Grecia. 2002. *Nuovi documenti dai territori tarantini.* (Tavola rotonda di Taranto, 7 giugno 2001; *Atti Taranto* 41.) Taranto.

Territory of Metaponto
Carter, J.C., ed. 1983. *The Territory of Metaponto 1981-1982.* Austin.

Torre di Mare I
Bertelli, G., and D. Roubis, eds. 2002. *Torre di Mare* I. *Ricerche archeologiche nell'insediamento medievale di Metaponto (1995-1999).* (*SIRIS* 2. *Studi e ricerche della Scuola di Specializzazione in Archeologia di Matera [1999-2000].*) Bari.

Torre di Satriano I
Osanna, M., and M.M. Sica, eds. 2005. *Torre di Satriano* I. *Il santuario lucano.* Venosa.

Uomo, acqua e paesaggio
Quilici Gigli, S., ed. 1997. *Uomo, acqua e paesaggio: Atti dell'incontro di studi sul tema regimentazione delle acque e trasformazione del paesaggio antico; S. Maria Capua Vetere 22–23 novembre 1996.* Rome: L'Erma di Bretschneider.

Valesio I
Yntema, D. 2001. *Pre-Roman Valesio. Excavations of the Amsterdam Free University at Valesio, Province of Brindisi, Southern Italy* I. *The Pottery.* Amsterdam.

Vecchi scavi
D'Amicis, A., ed. 1991. *Vecchi scavi, nuovi restauri.* Taranto.

Venusia
Marchi, M.L., and G. Sabbatini. 1996. Venusia. (*Forma Italiae* 37.) Florence.

Abbott, J.T. 1997. *Late Quaternary Alluviation and Soil Erosion in Southern Italy.* Doctoral dissertation, The University of Texas at Austin.
———. 1998. Geomorphic Context. In *Necropoleis*, 483–501.
Abbott, J.T., and S. Valastro, Jr. 1995. The Holocene Alluvial Records of the Chorai of Metapontum, Basilicata, and Croton, Calabria, Italy. In *Mediterranean Quaternary River Environments*, 195–205.
Abramowitz, M., and I.A. Stegun. 1968. *Handbook of Mathematical Functions.* Washington, D.C.
Abruzzese Calabrese, G. 1996. Taranto. La coroplastica votiva. In *Arte e artigianato*, 188–205.
Accorsi, C. A., M. Bandini Mazzanti, and L. Forlani, 1989. Segni palinologici, antraco-xilologici e carpologici dell'azione antropica sul paesaggio vegetale Olocenico in Emilia-Romagna. *Memoir Societa Geologica Italiana* 42: 95–108.
Ackerknecht, E.H. 1965. *History and Geography of the Most Important Diseases.* New York.
Acsádi, G., and J. Nemeskéri. 1970. *History of Human Life Span and Mortality.* Trans. K. Balás. Budapest.
Adam, J.-P. 2005. *Roman Building. Materials and Techniques.* London.
Adamesteanu, D. 1958a. L'opera di Timoleonte nella Sicilia centro-meridionale vista attraverso gli scavi e le ricerche archeologiche. *Kokalos* 4: 31–68.
———. 1958b. Manfria (Gela). Scavo di una fattoria officina. *NSc* Ser. 8, V. 1-42/43: 290–334.
———. 1963. La fotografia aerea e le vie della Magna Grecia. *Atti Taranto* 2: 39–58.
———. 1965. La documentazione archeologica in Basilicata. *Atti Taranto* 4: 121–143.
———. 1967a. Problèmes de la zone archéologique de Métaponte. *RA* 1967.1: 3–38.
———. 1967b. Attività delle Soprintendenze – Basilicata. *BdA* 52.1: 44–50.
———. 1967c. L'acropoli di Heraclea. Appunti di fotointerpretazione archeologica. In *Archäologische Forschungen in Lukanien* II: *Herakleiastudien*. Ed. B. Neutsch. 96–99. (*Mitteilungen des Deutschen Archäologischen Instituts, Römische Abteilung.* Erganzungsheft 11.) Heidelberg.
———. 1968. La documentazione archeologica in Basilicata. *Atti Taranto* 7: 243–263.
———. 1970a. L'attività archeologica in Basilicata. *Atti Taranto* 9: 215–237.
———. 1970b. ΑΡΓΟΙ ΛΙΘΟΙ a Metaponto. In *Adriatica Praehistorica et Antiqua. Miscellanea Gregorio Novak Dicata.* Eds. V. Mirosavljevic, D. Rendic-Miocevic and M. Suic. 307–324. Zagreb.
———. 1971. L'attività archeologica in Basilicata. *Atti Taranto* 10: 467–485.
———. 1972. L'attività archeologica in Basilicata. *Atti Taranto* 11: 445–459.
———. 1973a. Le suddivisioni di terra nel metapontino. In *Problèmes de la terre*, 49–61.
———. 1973b. *Metaponto.* Naples.
———. 1974a. *La Basilicata antica.* Cava dei Tirreni.
———. 1974b. Problemi topografici ed urbanistici metapontini. *Atti Taranto* 13: 153–186.
———. 1975. Tipi di insediamenti umani in Puglia e in Basilicata. *Posebna Izdanja* (Centar za Balkanoloska Ispitivanja) 24.6: 259–269.
———. 1978. Fotografia aerea e problemi topografici. Serra di Vaglio e Metaponto. In Un decennio di ricerche archeologiche. 2 vols. II. 585–596. (*Quaderni de La Ricerca Scientifica* 100.) Rome.
———. 1979. ΔΙΟΣ ΑΓΟΡΑ a Metaponto. *PP* 187: 296–312.
———. 1980. Il santuario di Apollo e urbanistica generale. In *Metaponto* I, 15–311.
———. 1987. Una porta metapontina. In *Saggi in onore di Guglielmo De Angelis d'Ossat. Quaderni dell'Istituto di Storia dell'Architettura* N.S. 1–10 (1983–1987). Eds. S. Benedetti and G. Miarelli Mariani. 15–18. Rome.
———. 1999. Le "poleis" della costa nel V secolo a.C. In *Storia della Basilicata* 1, 339–357.
Adamesteanu, D., and H. Dilthey. 1978. Siris – Nuovi contributi archeologici. *MEFRA* 90.2: 515–565.
———. 1992. *Macchia di Rossano: il santuario della Mefitis. Rapporto preliminare.* Galatina.
Adamesteanu, D., L. Lissi Caronna, R.R. Holloway, J.-P. Morel, and G. Tocco. 1971. *Popoli anellenici in Basilicata.* Potenza.
Adamesteanu, D., D. Mertens, and A. De Siena. 1975. Metaponto: Santuario di Apollo – Tempio D (Tempio Ionico). Rapporto preliminare. *BdA* 1–2: 26–49.
Adamesteanu, D., and C. Vatin. 1976. L'arrière-pays de Métaponte. *CRAI* 1976: 110–123.
Adams, W.Y., and E.W. Adams. 1991 *Archaeological Typology and Practical Reality. A Dialectical Approach to Artifact Classification and Sorting.* Cambridge.
Agresti, G. 1988a. San Pietro. In *Bari*, 217–220.
———. 1988b. Rione Sant'Angelo. In *Bari*, 340–348.
Alberti, L. 1596. *Descrittione di tutta l'Italia et isole pertinenti ad essa.* Venetia.
Alcock, S.E., J.F. Cherry, and J.L. Davis. 1994. Intensive Survey, Agricultural Practice, and the Classical Landscape of Greece. In *Classical Greece: Ancient Histories and Modern Archaeologies.* Ed. I. Morris. 137–170. Cambridge.
Alessio, A. 1988. La necropoli di Piazza d'Armi. Lo scavo 1911–1914. In *Museo di Taranto*, 325–370.
———. 2002. L'area a sud-est di Taranto. In *Territori tarantini*, 87–113.
Alessio, A., and P.G. Guzzo. 1989-1990. Santuari e fattorie ad est di Taranto. Elementi archeologici per un modello di interpretazione. *ScAnt* 3/4: 363–396.

Alexander, D.E. 1980. I *Calanchi*: Accelerated Erosion in Italy. *Geography* 65: 95–100.

———. 1981a. Disaster in Southern Italy, November 1980. *GeogM* 53: 555–561.

———. 1981b. Preliminary Assessment of Landslides Resulting from the Earthquake of 23 November, 1980, in Southern Italy. *Disasters* 5: 376–383.

———. 1982a. "Creeping Disaster" Speeds up. *Geotimes* 27.11: 17–20.

———. 1982b. Difference between "Calanchi" and "Biancane" Badlands in Italy. In *Badland Geomorphology and Piping*. Eds. R. Bryan and A. Yair. 71–87. Norwich.

Allen, E. 1969. *Stone Shelters*. Cambridge, MA: Massachusetts Institute of Technology Press.

Ambrosetti, P.L., and F. Carraro. 1980. Pliocène et quaternaire. In *Introduction a la géologie generale d'Italie*. (*Guide à l'excursion* 122A, XXVI Cong. Geol. Intern., Paris.) 77–82. Paris.

Ammerman, A. 1985. Plow-zone Experiments in Calabria, Italy. *JFA* 12: 33–40.

Ammerman, R.M. 1990. The Religious Context of Hellenistic Terracotta Figurines. In *The Coroplast's Art*, 37–45.

———. 2002. *The Sanctuary of Santa Venera at Paestum* II. *The Votive Terracottas*. Ann Arbor.

Amore, M.G. 1992–1993. *Aspetti di vita quotidiana nella chora metapontina. La documentazione delle fattorie*. Tesi di laurea in Archeologia della Magna Grecia, Università degli Studi di Torino.

Amouretti, M.C. 1986. *Le pain et l'huile dans la Grèce antique. De l'araire au moulin*. (*Annales Littéraires de l'Université de Besançon* 328. *Centre de Recherche d'Histoire Ancienne* 67.) Paris.

Ampolo, C. 1987. Organizzazione politica, sociale ed economica delle "poleis" italiote. In *Magna Grecia* II, 89–98.

Amyx, D.A. 1988. *Corinthian Vase-painting of the Archaic Period*. Berkeley.

Anderson-Stojanović, V.R. 1987. The Chronology and Function of Ceramic Unguentaria. *AJA* 91.1: 105–122.

Andreassi, G. 1979. Scavi a Gravina, Salentino ed Egnazia. *Atti Taranto* 28: 437–443.

———. 1988a. Piazza Aldo Moro. In *Bari*, 242.

———. 1988b. Via Francesco Saverio Abbrescia-Via Marcello Celentano. In *Bari*, 243–248.

———. 1988c. Via Giovanni Amendola. In *Bari*, 257–275.

———. 1988d. Via Trento. In *Bari*, 287–290.

———. 2001. L'attività archeologica in Puglia nel 2000. *Atti Taranto* 40: 1007–1033.

———. 2002. L'attività archeologica in Puglia nel 2001. *Atti Taranto* 41: 791–801.

Andreassi, G., V. Scattarella, and A. De Lucia. 1988. Ferrovie del Sud-Est. In *Bari*, 276–286.

Angel, J.L. 1972. Ecology and Population in the Eastern Mediterranean. *WorldArch* 4: 88–105.

Angelelli, C. 1998–1999. Amelia (Terni). Scavo all'interno della Torre del S. Uffizio. I materiali. *NSc* Ser. 9, Vol. 9–10: 12–55.

Annese, C. 2000. Le ceramiche tardoantiche della *domus* B. In *Ordona* X, 285–342.

Aprosio, M. 2008. *Archeologia dei paesaggi a Brindisi. Dalla romanizzazione al medioevo*. Bari.

Arangio-Ruiz, V., and A. Olivieri. 1965. *Inscriptiones graecae Siciliae et infirmae Italiae ad ius pertinentes*. Edizione anastatica. Rome: L'Erma di Bretschneider.

Arias, P.E., M. Hirmer, and B. Shefton. 1961. *A Thousand Years of Greek Vase-painting*. New York.

Arthur, P. 1986. Appunti sulla circolazione della ceramica medievale a Napoli. In *Ceramica medievale*, 545–554.

———. 1991. *Romans in Northern Campania: Settlement and Land-use around the Massico and the Garigliano Basin*. (*Archaeological Monographs of the British School at Rome* 1.) London.

———. 1999. Riflessioni intorno ad alcune produzioni di anfore tra la Calabria e la Puglia in età medievale. In *Atti del XXX Convegno Internazionale della Ceramica*. 9–18. Florence.

———. 2000a. La città in Italia meridionale in età tardoantica: riflessioni intorno alle evidenze materiali. *Atti Taranto* 39: 167–200.

———. 2000b. Le prime ceramiche invetriate monocrome in Terra d'Otranto. In *Ceramica invetriata*, 159–166.

———. 2003. I Balcani e il Salento nel Medioevo. In *L'Archeologia dell'Adriatico dalla Preistoria al Medioevo*. (Atti del Convegno, Ravenna 7–9 giugno 2001.) Ed. F. Lenzi. 654–665. Florence.

———. 2004. From Vicus to Village: Italian Landscapes, AD 400–1000. In *Landscapes of Change: Rural Evolutions in Late Antiquity and the Early Middle Ages*. Ed. N. Christie. 103–133. Aldershot.

Arthur, P., and H. Patterson. 1998. Local Pottery in Southern Puglia in the Sixth and Seventh Centuries. In *Ceramica in Italia*, 511–530.

Asheri, D. 1966. *Distribuzioni di terre nell'antica Grecia*. Turin: Accademia delle Scienze.

Ashmole, B. 1972. *Architect and Sculptor in Classical Greece*. London.

ASTER DEM Validation Team. 2009. ASTER Global DEM Validation Summary Report. Pasadena.

Auriemma, R. 2004. *Salentum a salo* 2. *Forma Maris Antiqui*. Lecce.

Avagliano, G. 1988. Intervento. *Atti Taranto* 27: 329–330.

Bacci, G.M. and G. Tigano, eds. 2002. *Da Zancle a Messina, un percorso archeologico attraverso gli scavi, II*. Messina.

Bagnasco Gianni, G. 1999. La ceramica depurata "acroma" e a "bande." In *Tarquinia. Scavi sistematici nell'abitato. Campagne 1982–1988. I materiali* 1. (Tarchna 2.) Ed. C. Chiaromonte Treré. 99–176. Rome.

———. 2007. Le ricerche sulle ceramiche depurate a Tarquinia: tra importazioni e produzioni locali. In *Ceramiche fini*, 93–99.

Bailo Modesti, G. 1980. *Cairano nell'età arcaica.* Naples.

Bailey, D. M. 1975. *A Catalogue of the Lamps in the British Museum.* I. *Greek, Hellenistic, and Roman Pottery Lamps.* Oxford.

Baldassare, G., and B. Radina. 1971. Sorgenti di una parte del medio bacino del fiume Bradano. *GAIG* 6: 137–159.

Barberis, V. 1995. Le fattorie della chora metapontina. Note sui culti. *BBasil* 11: 11–40.

———. 1997. I materiali 1. La ceramica arcaica. In *Pomarico Vecchio* I, 55–62.

———. 1999. I siti tra Sinni e Bradano dall'età arcaica all'età ellenistica: schede. In *Magna Grecia e Sicilia*, 59–105.

———. 2004. *Rappresentazioni di divinità e di devoti dall'area sacra urbana di Metaponto. La coroplastica votiva dalla fine del VII all'inizio del V sec. a.C.* Florence.

Barker, G. 1972. The Conditions of Cultural and Economic Growth in the Bronze Age of Central Italy. *PPS* 38: 170–208.

———. 1981. *Landscape and Society: Prehistoric Central Italy.* New York.

———. 1985. Landscape Archaeology in Italy. In *Human Landscape*, 1–19.

———. 1991. Approaches to Archaeological Survey. In *Roman Landscapes. Archaeological Survey in the Mediterranean Region.* (*Archaeological Monographs of the British School at Rome* No. 2.) Eds. G. Barker and J. Lloyd. 1–9. London.

Barker, G., and C. Hunt. 1995. Quaternary Valley Floor Erosion and Alluviation in the Biferno Valley, Molise, Italy: The Role of Tectonics, Climate, Sea Level Change, and Human Activity. In *Mediterranean Quaternary River Environments*, 145–157.

Barone, A., F. Lafage, and R. Cantilena. 1989. Area incrocio. In *Laos* I, 90–97.

Barone, G., V. Crupi, S. Galli, F. Longo, D. Majolino, P. Mazzoleni, and G. Spagnolo. 2002. Archaeometric Analyses on "Corinthian B" Transport Amphorae Found at Gela (Sicily, Italy). *Archaeometry* 46.4: 553–568.

Baroni, S., and V. Casolo. 1990. *Capua preromana. Terrecotte votive V: Piccole figure muliebri panneggiate.* Florence.

Barra Bagnasco, M. 1992. Le anfore. In *Locri* IV, 205–240.

———. 1996a. La coroplastica. In *Greci, Enotri e Lucani*, 219–223.

———. 1996b. La casa in Magna Grecia. In *Ricerche sulla casa*, 41–66.

———. 1997a. Le anfore. In *Pomarico Vecchio* I, 1.201–208.

———. 1997b. La coroplastica. In *Pomarico Vecchio* I, 1.215–226.

———. 1999a. Sistemi insediativi nella Basilicata dal Sinni al Bradano, tra il IV e il III sec. a.C. In *Magna Grecia e Sicilia*, 39–57.

———. 1999b. Il culto delle acque in Magna Grecia dall'età arcaica alla romanizzazione: documenti archeologici e fonti letterarie. In *Archeologia dell'acqua*, 53–72.

Barra Bagnasco, M., and A. Russo Tagliente. 1996. I culti. In *Greci, Enotri e Lucani*, 183–205.

Bartoccini, R. 1936. Taranto. Rinvenimenti e scavi (1933-34). *NSc* Ser. 6, Vol. 12.4-6: 107–232.

Bats, M. 1988. *Vaisselle et alimentation à Olbia de Provence (v. 350 – v. 50 av. J.-C.). Modèles culturels et catégories céramiques.* Paris.

———. 1996. Remarques finales. In *Céramiques communes*, 481–484.

Bayliss, D.D. 1969. The Distribution of *Hyalinea balthica* and *Globorotalia truncatuloides* in the Type Calabrian. *Lethaia* 2: 133–143.

Beckby, H. 1957. *Anthologia Graeca: Griechisch-Deutsch.* Munich.

Bell, M., III. 1981. *Morgantina Studies* I: *The Terracottas.* Princeton.

———. 1990. Hellenistic Terracottas of Southern Italy and Sicily. In *The Coroplast's Art*, 63–70.

Beneduce, P., V. Festa, R. Francioso, M. Schiattarella, and M. Tropeano. 2004. Conflicting Drainage Patterns in the Matera Horst Area, Southern Italy. *PCE* 29: 717–724.

Berger, S. 1992. *Revolution and Society in Greek Sicily and Southern Italy.* (*Historia* Einzelschriften 71.) Stuttgart.

Berggren, W.A., D. Kent, and J.A. Van Couvering. 1985. Cenozoic Geochronology. *BGSA* 96: 1407–1418.

Berggren, W.A., and J.A. Van Couvering. 1974. The Late Neogene: Bio-Stratigraphy, Geochronology, and Paleoclimatology of the Last 15 Million Years in Marine And Continental Sequences. *PPP* 16: 1–216.

Berlingò, I. 1986. La necropoli arcaica di Policoro in Contrada Madonnelle. In *Siris-Polieion*, 117–127.

Bernardi, M., L. Cenciaioli, D. Manconi, and G. Venturini. 1991. Indagini archeologiche all'interno del Palazzo Ducale di Gubbio (PG). Nuove conoscenze sulla dinamica insediativa e sull'urbanistica del sito. *ArchMed* 18: 429–476.

Bernardini, M. 1961. *Vasi dello stile di Gnathia, vasi a vernice nera (Museo Provinciale "S. Castromediano" di Lecce).* Bari.

Berry, A.C., and R.J. Berry. 1967. Epigenetic Variation in the Human Cranium. *JA* 101: 361–379.

Bertelli, G. 2002. Interpretazione e considerazioni. In *Torre di Mare* I, 159–168.

Berti, G. 1997. Pisa: ceramiche e commerci (2° metà X-1° metà XIV s). In *Atti del I Congresso Nazionale di Archeologia Medievale.* Ed. S. Gelichi. 346–351. Florence.

Berti, G., and S. Gelichi. 1995. Ceramiche, ceramisti e trasmissioni tecnologiche tra XII e XIII secolo nell'Italia centro-settentrionale. In *Miscellanea in memoria di Giuliano Cremonesi.* 409–445. Pisa.

Besques, S. 1972. *Musée national du Louvre. Catalogue raisonné des figurines et reliefs en terre-cuite grecs, étrusques et romains* III: *époques hellénistique et romaine. Grèce et Asia Mineure.* Paris.

———. 1986. *Musée national du Louvre. Catalogue raisonné des figurines et reliefs en terre-cuite grècs, étrusques et romains* IV.1: *époques hellénistique et romaine. Italie méridionale–Sicilie–Sardaigne.* Paris.
Bianco, O. 1997a. Ceramica da mescita. In *Pomarico Vecchio* I, 1.182–186.
———. 1997b. Ceramica da fuoco. In *Pomarico Vecchio* I, 1.186–192.
Bianco, S. 1978. I materiali del villaggio dell'età del Bronzo di S. Marco presso Metaponto (MT). *Atti XX RSIIPP*: 295–306.
———. 1981. Aspetti culturali dell'Eneolitico della prima età del bronzo sulla costa ionica della Basilicata. *StAnt* 2: 13–17.
———. 1983a. Il Neolitico in contrada Cetrangolo nella valle del Cavone (Matera). In *Studi in onore di Dinu Adamesteanu*, 1–16.
———. 1983b. Aspetti dell'età del bronzo sulla costa ionica della Basilicata. *Atti Taranto* 22: 131–137.
———. 1986. Aspetti della età del bronzo e del ferro sulla costa ionica della Basilicata. In *Siris-Polieion*, 17–26.
———. 1991–1992. Nuovi dati da Calabria, Basilicata, e Puglia meridionale. L'età del bronzo in Italia nei secoli dal XVI al XIV a.C. *RassAPiomb* 10: 509–522.
———. 1993. La chora di Herakleia: Tursi, S. Maria d'Anglona. In *Da Leukania a Lucania*, 197–207.
———. 1994. La necropoli di contr. S. Brancato di S. Arcangelo (PZ). *StAnt* 7: 111–136.
———. 1996. L'età del Ferro. In *Greci, Enotri e Lucani*, 31–50.
———. 1998. La prima Età del ferro nel Metapontino e nella Siritide. In *Siritide e Metapontino*, 15–30.
———. 2001. La chora di Siris-Herakleia. *Atti Taranto* 40: 807–818.
Bianco, S. and L. Cataldo. 1991–1992. L'insediamento di Civita di Paterno (Potenza). L'età del bronzo in Italia nei secoli dal XVI al XIV a.C. *RasArch* 10: 744–746.
Bianco, S. and A. De Siena. 1982. La Basilicata. Termitito (Montalbano Ionico, Matera). In *Magna Grecia e mondo miceneo. Nuovi documenti*. Ed. L. Vagnetti. 69–96. Taranto.
Bianco, S., and D.A. Marino. 1991–1992. L'insediamento di Capo Piccolo di Isola di Capo Rizzuto (Catanzaro). *RasArch* 10: 754–755.
Bietti Sestieri, A.M. 1984. Central and Southern Italy in the Late Bronze Age. In *Crossroads of the Mediterranean* (*Archaeologia Transatlantica* II). Eds. T. Hackens, N.G. Holloway, and R.R. Holloway. 55–122. Providence/ Louvain-la-Neuve.
Bintliff, J. 2002a. Going to Market in Antiquity. In *Zu Wasser und Zu Land. Verkehrswege in der antiken Welt. Stuttgart Kolloquien zu Historischen Geographie des Altertums, 1999.* Eds. E. Olshausen and H. Sonnabend. 244–247. Stuttgart.
———. 2002b. Settlement and Territory. In *Companion Encyclopedia of Archaeology, Vol. 1.* Ed. G. Barker. 505–545. London and New York.
Bintliff, J., and P. Howard. 2004. A Radical Rethink on Approaches to Surface Survey and the Rural Landscape of Central Greece in Roman Times. In *Chora und Polis*, 43–78.
Bintliff, J.L., and A.M. Snodgrass. 1985. The Cambridge/ Bradford Boeotia Expedition: The First Four Years. *JFA* 12.2: 123–161.
———. 1988. Off-site Pottery Distributions: A Regional and Interregional Perspective. *Current Anthropology* 29: 506–513.
Bitti, M.C. 1989. La ceramica a vernice nera. Il vasellame del II strato: vasi da mensa. In *Locri* II, 141–188.
Blasi, M. 2001. La ceramica comune. In *Dea lucana*, 63–74.
Blattmann D'Amelj, I. 1996. *La ceramica medioevale di contrada "Badia" in Cutrofiano (Le).* Galatina.
Bloesch, H. 1940. *Formen attischer Schalen von Exekias bis zum Ende des strengen Stils.* Bern.
Blume, F., K. Lachmann, and A. Rudorff. 1848. *Schriften der Römischen Feldmesser.* 2 vols. Berlin.
Boardman, J. 1989. *Athenian Red-figure Vases: the Classical period.* London.
———. 2001. *Greek Gems and Finger Rings.* London.
Boccaletti, M., M. Coli, and G. Napoleone. 1980. Landsat Lineation Pattern in the Apennines and Its Geodynamic Significance. *ModGeol* 7: 95–103.
Boccaletti, M., R. Nicolich, and L. Tortorici. 1984. The Calabrian Arc and the Ionian Sea in the Dynamic Evolution of the Central Mediterranean. *MarGeol* 55: 219–245.
Boemi, M.F. 2003a. L'Aerofototeca Nazionale: le raccolte aerofotografiche. In *Lo sguardo di Icaro*, 29–31.
———. 2003b. La gestione delle immagini ed i riferimenti catalografici. In *Lo sguardo di Icaro*, 34–42.
Boenzi, F. 1984. Episodi di sedimentazione ed erosione fluviale di epoca storica nell'alto valle del fiume Sinni. *BSGRI* 11.1: 509–517.
Boenzi, F., M. Caldara, and L. Pennetta. 1995. Il Metaponto e il Tavoliere: due piane dell'Italia meridionale nell'ambiente e nella storia. In *Assetto fisico e problemi ambientali delle pianure italiane.* (Atti delle giornate di studio della Società Geografica Italiana, Roma, 3–4 giugno 1993. *Memorie della Società Geografica Italiana* 53.) Eds. G.B. Castiglioni and P.R. Federici. 337–358. Rome.
Boenzi, F., C. Cherubini, and C. Giasi. 1987. Dati e considerazioni sull'evoluzione recente e sui caratteri idrogeologici della piana costiera metapontina compresa tra il f. Bradano ed il f. Basento (Basilicata). *GFDQ* 10: 34–46.
Boenzi, F., C. Cherubini, C. Giasi, and A. De Siena. 1988. Engineering Geology around the Ancient Urban Area of Metapontum (Lucania, Italy). In *The Engineering Geology of Ancient Works, Monuments and Historical Sites: Preservation and Protection.* (Proceedings of an international

symposium organized by the Greek National Group of IAEG, Athens, 19-23 September 1988.) Eds. P.G. Marinos and G.C. Koukis. 4 vols. 3.1633–1638. Rotterdam-Brookfield.

Boenzi, F., A. De Marco, F. Loiacono, and F. Balenzano. 1989. Osservazioni sugli ambienti fluviali olocenici della media e bassa valle di alcuni fiumi Lucani. *MSGLI* 42: 191–196.

Boenzi, F., M.A. di Gennaro, and L. Pennetta. 1978. I terrazzi della valle del Basento (Basilicata). *RGI* 85: 396–418.

Boenzi, F., G. Palmentola, and A. Valduga. 1976. Caratteri geomorfologici dell'area del Foglio 'Matera' *BSGLI* 95: 527–566.

Boersma, J., and D.G. Yntema. 1989. The Valesio Project. Fourth Interim Report (Campaigns of 1987 and 1988). *BABesch* 64: 134–159.

Boitani, F. 1990. Le ceramiche laconiche a Gravisca. In *Lakonikà* I. *Ricerche e nuovi materiali di ceramica laconica*. Eds. P. Pelagatti and C.M. Stibbe. *BdA*, Suppl. al n. 64: 19–67.

Bökönyi, S., and E. Gál. 2010. *The Chora of Metaponto* 2. *Archaeozoology at Pantanello and Five Other Sites*. Ed. L. Bartosiewicz. Austin.

Bolton, J.D.P. 1962. *Aristeas of Proconnesus*. Oxford.

Bonanno, C. 1998. *I sarcofagi fittili della Sicilia*. Rome.

Bonfiglio, L. 1981. Terrazzi marini e depositi continentali quaternari di Taormina (Sicilia). *Quaternaria* 23: 81–98.

Bonifacio, R. 2004. Le tombe 89-90 della necropolis di Carife. In *Safinum. Studi in onore di Adriano La Regina per il premio I Sanniti*. Ed. D. Caiazza. 237–259. Benevento.

Bonnechere, P. 2003. Trophonius of Lebadea: Mystery Aspects of an Oracular Cult in Boeotia. In *Greek Mysteries: The Archaeology and Ritual of Ancient Greek Secret Cults*. Ed. M.B. Cosmopoulos. 169–192. London.

Borca, F. 1996. Stagna, paludese e presenza antropica: Il caso dell' alto Adriatico; Un *unicum* nell' antichità classica? *Quaderni di storia* 44: 115–45.

Bortolin, R. 2008. *Archeologia del miele*. Mantova.

Borza, E.N. 1979. Some Observations on Malaria and the Ecology of Central Macedonia in Antiquity. *AJAH* 4: 102–124.

Bosio, L. 1984. Capire la terra: la centurazione romana nel Veneto. In *Misurare la terra. Centurazione e coloni nel mondo romani: il caso Veneto*. 15–21. Modena.

Bottini, A. 1981. Ruvo del Monte (Potenza). Necropoli in contrada S. Antonio: scavi 1977. *NSc* Ser. 8, Vol. 35: 183–288.

———. 1982. *Principi guerrieri della Daunia del VII secolo: le tombe principesche di Lavello*. Bari.

———. 1983a. L'attività archelogica in Basilicata In *Atti Taranto* 22: 451–473.

———. 1983b. Cinturoni a placche dell'area melfese. *AnnAStorAnt* 5: 33–63.

——— 1983–1984. Tombe di età romana a Venosa. *AttiMGrecia* 23–24: 215–231.

——— 1987a. Ambre a protome umana dal Melfese. *BdA* 41: 1–16.

——— 1987b. I Lucani. In *Magna Grecia* II, 259–280.

——— 1988a. Gli italioti fino alla conquista romana. In *Popoli e civiltà dell'Italia antica* 8. *Greci e indigeni nel sud dall'VIII secolo alla conquista romana*. Eds. C. Ampolo, A. Bottini, and P.G. Guzzo. 255–341. Rome.

———. 1988b. Elena in Occidente: una tomba della chora di Metaponto. *BdA* 73, Ser. 6, Vol. 50–51: 1–18.

———. 1990. L'attività archeologica in Basilicata – 1989. *Atti Taranto* 29: 553–569.

———. 1992. *L'archeologia della salvezza. L'escatologia greca nelle testimonianze archeologiche*. Milan.

———. 1993a. L'attività archeologica in Basilicata: 1992. *Atti Taranto* 32: 755–768.

———, ed. 1993b. *Armi. Gli strumenti della guerra in Basilicata*. Bari.

———. 1994. L'attività archeologica in Basilicata. *Atti Taranto* 33: 695–709.

———. 1995. L'attività archeologica in Basilicata nel 1994. *Atti Taranto* 34: 759–770.

Bottini, A., and E. Setari. 2003. *La necropoli italica di Braida di Vaglio in Basilicata. Materiali dallo scavo del 1994*. (*Monumenti Antichi* 60, *Serie Miscellanea* VII.) Rome.

Bottini, P. 1986. Il lagonegrese in età arcaica. In *Siris-Polieion*, 199–203.

———. 1988. La Lucania meridionale tirrenica tra VI e l'inizio del V secolo. In *Archeologia arte e storia alle sorgenti del Lao*. Ed. P. Bottini. 93–139. Matera.

———, ed. 1997. *Il Museo archeologico nazionale dell'Alta Val D'Agri*. Lavello.

———, ed. 1998. *Greci e Indigeni tra Noce e Lao*. Lavello.

Bowden, M.C.B., S. Ford, V.L. Gaffney, and M. Tingle. 1991 Skimming the Surface or Scraping the Barrel: A Few Observations on the Nature of Surface and Subsurface Archaeology. In *Ploughzone Archaeology*, 107–113.

Boyd, T.D., and M.H. Jameson. 1981. Urban and Rural Land Division in Ancient Greece. *Hesperia* 50.4: 327–342.

Bradford, J. 1957. *Ancient Landscapes: Studies in Field Archaeology*. London: Bell & Sons, Ltd.

Bradley, R.S. 1985. *Quaternary Paleoclimatology: Methods of Paleoclimatic Reconstruction*. London.

Bragantini, I., E. Setari, and A. De Filippis. 1991. Ceramica comune. In *Ricerche archeologiche a Napoli. Lo scavo di Palazzo Corigliano*. 55–88. Naples.

Brandt, E. 1968. *Antieke Gemmen in deutschen Sammlungen. Bd I Teil 1*. Staatliche Münzsammlung München. Munich.

Brandt, R. 1985. Ostia, Minturnae, Pyrgi. In *Acta ad archaeologiam et artium historiam pertinentia* 5: 25–87.

Braudel, F. 1972. *The Mediterranean and the Mediterranean*

World in the Age of Phillip II. 2nd ed. Trans. S. Reynolds. 2 vols. New York.

Brauer, G.C. 1985. *Taras: its history and coinage.* New Rochelle, NY.

Brecciaroli Taborelli, L. 2005. Ceramiche a vernice nera. In *La ceramica e i materiali di età romana. Classi, produzioni, commerci, e consumi.* Ed. D. Gandolfi. 59–103. Bordighera.

Brehob, J. 1983. A Note on the Amphorae. In *Territory of Metaponto*, 39–40.

Breitenstein, N.J. 1941. *Danish National Museum. Catalogue of the Terracottas, Cypriote, Greek, Etrusco-Italian, and Roman.* Trans. W.E. Calvert. Copenhagen.

Briant, P. 1994. Travaux hydrauliques et contrôle de l'eau dans l'Empire Achéménide. In *Les problèmes institutionnels de l'eau en Égypte ancienne et dans l'Antiquité méditerranéenne.* Ed. B. Menu. 91–101. Cairo.

Brizzi, M. 1999a. Le anfore da trasporto. In *Oppido Mamertina*, 299–309.

———. 1999b. La ceramica comune. In *Oppido Mamertina*, 310–327.

———. 1999c. L'opus doliare. In *Oppido Mamertina*, 328-331.

Brogan, G.E., L.S. Cluff, and C.L. Taylor. 1975. Seismicity and Uplift of Southern Italy. *Tectonophysics* 29: 323–330.

Brogiolo, G.P., and A. Chavarria Arnau. 2005. *Aristocrazie e campagne nell'Occidente da Costantino a Carlo Magno.* Florence.

Brogiolo, G.P., and S. Gelichi. 1986. La ceramica grezza medievale nella pianura padana. In *Ceramica medievale*, 293–313.

Brooks, C.E.P. 1949. Post-glacial Climatic Changes in the Light of Recent Glaciological Research. *GeogAnn* 31: 21–24.

Bruce-Chwatt, L.J. 1965. Paleogenesis and Paleoepidemiology of Primate Malaria. *BWHO* 32: 363–387.

Brückner, H. 1980. Flussterrassen und Flusstäler im Küstentiefland von Metapont (Süditalien) und ihre Beziehung zu Meeresterrassen. *DGS* 15: 5–32.

———. 1982. Ausmass von Erosion und Akkumulation im Verlauf des Quartärs in der Basilicata (Süditalien). *ZG*, N.F. Suppl. 43: 121–137.

———. 1983. Holozäne Bodenbildungen in den Alluvionen süditalienischer Flüsse. *ZG*, N.F. Suppl. 48: 99–116.

———. 1986. Man's Impact on the Evolution of the Physical Environment in the Mediterranean Region in Historical Times. *GeoJournal* 13: 7–17.

———. 1990. Changes in the Mediterranean Ecosystem During Antiquity: A Geomorphological Approach As Seen in Two Examples. In *Man's Role*, 127–137.

Brunet, J. 1971. La Ve Olympique et la reconstruction de Camarine en 461/460. *Revue des Études Anciennes* 73: 327–342.

Brunet, M. 1992. Campagnes de la Grèce antique: les dangers du prisme athénien. *Topoi* 2: 33–51.

———. 2005. Le territoire de Thasos: modèles et interpretation des donées. In *Chora und Polis*, 79–87.

Brunt, P.A. 1971. *Italian Manpower, 225 BC–AD 14.* Oxford.

Bruscella, A., and T. Virtuoso. 2005. Ceramica a vernice nera. In *Torre di Satriano* I, 261–298.

Bubbico, L. 1996. Le dipendenze dell'abbazia di Montescaglioso: gli insediamenti nel territorio. In *Monasteri italogreci e benedettini* II, 160–165.

Buck, R.J. 1975. The Ancient Roads of Southeastern Lucania. *PBSR* 43: 98–117.

Büdel, J. 1982. *Climatic Geomorphology.* Trans. L. Fischer and D. Busche. Princeton.

Buerger, J.E. 1973. Ceramica smaltata importata dagli scavi presso il Palazzo di Diocleziano a Spalato. In *Notiziario di Archeologia Medievale*, 5–7.

———. 1974. Ceramica smaltata tardo medievale della costa adriatica. In *Atti del VII Convegno Internazionale della Ceramica.* 243–259. Albisola.

———. 1979. The Medieval Glazed Pottery. In *Diocletian's Palace. American-Yugoslav Joint Excavations* III. 5–124. Split.

Buikstra, J.E., and D.H. Ubelaker, eds. 1994. *Standards for Data Collection from Human Skeletal Remains.* (*Arkansas Archeological Survey Research Series* 44.) Fayetteville, AR.

Bull, W.B. 1991. *Geomorphic Responses to Climatic Change.* New York.

Burford, A. 1993. *Land and Labor in the Greek World.* Baltimore/London.

Burgers, G.-J. 1991. Pizzica-Pantanello Kiln Deposit: Grey Ware. Unpublished report, Institute of Classical Archaeology, University of Texas at Austin.

Burke, P.F. 1981. *Malaria in the Ancient World: Prolegomenon to an Ecological and Environmental Study.* Unpublished manuscript.

Burn, L. 1987. *The Meidias Painter.* Oxford.

———. 1998. Figured Vases. In *Necropoleis*, 592–640.

Burn, L., and R.A. Higgins. 2001. *Catalogue of Greek Terracottas in the British Museum* III. London.

Burow, J. 2000. Attisch schwartzfigurige keramik. In Kunze-Götte, E., J. Heiden, and J. Burow, *Olympische Forschungen* XXVIII. Archaische Keramik aus Olympia. 199–316. Berlin/New York.

Butzer, K.W. 1969. Changes in the Land: Review of C. Vita-Finzi, *The Mediterranean Valleys. Science* 165: 52–53.

———. 1975. Pleistocene Littoral-sedimentary Cycles in the Mediterranean: A Mallorquin View. In *After the Australopithecines: Stratigraphy, Ecology, and Culture Change in the Middle Pleistocene.* Eds. K.W. Butzer and G.L. Isaac. 25–71. The Hague.

———. 1982. *Archaeology as Human Ecology: Method and Theory for a Contextual Approach.* Cambridge.

Buyskikh, A., and S. Buyskikh. 2001. I siti di Olbia e della sua chora nel periodo della colonizzazione greca. *Atti Taranto* 40: 667–682.

Buzzi Auxilia, M.C. 1989. La ceramica a vernice nera. Il vasellame del II strato. In *Locri* II, 107–141.

Cabona, D., A. Gardini, and O. Pizzolo. 1986. Nuovi dati sulla circolazione delle ceramiche mediterranee dallo scavo di Palazzo Ducale a Genova (sec. XII–XIV). In *Ceramica medievale*, 453–482.

Cahill, N. 2002. *Household and City Organization in Olynthus*. New Haven.

Caire, A. 1975. Italy in its Mediterranean Setting. In *Geology of Italy*, I.11–74.

Calabria, E. 2005. Coroplastica votiva dal santuario urbano di Metaponto: nuove attestazioni di culto di età classica ed ellenistica. In *Spazio del rito*, 69–82.

Calderone, S. 1976. La conquista romana della Magna Grecia. *Atti Taranto* 15: 33–81.

Califano Ascenti, C. 1998. I materiali rinvenuti in Basilicata al Museo di Reggio Calabria. *BBasil* 14: 79–114.

Callu, J.-P., J.-P. Morel, R. Rebuffat, and G. Hallier. 1965. Thamusida. Fouilles du Service des Antiquités du Maroc I. *MEFRA* Suppl. 2, 84–85.

Cambi, F. 2002. La ricognizione. In *Paesaggi d'Etruria. Valle dell'Albegna, valle d'Oro, valle del Chiarore, valle del Tafone*. Eds. A. Carandini and F. Cambi. 345–374. Rome.

———. 2003. *Archeologia dei paesaggi antichi: fonti e diagnostica*. Rome.

Cambi, F., and N. Terrenato. 1994. *Introduzione all'archeologia dei paesaggi*. Rome.

Cambitoglou, A., and J. Chamay. 1997. *Céramique de la Grande Grèce. La collection de fragments Herbert H. Cahn*. Zurich.

Cambitoglou, A., and M. Harari. 1997. *The Italiote Red-figured Vases in the Museo Camillo Leone at Vercelli*. Rome.

Camilli, A. 1999. *Ampullae. Balsamari ceramici di età ellenistica e romana*. Rome.

Campagna, L. 1995. Cisterne e buca di scarico di età repubblicana a Vaste (LE). Scavi di fondo S. Antonio. *StAnt* 8.2: 215–288.

Campbell, B. 2000. *The Writings of the Roman Land Surveyors: Introduction, Text, Translation and Commentary*. London.

Campenon, C. 1994. *La céramique attique à figures rouges autour de 400 avant J.-C. Le principales formes, evolution et production*. Paris.

Canosa, M.G. 1986. Il Materano. In *Siris-Polieion*, 171–182.

———. 2007. *Una tomba principesca da Timmari*. Rome.

Cantilena, R. 1995. Un obolo per Caronte? In *Caronte. Un obolo per l'aldilà*. Ed. R. Cantilena. PP L: 165–177.

Capelli, C., and G. Di Gangi. 2000. La ceramica invetriata in Calabria (sec. XI-XII). Dati archeologici e minero-petrografici. In *Ceramica invetriata*, 197–205.

Capogrossi Colognesi, L. 2002. *Persistenza e innovazione nelle strutture territoriali dell'Italia romana*. Naples.

Caputo, C., L. D'Alessandro, G.B. La Monica, B. Landini and E. Lupia Palmieri. 1991. Present Erosion and Dynamics of Italian Beaches. *ZG* Suppl. 81: 31–39.

Caputo, M., and D. Postpischl. 1975. Seismicity of the Italian Region. In *Structural Model of Italy*, 491–494.

Caputo, P. 1997. Opere idrauliche di età arcaica presso Cuma. In *Uomo, acqua e paesaggio*, 331–334.

Carandini, A. 1981. Terra sigillata dell'Asia Minore (Late Roman C o di Costantinopoli). In *Atlante delle Forme Ceramiche* I. (Supplemento alla Enciclopedia dell'Arte Antica.) 231–232. Rome.

———. 1993. Paesaggi agrari meridionali e etruschi a confronto. In *Atti del Convegno per il Bimillenario della morte di Q. Orazio Flacco* (Venosa 1992). 239–245. Venosa.

———. 1994. I paesaggi agrari dell'Italia romana visti a partire dall'Etruria. In *L'Italie d'Auguste à Dioclétien*. (Actes du Colloque International, Rome 1992.) 167–174. Rome.

Carandini, A., and F. Cambi, eds. 2002. *Paesaggi d'Etruria. Valle dell'Albegna, valle d'Oro, valle del Chiarore, valle del Tafone*. Rome.

Carando, E. 1997. La necropoli sud-orientale. In *Pomarico Vecchio* I, 1.277–306.

———. 1999. Sibari-Thura: nota per una revisione dei dati. *AnnAStorAnt* NS6: 165–176.

Carettoni, G. 1959. Tusa (Messina) – Scavi di Halaesa (prima relazione). *NSc* Ser. 8, Vol. 13: 293–349.

Carissimo, L., O. D'Agostino, C. Loddo, and M. Pieri. 1963. Petroleum Exploration by AGIP Mineraria and New Geological Information in Central and Southern Italy from the Abruzzi to the Taranto Gulf. *Sixth World Petroleum Congress*. 207–292. Frankfurt am Main.

Carter, J.C. 1974. *Excavations in the Territory of Metaponto*. Austin.

———. 1977. Scavi dell'Università del Texas nel territorio di Metaponto, 1976. *Atti Taranto* 16: 845–853.

———. 1978. Scavi a Pizzica e Incoronata nei dintorni di Metaponto. *Atti Taranto* 17: 397–407.

———. 1979. Rural Architecture and Ceramic Industry at Metaponto, Italy, 350–50 BC. In *Roman Brick and Tile. Studies in Manufacture, Distribution and Use in the Western Empire*. (*BAR-IS* 68.) Ed. A. McWhirr. 45–64. Oxford.

———. 1980a. *Excavations in the Territory, Metaponto, 1980*. Austin.

———. 1980b. A Burying Ground in the Territory of Metaponto. In *Attività archeologica*, 147–174.

———. 1981. Rural Settlement at Metaponto. In *Papers in Italian Archaeology* II. Archaeology and Italian Society. (*BAR-IS* 102.) Eds. G. Barker and R. Hodges. 167–178. Oxford.

———. 1983a. Preliminary Report on the Excavation at Pizzica Pantanello (1974-1976). In *Metaponto* II, 407–490.

———. 1983b. Excavations in the Territory. In *Territory of Metaponto*, 13–18.
———. 1983c. Excavation in the Sanctuary Area, Pizzica-Pantanello, 1982. In *Territory of Metaponto*, 19–28.
———. 1984. Pantanello (Metaponto): Scavi 1983. *Atti Taranto* 23: 469–472..
———. 1985a. Ricerca archeologica nella chora metapontina—la campagna del 1985. *Atti Taranto* 25: 477–491.
———. 1985b. Ricognizioni topografiche a Capo Colonna—la campagna del 1985. *Atti Taranto* 25: 448–453.
———. 1987. Agricoltura e pastorizia in Magna Grecia (tra Bradano e Basento). In *Magna Grecia* II, 173–212.
———. 1990. Metapontum: Land, Wealth, and Population. In *Greek Colonists and Native Populations*, 405–441.
———. 1992. The Decline of Magna Graecia in the Age of Pyrrhos? New Evidence from the Chora. In *The Age of Pyrrhus*. (Proceedings of an International Conference held at Brown University, April 8th-10th, 1988. *Archaeologia Transatlantica* 11.) Eds. T. Hackens, N.D. Holloway, R.R. Holloway, and G. Moucharte. 97–145. Louvain-la-Neuve/Providence.
———. 1994a. Sanctuaries in the Chora of Metaponto. In *Placing the Gods. Sanctuaries and Sacred Spaces in Ancient Greece*. Eds. S.E. Alcock and R. Osborne. 161–198. Oxford
———. 1994b. Risorse agricole della costa ionica (Metaponto e Crotone) in età romana. In *Ravitaillement en blé*, 177–196.
———. 1998a. Vent'anni di ricerca nel territorio di Metaponto. In *Siritide e Metapontino*, 237–259.
———. 1998b. Topography. In *Necropoleis*, 25–55.
———. 1998c. Burial Rites and Tomb Types. In *Necropoleis*, 57–111.
———. 1998d. Rites at the Tomb and Grave Markers. In *Necropoleis*, 114–140.
———. 1998e. Family Groups. In *Necropoleis*, 142–162.
———. 1998f. Historical Development. In *Necropoleis*, 167–236.
———. 1998g. Miniatures. In *Necropoleis*, 691.
———. 1998h. Horse Burial and Horsemanship in Magna Grecia. In *Man and the Animal World. Studies in Archeozoology, Archaeology, Anthropology and Paleolinguistics in Memoriam Sándor Bökönyi*. Eds. P. Anreiter, L. Bartosiewicz, E. Jerem, and W. Meid. 131–146. Budapest.
———. 2000. The Chora and the Polis of Metaponto. In *Die Ägäis und das westliche Mittelmeer*, 81–94.
———. 2001. La chora di Metaponto. Risultati degli ultimi 25 anni di ricerca archeologica. *Atti Taranto* 40: 771–792.
———. 2002. Ambiente e paesaggio nel Metapontino. *Atti Taranto* 42: 491–509.
———. 2004. The Greek Identity at Metaponto. In *Greek Identity in the Western Mediterranean*. (An International Conference in Honour of Prof. B.B. Shefton, University of Newcastle-upon-Tyne 1999; *Mnemosyne* Supplement 246.) Ed. K. Lomas. 363–390. Leiden.
———. 2006a. *Discovering the Greek Countryside at Metaponto*. Ann Arbor.
———. 2006b. A Bronze Mirror from Pantanello. In *Across Frontiers: Etruscans, Greeks, Phoenicians and Cypriots. Studies in Honour of David Ridgway and Francesca Romana Serra Ridgway*. Eds. E.J. Herring, I. Lemos, F. Lo Schiavo, L. Vagnetti, R. Whitehouse, and J. Wilkins. 133–145. London.
———. 2006c. The Contribution of the Sciences to the Discovery of Ancient Rural Civilization. *PAPS* 150.1: 15–61.
———. 2008. *La scoperta del territorio rurale Greco di Metaponto*. Venosa.
———. Forthcoming 2011. Forme e processi di territorializzazione. *Atti Taranto* 51.
Carter, J.C., and L. Costantini. 1994. Settlement Density, Agriculture, and the Extent of Productive Land Cleared from Forest in the Time of the Roman Empire in Magna Grecia. In *Evaluation of Land Surfaces Cleared from Forests in the Mediterranean Region During the Time of the Roman Empire. (Palaeoclimate Research 5. European Palaeoclimate and Man 5.)* Ed. B. Frenzel. 101–118. Stuttgart/New York.
Carter, J.C., L. Costantini, C. D'Annibale, J.R. Jones, R.L. Folk, and D. Sullivan. 1985. Population and Agriculture: Magna Grecia in the Fourth Century BC. In *Human Landscape*, 281–312.
Carter, J.C., M. Crawford, P. Lehman, G. Nikolaenko, and J. Trelogan. 2000. The Chora of Chersonesos in Crimea, Ukraine. *AJA* 104.4: 707–741.
Carter, J.C., and C. D'Annibale. 1985. Metaponto and Croton. In *Archaeological Field Survey in Britain and Abroad. (Society of Antiquaries of London Occasional Papers*, N.S. 6.) Eds. S. Macready and F.H. Thompson. 146–157. London.
———. 1993. Il territorio di Crotone. Ricognizioni topografiche 1983–1986. In *Crotone e la sua storia tra IV e III secolo a.C.* Ed. M.L. Napolitano. 93–99. Naples.
Carter, J.C., and J. Hall. 1998. Burial Descriptions. In *Necropoleis*, 237–447.
Carter, J.C., and G.R. Mack. 2003. *Crimean Chersonesos: City, Chora, Museum, and Environs*. Austin.
Carter, J.C., S.M. Thompson, and J. Trelogan. 2004. Dividing the Chora. In *Chora und Polis*, 127–145.
Carter, J.C., and A.P. Toxey. 1998a. Banded-ware and Dipped Pottery. In *Necropoleis*, 694–717.
———. 1998b. Alabastra. In *Necropoleis*, 757–769.
Casagrande, M. 2002. I materiali. In *Eraclea lucana*, 113–223.
Cassano, R. 1982a. Rinvenimenti del 1904. In *Ceglie Peuceta* I, 65–76.

———. 1982b. Scavi del 1930-1931. In *Ceglie Peuceta* I, 158–183.

———. 1988. Stazione Ferrovie del Sud-Est. In *Bari*, 349–360.

———. 1989. Società e cultura figurativa. In *Storia di Bari dalla preistoria al Mille*. Ed. F. Tateo. 131–164. Bari.

———. 1992a. Ceramica a figure rosse. In *Principi imperatori vescovi*, 261–301.

———. 1992b. Ipogeo di Via Legnano. In *Principi imperatori vescovi*, 385–402.

———. 1992c. Ipogeo di Via Molise. In *Principi imperatori vescovi*, 491–505.

Castagnoli, F. 1958. La pianta di Metaponto. Ancora sull'urbanistica ippodamea. *RendAccLinc* Ser. 8, Vol. 13: 49–55.

Castoldi, M. 1995. Oggetti diversi. In *Incoronata* 3, 109–113.

———. 2008. Oltre la chora: nuove indagini archeologiche nell'entroterra di Metaponto. In *Nova vestigia antiquitatis*. (Atti dei seminari del Dipartimento di Scienze dell'Antichità, Università degli Studi. *Quaderni di Acme* 102.) Eds. G. Zanetto, S. Martinello Tempesta, and M. Ornaghi. 143–160. Cisalpino.

Castronovi, C. 1996. Reperti ceramici medievali dell'abbazia di S. Michele Arcangelo di Montescaglioso. In *Monasteri italo greci e benedettini* II, 220–223.

———. 1999. La ceramica postmedievale. In *Da Apigliano a Martano. Tre anni di archeologia medioevale (1997-1999)*. Ed. P. Arthur. 68–72. Galatina.

Catenacci, V. 1992. *Il dissesto geologico e geoambientale in Italia dal dopoguerra al 1990*. (*Memoria Descrittiva della Carta Geologica d'Italia* Vol. 47.) Rome.

Cavagnera, L. 1995. Ceramica protocorinzia dall'Incoronata presso Metaponto (scavi 1971–1993). *MEFRA* 107.2: 869–936.

Cavagnera, L., and M. Pizzo. 1995. Pithoi. In *Incoronata* 3, 105–108.

Cavazza, S. 1962. Sulla erodibilità dei terreni di alcuni bacini calabro-lucani. In *Aspetti geografici dell'erosione del suolo in Italia*. (*L'erosione del suolo in Italia* II.) Ed. G. Morandini. 177–196. Padua.

Ceraudo, G. 2003. Elementi di fotogrammetria. In *Lo sguardo di Icaro*, 94–96.

Cerchiai, L. 2000. "Il cerchio di Aristodemo." *Annali dell'Istituto Universitario Orientale di Napoli, Sezione Archeologia e Storia Antica*, n.s., 7: 115–16.

———. 2007. Introduzione. In *Ceramiche fini*, 1–5.

Champion, T., C. Gamble, S. Shennan, and A. Whittle. 1984. *Prehistoric Europe*. New York.

Channell, J.E.T., B. D'Argenio, and F. Horvath. 1979. Adria, the African Promontory, in Mesozoic Mediterranean Paleogeography. *ESR* 15: 213–292.

Cherry, J.F. 1983. Frogs round the Pond: Perspectives on Current Archaeological Survey Projects in the Mediterranean Region. In *Archaeological Survey*, 375–416.

———. 1994. Regional Survey in the Aegean: The "New Wave" (and After). In *Beyond the Site: Regional Studies in the Aegean Area*. Ed. P.N. Kardulias. 91–112. Lanham, MD.

Cherry, J.F., J.L. Davis, A. Demitrack, E. Mantzourani, T.F. Strasser, and L.E. Talalay. 1988. Archaeological Survey in an Artifact-Rich Landscape: A Middle Neolithic Example from Nemea, Greece. *AJA* 92: 159–176.

Chester, D.K., and P.A. James. 1991. Holocene Alluviation in the Algarve, Southern Portugal: The Case for an Anthropogenic Cause. *JAS* 18: 73–87.

Chevallier, R. 1971. Mission archéologique de la Faculté des Lettres de Tours à Métaponte. *RA* 1971.2: 309–326.

Chiartano, B. 1983. La necropoli dell'Età del Ferro dell'Incoronata e di S. Teodoro (scavi 1970–1974). In *Metaponto* II, 9–190.

———. 1994. *La necropoli dell'Età del Ferro dell'Incoronata e di S. Teodoro* I *and* II *(Scavi 1978–1985)*. Galatina.

———. 1996. *La necropoli dell'Età del Ferro dell'Incoronata e di S. Teodoro* III *(Scavi 1986–1987)*. Galatina.

Chieco Bianchi Martini, A.M. 1964. Conversano (Bari). Scavi in via T. Pantaleo. *NSc* Ser. 8, Vol. 18: 100–76.

Chisholm, M. 1970. *Rural Settlement and Land Use*. Chicago/New York.

Chorley, R.J., S.A. Schumm, and D.E. Sugden. 1984. *Geomorphology*. London.

Chtcheglov (Shcheglov), A. 1992. *Polis et chora: cité et territoire dans le Pont-Euxin*. *(Annales Littéraires de l'Université de Besançon* 476. *Centre de Recherches d'Histoire Ancienne*, Vol. 118. *Espaces et paysages* 1.) Besançon.

Ciaceri, E. 1931. La cooperazione navale delle città italiote alla vittoria romana nella prima guerra punica. *RendNap* N.S. 12: 41–59.

Ciancio, A. 1985. Tombe arcaico-classiche nei territori di Noicattaro e di Valenzano – Bari (Scavi 1978–1981). *Taras* 5.1: 45–107.

———. 1989. Fase IIa. La fase protourbana di età arcaica (VI sec. a.C.). In *Monte Sannace*, 107–131.

———. 1997. *Silbíon. Una città tra greci e indigeni*. Bari.

———. 2005. *I fili della meraviglia. L'abbigliamento di Greci e Apuli tra funzionalità e comunicazione*. Catalogo della Mostra, Gioia del Colle.

———. 2006. Ceramica a figure rosse apula (antica e media). In *Museo Taranto* II.2, 417–430.

Ciancio, A., and F. Radina. 1983. Madonna delle Grazie (Rutigliano): campagna di scavo 1979. *Taras* 3.1–2: 7–61.

Ciancio, A., A. Dell'Anna, and R. Laviano. 1994. Indagini chimico-mineralogiche su ceramica a pasta grigia proveniente dalla Puglia centrale. In *Ceramica romana e archeometria*, 261–276.

Ciarallo, A. 2006. The *Saltus metapontinus* e le tavole di Eraclea. *Rivista di Studi Pompeiani* 17: 139–142.

Ciaranfi, N., M. Guida, G. Iaccarino, T. Pescatore, P. Pieri, L. Rapisardi, G. Ricchetti, et al. 1983. Elementi sismotettonici dell'Appennino meridionale. *BSGLI* 102: 201–222.

Cicala, L. 2002. *L'edilizia domestica tardoarcaica di Elea*. Pozzuoli.

Ciminale, D. 2000. Lecce nel XVI secolo e l'Isola del Governatore: i materiali ceramici. In *Lecce. Frammenti di storia urbana. Tesori archeologici sotto la Banca d'Italia*. Eds. L. Giardino, P. Arthur, and G.P. Ciongoli. 93–102. Bari.

Cipollone Sampò, M. 1999. L'Eneolitico e l'età del Bronzo. In *Storia della Basilicata* 1, 67–136.

Cipriani, M. 1989. *S. Nicola di Albanella. Scavo di un santuario campestre nel territorio di Poseidonia-Paestum*. Rome.

———. 1996. I santuari rurali: Albanella. In *I Greci in Occidente: Poseidonia e i Lucani*. Eds. M. Cipriani and F. Longo. 233–236. Naples.

Clark, R.H., and A.J. Schofield. 1991. By Experiment and Calibration: An Integrated Approach to Archaeology of the Ploughsoil. In *Ploughzone Archaeology*, 93–105.

Clarke, D.L. 1978. *Analytical Archaeology*. 2nd ed. London.

Clarke, N.G., and R.S. Hirsch. 1991. Physiological, Pulpal, and Periodontal Factors Influencing Alveolar Bone. In *Advances in Dental Anthropology*. Eds. M.A. Kelley and C.S. Larsen. 241–266. New York.

Coarelli, F. 1998. Problemi e ipotesi sulle tavole greche di Eraclea. In *Siritide e Metapontino*, 281–290.

Cocchiaro, A. 1981. Contributo per la carta archeologica del territorio a Sud-Est di Taranto. *Taras* 1.1: 53–75.

Cocchiaro, A., and G. Andreassi, eds. 1988. *La necropoli di via Cappuccini a Brindisi*. Brindisi.

Cocco, E. 1975. Interpretazione aereofotografica delle variazioni della linea di costa lungo alcune zone del litorale alto ionico (Golfo di Taranto). *GAIG* 10.1: 277–28?.

———. 1976. The Italian North Ionian Coast: Tendency Towards Erosion. *MarGeol* 21: M49–M57.

Cocco, E., E. Cravero, S. di Geronimo, G. Mazzadri, G.C. Parea, T. Pescatore, R. Valloni, and A. Vinci. 1975. Lineamenti geomorfologici e sedimentologici del litorale alto ionico (Golfo di Taranto). *BSGLI* 94: 993–1051.

Cocco, E., M.A. De Magistris, and T. De Pippo. 1978. Studi sulle cause dell'arretramento della costa lucana ionica (I). L'estrazione degli inerti lungo le aste fluviali. *MSGLI* 29: 421–428.

Colangelo, L. 2001. La ceramica con decorazione a bande. In *Dea lucana*, 74–76.

———. 2005. Ceramica a bande. In *Torre di Satriano* I, 251–260.

Coleman J.E., K. O'Neill, M. Pomeroy, K.E. Carr, and A. Heafitz. 1992. Excavations at Halai 1990–1991. *Hesperia* 61.3: 265–289.

Coles, J.M., and A.F. Harding. 1979. *The Bronze Age in Europe: An Introduction to the Prehistory of Europe c. 2000–700 BC*. London.

Colivicchi, F. 2006. Kantharoi attici per il vino degli Apuli. In *Il greco, il barbaro e la ceramica attica, Immaginario del diverso, processi di scambio e autorappresentazione degli indigeni*. Eds. F. Giudice and R. Panvini. *(Atti Convegno Catania, Caltanissetta, Gela, Camarina, Vittoria, Siracusa 2001.)* 117–130. Rome.

Colli, G. 1981. *La sapienza greca*. Milan.

Collin Bouffier, S. 2002. Eau et Campagne en Sicile Grecque: Observations préliminaire à l étude. In *Binos Actus Lumina: Rivista di studi e ricerche sull'idraulica storica e la storia della tecnica. 1, Atti del Convegno Internazionale di Studi su Metodologie per lo studio della scienza idraulica antica: Ravenna, 13–15 maggio 1999*, 27–35. La Spezia.

Colton, G.W. 1967. Orientation of Carbonate Concretions. *U.S. Geological Survey Professional Paper* 575B: 57–59.

Conti, M.C. 1989a. La ceramica comune. Criteri di classificazione. In *Locri* II, 257–260.

———. 1989b. La ceramica comune. Ceramica da fuoco. In *Locri* II, 262–294.

———. 1989c. La ceramica comune. Vasellame da cucina e da dispensa. In *Locri* II, 294–312.

———. 1989d. La ceramica comune. Vasellame da mensa. In *Locri* II, 312–327.

———. 1989e. La ceramica comune. Note conclusive. In *Locri* II, 344–345.

———. 1992. La ceramica comune. In *Locri* IV, 241–255.

Coppa, A., A. Cucina, D. Mancinelli, R. Vargiu, and J.M. Calcagno. 1998. Dental Anthropology of Central-Southern, Iron Age Italy: The Evidence of Metric versus Non-metric Traits. *AJPA* 107: 371–386.

Cordano, F. 1995. La forzata partenza dei fondatori di Taranto nell'attualità del V secolo a.C. In *Coercizione e mobilità umana nel mondo antico*. Ed. M. Sordi. 51–53. Milan.

Cordano, F., and G. Di Stefano. 1997. Il fiume e la città nella Sicilia meridionale: Il caso di Camarina. In *Uomo, acqua e paesaggio*, 289–300.

Corrado, M. 2003. Appunti per una prima carta delle ceramiche invetriate bassomedievali nel medio Ionio calabrese. In *III Congresso Nazionale di Archeologia Medievale*. Ed. G.P. Brogiolo. 159–164. Florence.

Corrente, M. 1992. La tomba degli ori. In *Principi imperatori vescovi*, 337–345.

———. 2003. *Canusium. L'ipogeo dei serpenti piumati*. Canosa.

Corrente, M., and M. Labellarte. 1992. Tombe a fossa di Vico San Martino. In *Principi imperatori vescovi*, 435–481.

Costamagna, L. 1983. Taranto: indagini archeologiche nell'area di Piazza Garibaldi. *Taras* 3.1–2: 101–123.

Costamagna, L., and C. Sabbione. 1990. *Una città in Magna Grecia. Locri Epizefiri: guida archeologica*. Reggio Calabria.

Costantini, L. 1983. Bioarchaeological Research at Pizzica-Pantanello. In *Territory of Metaponto*, 30–34.

Costantini, L., and L. Biasini Costantini. 2003. Agriculture and Diet in the Chora of Metaponto: The Paleobotanical Evidence from Pantanello. In *Living off the Chora: Diet and Nutrition at Metaponto*. Ed. J.C. Carter. 3–12. Austin.

Cotecchia, V., G. Dai Pra, and G. Magri. 1969. Oscillazioni tirreniane e oloceniche del livello mare nel Golfo di Taranto, correlate da datazione col metodo del radiocarbonio. *GAIG* 4: 93–148.

———. 1971. Morfogenesi litorale olocenica tra Capo Spulico e Taranto nella prospettiva della protezione costiera. *GAIG* 6: 65-78.

Cotton, M.A. 1971. The Finds. In Cherry, J.F., D. Whitehouse, and M. Aylwin Cotton. *A Trial Excavation at Monte d'Irsi, Basilicata*. *PBSR* 39: 145–169.

———. 1979. *The Late Republican Villa at Posto, Francolise*. (Supplementary Publications of the British School at Rome.) London.

———. 1983. Coarse Wares. In *Cozzo Presepe*, 364–379.

———. 1992a. Hellenistic and Roman Red Slipped Wares. In *Gravina*, II.157–161.

———. 1992b. Plain Light-surfaced Wares. In *Gravina*, II.162–179.

———. 1992c. Dark-surfaced or Cooking Ware. In *Gravina*, II.179–194.

———. 1992d. Mortaria. In *Gravina*, II.194–197.

———. 1992e. Amphorae. In *Gravina*, II.197–200.

———. 1992f. Pithoi and Dolia. In *Gravina*, II.200–202.

Cotton, M.A., and G.P.R. Métraux. 1985. *The San Rocco Villa at Francolise*. (Supplementary Publications of the British School at Rome.) London/New York.

Courty, M.A., P. Goldberg, and R. MacPhail. 1989. *Soils and Micromorphology in Archaeology*. Cambridge.

Cracolici, V. 2001. Le ceramiche. In *Metaponto/Archeologia*, 103–114.

———. 2004a. *I sostegni di fornace dal kerameikos di Metaponto*. (*Beni Archeologici – Conoscenza e Tecnologie* Quaderno 3.) Bari.

———. 2004b. Il sito in Località Casalini Sottana. In *Nobili e guerrieri*, 7–12.

———. 2004c. Catalogo. In *Nobili e guerrieri*, 45–60.

Cracolici, V., and L. Padalino. 2004. La necropoli. In *Nobili e guerrieri*, 13–30.

Crawford, M.H. 2003. Brave new world: Metapontum after Metapontum. In *Les élites et leurs facettes. Les elites locales dans le monde hellénstique et romain*. Eds. M. Cébeillac-Gervasoni and L. Lamoine. 15–30. Rome and Clermont-Ferrand.

———. 2005. Transhumance in Italy: Its History and Historians. In *Noctes campanae: Studi di storia antica e archeologia dell'Italia preromana in memoria di M.W. Frederiksen*. Eds. W.V. Harris and E. Lo Cascio. 159–179. Naples.

Crescenti, U. 1975. Sul substrato pre-pliocenico dell' avanfossa appenninica dalle Marche allo Ionio. *BSGLI* 94: 564–583.

Cubberly, A., J. Lloyd, and P. Roberts. 1988. Testa and clibani. *PBSR* 56: 98–119.

Cuomo di Caprio, N. 1985. *La ceramica in archeologia. Antiche tecniche di lavorazione e moderni metodi d'indagine*. Rome.

——— 1988. *Fornaci e officine da vasaio tardo-ellenistiche a Morgantina*. Milan.

——— 1992. *Fornaci e officine da vasaio tardo-ellenistiche*. (*Morgantina Studies* 3.) Princeton.

Cuozzo, M. 2003. *Reinventando la tradizione. Immaginario sociale, ideologie e rappresentazione nelle necropoli orientalizzanti di Pontecagnano*. Paestum.

Cuozzo, M., and A. D'Andrea. 1991. Proposta di periodizzazione del repertorio locale di Pontecagnano tra la fine de VII e la metà del V sec. a.C. alla luce della stratigrafia delle necropoli. *AION ArchStAnt* 13: 47–114.

Curti, F. 1998. *La céramique de Gnathia du Musée d'art et d'histoire de Genève*. Geneva.

Cuteri, F. 1993. Il paiolo in ceramica acroma grezza e la sua diffusione nella Toscana medievale. *ResArch* 11: 327–348.

——— 2002a. Resti di attività siderurgica a Torre di Mare. Analisi di alcuni reperti metallurgici dall'area delle strutture abitative (saggi I-II e III). In *Torre di Mare* I, 285–290.

——— 2002b. Appendice I: Cenni sulla produzione di ferro tra Calabria e Basilicata. In *Torre di Mare* I, 285–290.

D'Agostino, B. 1968. Pontecagnano-Tombe orientalizzanti in contrada S. Antonio. In *NSc* Ser. 8, Vol. 22: 75–196.

Dai Pra, G., and P.J. Hearty. 1988. I livelli marini pleistocenici del Golfo di Taranto. Sintesi geocronostratigrafica e tettonica. *MSGLI* 41: 637–644.

Dalena, P. 2002. Torre di Mare nel medioevo: habitat-viabilità-porto. In *Torre di Mare* I, 53–57.

D'Amicis, A. 1984. Taranto. Rinvenimenti in via Alto Adige *Taras* 4,1–2: 47–84.

——— 1995. La ceramica di Gnathia. In *Collezione Chini*, 45–90.

Dana, Peter H. 2004. Incorporating the Advice of Experts in GIS Analysis. In *Ancient Territories 2004*, 77–81.

D'Andria, F. 1969. Un piatto in ceramica pergamena. In *Missione archeologica a Malta. Rapporto preliminare della campagna 1968*. 81–89. Rome.

——— 1972. Catalogo della ceramica greca, italiota e a vernice nera. In *Missione archeologica italiana a Malta. Rapporto preliminare della campagna 1969*. 77–95. Rome.

——— 1975. Metaponto: scavi nella zona del *kerameikos* (1973). *NSc* Ser.8, Vol. 29 suppl.: 355–452.

——— 1976. Metaponto romana. *Atti Taranto* 15: 539–544.

——— 1977. Scavi a Metaponto (1976). *Atti Taranto* 16: 855–859.

———. 1978a. La ceramica japigia e messapica. In *Leuca*, 48–90.

———. 1978b. La documentazione archeologica negli insediamenti del materano tra tardo-antico e altomedioevo. In *Habitat, strutture, territorio. Atti del III Convegno Internazionale di Studio sulla Civiltà Rupestre Medioevale nel Mezzogiorno d'Italia* (Taranto-Grottaglie, 24–27 settembre 1975). Ed. C.D. Fonseca. 157–162. Galatina.

———. 1980a. Scavi nella zona del *kerameikos* (1973). In *Metaponto* I, 355–452.

———. 1980b. I materiali del V secolo a.C. nel ceramico di Metaponto e alcuni risultati delle analisi sulle argille. In *Attività archeologica*, 117–146.

———, ed. 1990. *Archeologia dei Messapi.* (Catalogo della mostra, Lecce, Museo Provinciale "Sigismondo Castromediano," 7 ottobre 1990 – 7 gennaio 1991.) Bari.

D'Andria, F., and G. Semeraro, eds. Forthcoming. *Contestualizzare la ceramica greca.* (Atti della Summer School, Cavallino 2001). Bari.

D'Angela, C., and G. Volpe. 1994. Aspetti storici e archeologici dell'Alto Medioevo in Puglia. In *Storia dell'alto Medioevo*, 299–332.

D'Angelo, F. 1984. Ceramica locale e ceramica d'importazione a Brucato. In *Brucato*, 451–471.

D'Anisi, M.C. 2005. Nuovi dati sui culti lucani: un deposito votivo inedito da Accettura. In *Spazio del rito*, 167–178.

Danner, P. 1993. *Westgriechische Akrotere.* Mainz.

D'Annibale, C. 1983a. Field Survey of the Chora of Metaponto, 1981–82. In *Territory of Metaponto*, 4–9.

———. 1983b. Field Survey of the Chora of Metaponto. In *Archaeological Survey*, 191–194.

D'Argenio, B., T. Pescatore, and P. Scandone. 1975. Structural Pattern of the Campania-Lucania Apennines. In *Structural Model of Italy*, 313–327.

Dark, K.R. 1995. *Theoretical Archaeology.* London.

Davidson, D.A. 1980. Erosion in Greece During the First and Second Millennia BC. In *Timescales in Geomorphology*. Eds. R.A. Cullingford, D.A. Davidson, and J. Lewin. 143–158. London.

Daverio Rocchi, G. 1973. Aristocrazio genetica ed organizzazione politica arcaica. *PP* 148–149: 92–116.

Davidson, G.R., D.B. Thompson, and H.A. Thompson. 1943. *Small Objects from the Pnyx* I. (*Hesperia Supplement* 7.) Princeton.

Davis, D. 2003. 2003 Gradiometer Survey in the Chora of Metaponto. In *Ancient Territories 2003*, 77–80.

Davis, J.C. 1986. *Statistics and Data Analysis in Geology.* New York.

Davis, J.G. 1965. *Cheese.* 2 vols. London.

De Boe, G., and M. Vanderhoeven. 1979. Un lot de céramique du troisième quart du Ier siècle avant J.-C. In *Ordona* VI, 107–127.

De Castro Coppa, M.G., M. De Nucci, F.M. Guadagno, M. Moncharmont Zei, L. Muoio, B. Placella, F. Sgarrella, E. Taddei Ruggiero, and E. Turco. 1979. Biostratigrafia e paleoecologia delle argille grigio-azzurre bradaniche tra i fiumi Basento e Cavone. In *Giornate di studio*, 503–534.

De Cazanove, O. 2008. *Civita di Tricarico* 1. *Quartier de la maison du monolithe et l'enceinte intermédiaire.* Rome.

De Coronei, N.J. 1880. *Sinodo materse del 1567 pubblicato ed annotato per cura del Cavaliere Niccolò Jeno de Coronei con appendice di 23 documenti inediti e fac simili di autografi.* Naples.

De Crescenzo, A. 2002. I manufatti ceramici. In *Torre di Mare* I, 171–234.

Deferrari, G., P. Palazzi, and L. Parodi. 2001. La ceramica medievale e moderna. In *S. Antonino dei Perti: un insediamento fortificato nella Liguria bizantina.* Eds. T. Mannoni and G. Murialdo. 705–716. Florence.

Degrassi, N. 1965. Il Pittore di Policoro e l'officina di ceramica protoitaliota di Eraclea Lucana. *Bollettino d'Arte Serie V*, L, 1–2: 5–37.

De Juliis, E. 1973. Ordona (Foggia). Scavi nella necropoli. *NSc* Ser. 8, Vol. 27: 285–399.

———. 1990. *L'ipogeo dei vimini di Canosa.* Bari.

———. 1992. L'ipogeo dei vimini. In *Principi imperatori vescovi*, 350–382.

———. 2001. *Metaponto.* Bari.

———. 2002. *La ceramica sovraddipinta apula.* Bari.

———. 2006. Ceramica sovraddipinta. In *Museo Taranto* II.2, 527–540.

De La Genière, J. 1968. *Recherches sur l'âge du Fer en Italie méridionale. Sala Consilina.* (*Publications du Centre Jean Bérard* 1.) Naples.

———. 1971. Amendolara (Cosenza). Campagne del 1967 e 1968 (relazione preliminare). *NSc* Ser. 8, Vol. 25.II: 439–475.

———. 1979. The Iron Age in Southern Italy. In *Italy before the Romans.* Eds. D. Ridgway and F.R. Ridgway. 59–93. New York.

———. 1989. Épire et Basilicate. À propos de la couronne d'Armento. *MEFRA* 101.2: 691–698.

Delano Smith, C. 1978. Coastal Sedimentation, Lagoons and Ports in Italy. In *Papers in Italian Archaeology* I, Part i: *Recent Research in Prehistoric, Classical and Medieval Archaeology.* (*BAR Supplementary Series* 41.i.) Eds. H.M. Blake, T.W. Potter, and D.B. Whitehouse. 24–33. Oxford.

———. 1979. *Western Mediterranean Europe: A Historical Geography of Italy, Spain and Southern France since the Neolithic.* London/New York.

———. 1981. Climate or Man? The Evidence of Sediments and Early Maps for the Agents of Change in the Post-Medieval Period. In *Consequences of Climate Change.* Eds. C. Delano-Smith and M. Parry. 88–105. Nottingham.

Delatte, A. 1921. *Essai sur le politique pythagoricienne.* Liège.

De Ligt, L. 2006. Agrarian change during the second century. In *A Companion to the Roman Republic.* Eds. N. Rogenstein and R. Morstein-Marx. 590–605. Oxford.

Dell'Aglio, A. 1991a. Taranto, la necropoli arcaica. In Vecchi scavi, 17–34.

———. 1991b. Taranto, la necropoli tardo-classica ed ellenistica. Via Campania angolo Corso Italia. In *Vecchi scavi*, 87–94.

———. 1996a. Taranto. In *Arte e artigianato*, 51–79.

———. 1996b. La ceramica a vernice nera. Taranto. In *Arte e artigianato*, 323–328.

———. 2002. La proschoros tarentina. In *Territori tarantini*, 19–42.

Dell'Anna, L., and A. De Marco. 1977. Ulteriori dati sulla composizione mineralogica e granulometrica delle sabbie del litorale ionico tra Metaponto e Taranto. *BSGLI* 96: 569–585.

Delli Ponti, G. 1981. Gli scavi di Rocavecchia dal 1969 al 1970 (tombe). In *Momenti e figure di storia pugliese. Studi in memoria di M. Viterbo (Peucezio).* 51–84. Galatina.

———. 1996. Vaste-Poggiardo (LE). La necropoli di Fondo Aia. *StAnt* 9: 99–214.

Delogu, P. 1994. La fine del mondo antico e l'inizio del medioevo: nuovi dati per un vecchio problema. In *Storia dell'alto Medioevo*, 7–29.

Delplace, C. 1974. Présentation de l'ensemble des lampes trouvées de 1962 à 1971 à Ordona. In *Ordona* IV, 7–101.

Del Prete, M., and G. Valentini. 1971. Le caratteristiche geotecniche delle argille azzurre dell'Italia sud-orientale in relazione alle differenti situazioni stratigrafiche e tettoniche. *GAIG* 6: 197–215.

De Maio, A., L. Mirabile, M. Moretti, T. Pescatore, E. Sansone, G. Spezie, M. Tramutoli, and M. Vultaggio. 1979. Primi risultati degli studi eseguiti sulla piattaforma continentale del Golfo di Taranto. In *Atti del Convegno Scientifico Nazionale Progetto Finalizzato Oceanografia e Fondi Marini.* 1333–1347. Rome.

De Marinis, G. 1979. Un piccolo castrum altomedievale presso Pomino in Val di Sieve. *ArchMed* 6: 275–288.

De Miro, E. 1963. Agrigento. Scavi nell'area a sud del Tempio di Giove. *MonAnt* XLVI: 81–198.

———. 2000. Agrigento I. *I santuari urbani*. 2 vols. Rome.

Denoyelle, M. 1992. Du peinture de Pisticci au peintre du Cyclope. Quelques problèmes de style au sein du premier atelier à figures rouges de Métaponte (deuxième moitié du V siècle av. J.-C.). *Revue du Louvre* 42.4: 21–29.

———. 1997. Attic or Non-Attic? The Case of the Pisticci Painter. In *Athenian Potters and Painters*, 395–405.

———. 2002a. Il mito greco in Occidente: Metaponto ed Herakleia. In *Immagine e mito*, 104–112.

———. 2002b. Style individuel, style local et centres de production: rertour sur le cratère des "Karneia." *MEFRA* 114: 587–609.

———. 2002c. Some Little Vases by the Creusa and Dolon Painter. In *Essays in Honour of Dietrich von Bothmer.* Eds. A.J. Clarke, J. Gaunt, and B. Gilman. 107–112. Amsterdam.

———. 2008. La ceramica: appunti sulla nascita delle produzioni italiote. *Atti Taranto* 47: 339–350.

Denton, G.H., and S.C. Porter. 1970. Neoglaciation. *SciAm* 222.6: 100–110.

Deodato, A., and O. Bianco. 1997. La ceramica comune. In *Pomarico Vecchio* I, 1.175–199.

De Palma, G. 1989. La ceramica dorata in area apula. Contributo al problema delle ceramiche di imitazione metallica. *Taras* 9: 7–94.

Depalo, M.R. 1984. Alcuni rinvenimenti tombali a Bitetto (Bari). *Taras* 4: 175–88.

———. 1997. *La collezione Loiudice.* Bari.

De Polignac, F. 1991. *La nascita della città greca. Culti, spazio e società nei secoli VIII e VII a.C.* Trans. E. Robberto. Milan. (Enlarged and updated translation of *La naissance de la cité grecque: cultes, espace et société VIIIe–VIIe siècles avant* J.-C. 1984. Paris.)

Deschenes, G. 1977. Plain Arretine. In *Monte Irsi*, 176–188.

De Sensi Sestito, G. 2004. Alessandro e le popolazioni della Lucania e del Bruzio. *Atti Taranto* 43: 519–560.

De Siena, A. 1980. Note stratigrafiche sul santuario di Apollo Licio a Metaponto. In *Attività archeologica*, 83–116.

———. 1986a. Termitito e S. Vito di Pisticci: dati archeologici e problemi interpretativi. In *Siris-Polieion*, 27–34.

———. 1986b. Metaponto. Nuove scoperte in proprietà Andrisani. In *Siris-Polieion*, 135–156.

———. 1986c. Termitito. In *Traffici micenei nel Mediterraneo: i problemi storici e la documentazione archeologica.* (Atti del convegno di Palermo, 11–12 maggio e 3-6 dicembre 1984). Eds. M. Marazzi, S. Tusa, and L. Vagnetti. 41–54. Taranto.

———. 1986d. Scavi in località Incoronata ed a Metaponto: nuove scoperte. In *Greci sul Basento*, 199–212.

———. 1990a. Contributi archeologici alla definizione della fase protocoloniale del metapontino. *BBasil* 6: 71–88.

———. 1990b. Il castro romano di Metaponto. In *Basilicata/espansionismo romano*, 301–314.

———. 1993. Metaponto e il metapontino. In *Da Leukania a Lucania*, 114–135.

———. 1996. Metapontino: strutture abitative ed organizzazione territoriale prima della fondazione della colonia achea. In *Ricerche sulla casa*, 161–195.

———. 1998. Metaponto: problemi urbanistici e scoperte recenti. In *Siritide e Metapontino*, 141–170.

———. 1999a. Il Metapontino: insediamenti antichi e bonifiche. In *Archeologia dell'acqua*, 53–72.

———. 1999b. La colonizzazione achea del Metapontino. In *Storia della Basilicata* 1, 211–245.

———. 2001a. Il territorio di Metaponto. *Atti Taranto* 40: 757–769.

———. 2001b. Profilo storico archeologico. In *Metaponto/Archeologia*, 7–43.

———. 2002. Appunti di topografia metapontina. In *Torre di Mare* I, 25–40.

———. 2003. Ambiente e paesaggio nella Magna Grecia: Metaponto. *Atti Taranto* 42: 511–513.

———. 2005. Tramonto della Magna Grecia: la documentazione archeologica dai territori delle colonie greche di Metaponto ed Herakleia. *Atti Taranto* 44: 433–458.

———. 2007. L'attività archeologica in Basilicata nel 2006. *Atti Taranto* 46: 407–463.

———. 2008. Atene e l'area ionica della Basilicata. *Atti Taranto* 47: 581–596.

———. 2010. Il santuario extraurbano delle Tavole Palatine: ruolo e tradizione. In *Il temptio di Hera (Tavole Palatine) di Metaponto: archeologia, archeometria, conservazione*. Ed. L. Lazzarini. 13–15. Pisa-Rome.

De Siena, A. and S. Bianco. 1982. La Basilicata, Termitito (Montalbano Ionico, Matera). In *Magna Grecia e mondo Miceneo. Nuovi documenti*. Ed. L. Vagnetti. 69–94. Taranto.

De Siena, A., A. Di Tursi, E. Giannichedda, E. Lapadula, A. Patrone, and L. Trivigno. Forthcoming. *Metallurghi e ceramisti a Montescaglioso (MT) tra Medioevo e Postmedioevo. Scavi archeologici nell'area dell'abbazia di San Michele Arcangelo.*

De Siena, A., and L. Giardino. 1994. Herakleia e Metaponto. Trasformazioni urbanistiche e produzione agricola tra tarda repubblica a primo impero: i nuovi dati archeologici. In *Ravitaillement en blé*, 197–211.

———. 2001. Trasformazioni delle aree urbane e del paesaggio agrario in età romana nella Basilicata sudorientale. In *Modalità insediative*, 129–168.

Desio, A., ed. 1973. *Geologia dell'Italia*. Turin.

Desy, P. 1989. *Les timbres amphoriques de l'Apulie républicaine. Documents pour une histoire économique et sociale.* (*BAR-IS* 554.) Oxford.

Desy, P. and P. De Paepe. 1990. Torre San Giovanni (Ugento). Les amphores commerciales hellénistiques et républicaines. *StAnt* 6: 187–234.

Deussen, P.W. 1994. The Granaries of Morgantina and the *Lex Hieronica*. In *Ravitaillement en blé*, 231–235.

Devereaux, C.M. 1982. Climate Speeds Erosion of the Algarve's Valleys. *TGM* 54: 10–17.

D'Henry, G. 1981. Una tomba a Nocera della seconda metà del quinto secolo. Problemi di inquadramento. *AnnASorAnt* 3: 159–174.

Di Gangi, G. 1997. Status quaestionis e spunti per una riflessione sulla protomaiolica in Calabria. Materiali, insediamenti, distribuzione, commerci alla luce degli scavi stratigrafici di Troppa. In *La protomaiolica*, 157–184.

Di Gangi, G., and C. Lebole. 1998. Aspetti e problemi dell'età normanna in Calabria alla luce dell'archeologia. In *Società ed insediamenti dell'età dei Normanni in Italia meridionale*. (Atti del Seminario di studi, Roccelletta 12-13 novembre 1994.) *MEFRM* 110.1. 397–424. Rome.

Di Giuseppe, H. 1996. Insediamenti rurali della Basilicata interna tra la romanizzazione e l'età tardoantica: materiali per una tipologia. In *Epigrafia e territorio. Politica e società. Temi di antichità romane* IV. Ed. M. Pani. 189–252. Bari.

———. 1998. La fornace di Calle di Tricarico: produzione e diffusione. In *Ceramica in Italia*, 735–752.

Di Lieto, M. 2001. Thymiateria. In *Dea lucana*, 57–61.

———. 2005a. I thymiateria. In *Spazio del rito*, 151–154.

———. 2005b. Thymiateria. In *Torre di Satriano* I, 357–387.

Dilke, O.A.W. 1971. *The Roman Land Surveyors*. Newton Abbot.

———. 1987. Roman Large-scale Mapping in the Early Empire. In *History of Cartography*. Eds. J.B. Harley and D. Woodward. 2 vols. I.212–233. Chicago.

Dillon, J., and J. Hershbell. 1991. *Iamblichus. On the Pythagorean Way of Life. Text, Translation, and Notes*. Atlanta.

Dilthey, H. 1980. Sorgenti acque luoghi sacri in Basilicata. In *Attività archeologica*, 539–560.

Di Noia, A. 2005. Ceramica miniaturistica. In *Torre di Satriano* I, 342–347.

Di Stefano, G. 2001. La chora di Camarina. *Atti Taranto* 40: 689–705.

Djindjian, F. 1991. *Méthodes pour l'archéologie*. Paris.

Dobkins, J.E., Jr., and R.L. Folk. 1970. Shape Development on Tahiti-Nui. *JSP* 40: 1167–1203.

Dobney, K., and D. Brothwell. 1986. Dental Calculus: Its Relevance to Ancient Diet and Oral Ecology. In *Teeth and Anthropology*. (*BAR-IS* 291.) Eds. E. Cruwys and R. A. Foley. 55–82. Oxford.

Doepner, D. 2002. *Steine und Pfeiler für die Götter. Weihgeschenkgattungen in westgriechischen Stadtheiligtümern*. (*Palilia* 10.) Wiesbaden.

Doglioni, C., F. Mongelli, and P. Pieri. 1994. The Puglia Uplift (SE Italy): An Anomaly in the Foreland of the Apenninic Subduction due to Buckling of a Thick Continental Lithosphere. *Tectonics* 13.5: 1309–1321.

Donnarumma, R., and L. Tomay. 1990. I corredi di VI e V secolo a.C. In *Fratte*, 207–275.

———. 1999. La necropoli di San Brancato di Tortora. In *Terra degli Enotri*, 49–59.

Donvito, A. 1988. *Egnazia dalle origini alla riscoperta archeologica*. Fasano.

———. 1991. Santo Mola. Un insediamento peuceta inedito in territorio di Gioia. In *Gioia. Una città nella storia e nella civiltà di Puglia* III. Ed. M. Girardi. 23–126. Fasano.

Dubois, L. 2002. *Inscriptions grecques dialectales de Grande Grèce* II. *Colonies achéennes*. Geneva.

Dufaure, J.J. 1976. La terasse holocene d'Olympie et ses equivalents Mediterraneens. *BAGF* 433: 85–94.
Dufkova, M., and J. Pečirka. 1970. Excavations of Farms and Farmhouses in the Chora of Chersonesos in the Crimea. *Eirene* 8: 123–174.
Dufournier, D., A.M. Flambard, and G. Noyé. 1986. A propos de céramique «RMR»: problemes de définition et de classement, problemes de répartition. In *Ceramica medievale*, 215–277.
Dumas, B., P. Gueremy, P.J. Hearty, R. Lhenaff, and J. Raffy. 1988. Morphometric Analysis and Amino-acid Geochronology of Uplifted Shorelines in a Tectonic Region near Reggio Calabria, South Italy. *PPP* 68: 273–289.
Dunbabin, T.J. 1948. *The Western Greeks*. Oxford.
du Plat Taylor, J. 1983. Wheelmade Painted Pottery. In *Cozzo Presepe*, 331–346.
du Plat Taylor, J., and A.M. Small. 1992. Greek and Ionian Type Cups. In *Gravina*, II.65–66.
Dyson, S.L. 1976. *Cosa: The Utilitarian Pottery*. Rome.
———. 1983. *The Roman Villas of Buccino. Wesleyan University Excavation in Buccino, Italy 1969–1972*. (*BAR-IS* 187.) Oxford.
Ebanista, C., and F. Fusaro. 2000. La ceramica invetriata del castello di Montella. Nota preliminare. In *Ceramica invetriata*, 113–133.
Edlund, I.E.M., ed. 1979. *Excavations at Metaponto, 1979*. Austin.
———. 1984. Sacred and Secular: Evidence of Rural Shrines and Industry among Greeks and Etruscans. In *Crossroads of the Mediterranean*. (*Archaeologia Transatlantica* II.) Eds. T. Hackens, N.D. Holloway, and R.R. Holloway. 277–290. Providence/Louvain-la-Neuve.
———. 1986. Scavi nella zona di Metaponto: Sant'Angelo Vecchio. *BdA* 71: 119–122.
———. 1987. *The Gods and the Place: Location and Function of Sanctuaries in the Countryside of Etruria and Magna Graecia (700–400 BC)*. (*Acta Instituti Romani Regni Sueciae* Ser. 3, Vol. 43.) Stockholm.
———. 1988. Classical Themes in South Italian Votive Terracottas of the Fifth and Fourth Centuries BC: Continuity and Development in Cults and Iconography. In *Acts of the XIIth International Congress of Classical Archaeology* (Athens, September 4-10, 1983). 2 vols. 2.237–242. Athens.
Eisma, D. 1964. Stream Deposition in the Mediterranean Area in Historical Times. *Nature* 203: 1061.
Elia, D. 1997. La ceramica a figure rosse. In *Pomarico Vecchio* I, 1.111–125.
———. 2003. L'offerta di sostanze alimentari liquide presso la tomba e l'uso rituale del cratere nelle necropoli greche d'Occidente. *Orizzonti* 4: 146–154.
———. 2006. Ceramica fine dei secoli IV e III a.C. In *Da Terravecchia di Grammichele a Occhiolà. Archeologia di un insediamento dell Sicilia centro-orientale: campagne di scavo 2000–2001*. Ed. M. Barra Bagnasco. 157–214. Alessandria.
Elia, D., and V. Meirano. 1997. La ceramica a vernice nera: forme chiuse, crateri, coperchi. In *Pomarico Vecchio* I, 1.83–92.
Elias, M.K. 1931. *The Geology of Wallace County, Kansas*. (*Kansas Geological Survey Bulletin* 18.) Lawrence, KA.
Elliott, M. 1998. Black-glazed Pottery. In *Necropoleis*, 642–693.
Embleton, C. 1984. Structural and Tectonic Framework of the Continent of Europe: Introduction. In *Geomorphology of Europe*. Ed. C. Embleton. 1. New York.
Empereur, J.-Y., and A. Hesnard. 1987. Les amphores hellénistiques. In *Céramiques hellénistiques et romaines* II, 9–71.
Esteva, M., J. Trelogan, A. Rabinowitz, D. Walling, and S. Pipkin. Forthcoming. From the Site to Long-term Preservation: A Reflexive System to Manage and Archive Digital Archaeological Data. *Proceedings of the IS&T Archiving 2010 Conference, The Hague, Netherlands 1–4 June 2010*.
Euwe, J. 1988. The Shape of the Early Nolan Amphorae, 490–480 BC: The Origin and the Relation with Other Small Neck-amphorae. In *Proceedings of the 3rd Symposium of Ancient Greek and Related Pottery* (Copenhagen, August 31–September 4, 1987). 144–151. Copenhagen.
Fabbri, M., and M. Osanna. 2005. Aspetti del sacro nel mondo apulo: rituali di abbandono tra area sacra e abitato nell'antica Ausculum. In *Spazio del rito*, 215–233.
Fabbricotti, E. 1970. Ceramica campana. In Ostia II. *Le terme del nuotatore: scavo dell'ambiente* I. (*Studi Miscellanei* 16.) 57–58. Rome.
———. 1976. Cancellara (Potenza). Scavi 1972. *NSc* Ser. 8, Vol. 30: 327–358.
———. 1979. Ruoti (Potenza). Scavi in località Fontana Bona, 1972. *NSc* Ser. 8, Vol. 33: 347–413.
Fagles, R., trans. 1961. *Bacchylides: Complete Poems*. Westport, CT.
Fairbridge, R.W. 1972. Quaternary Sedimentation in the Mediterranean Region Controlled by Tectonics, Paleoclimates, and Sea Level. In *The Mediterranean*. Ed. D.J. Stanley. 93–113. Stroudsburg, PA.
Fallico, A.M. 1971. Siracusa. Saggio di scavo nell'area della Villa Maria. *NSc* Ser. 8, Vol. 25.II: 581–639.
Farrington, I.S. 1980. The archaeology of irrigation canals, with special reference to Peru. *World Archaeology* 11: 287–305.
Favia, P., R. Giuliani, A.M. Small, and C. Small. 2006. La valle del Basentello e l'insediamento rurale di Vagnari in età tardoantica. In *Paesaggi e insediamenti rurali*, 193–222.
Fedele, B., A. Alessio, and O. Del Monaco. 1992. *Archeologia, civiltà e culture nell'area ionico-tarantina. Origini e sviluppo dell'artigianato ceramico*. Taranto
Fentress, E. 2000. What Are We Counting For? In *Ploughsoil Assemblages*, 44–52.
———. 2005. Toynbee's Legacy. *JRA* 18: 482–438.

Ferri, S. 1962. L'ex-volo metapontino di Theayes. *RendAccLine* ser. VIII, 17: 3–10.

———. 2005. Toynbee's Legacy. *JRA* 18: 482–488.

Filangieri, R., ed. 1950. *I registri Cancelleria angioina: Testi e documenti di storia napoletana*. Naples.

Finley, M.I. 1979. *Ancient Sicily*. London.

———. 1999. *The Ancient Economy*. Updated ed. Berkeley.

Finnegan, M. 1978. Non-metric Variation of the Infracranial Skeleton. *JA* 125: 23–37.

Fiorilla, S. 1996. *Gela. Le ceramiche medievali dai pozzi di Piazza S. Giacomo*. Agrigento.

Fiorillo, R. 2005. *La tavola dei d'Angiò. Analisi archeologica di una spazzatura reale. Castello di Lagopesole (1266–1315)*. Florence.

Fischer-Hansen, T., J. Christiansen, M. Moltesen, and J.S. Ostergaard. 1992. Catalogue of the Ny Carslberg Glyptotek. Campania, South Italy and Sicily. Copenhagen.

Flambard, A.M. 1983. Le chateau de Scribla et la basse vallée du Crati du XIII au XV siècle. Thèse de 3° cycle, Université de Caen.

Flambard, A.M., and G. Noyé. 1984. La ceramica invetriata rinvenuta nello scavo del castello di Scribla (Calabria). XII-XV sec. In *San Lorenzo Maggiore*, 451–479.

Flemming, N.C. 1969. *Archaeological Evidence for Eustatic Change of Sea Level and Earth Movements in the Western Mediterranean During the Last 2,000 Years*. (*Geological Society of America Special Publication* 109.) Boulder.

Folk, R.L. 1980. *Petrology of Sedimentary Rocks*. Austin.

———. 1983. A Note on the Analysis of Metapontine Roof Tiles, 1982. In *Territory of Metaponto*, 9.

———. 1988. Crotone and Environs. Unpublished report, Institute of Classical Archaeology, The University of Texas at Austin.

———. 1998. Appendix 3A.1. Burial Rites and Tomb Types: Stones in Use. In *Necropoleis*, 112–113.

Folk, R.L., and J.C. Carter. 1981. Geological Factors Controlling Settlement Patterns on the Ionian Coast of Basilicata, Italy (abstr.). *Program of the Annual Meeting of the Geological Society of America*. 453. Cincinnati.

Fonseca, C.D. 2002. Dalla civitas Sanctae Trinitatis alla Turris ad Mare (secc. XI-XIV). In *Torre di Mare* I, 41–43.

Fornaro, A. 1988. Santa Scolastica. In *Bari*, 176–216.

Forti, L. 1962. Gli unguentari del primo periodo ellenistico. *RendNap* 30: 143–157.

———. 1965. *La ceramica di Gnathia*. Naples.

———. 1972. Questioni di ceramica messapica. *ASP* 25: 3–27.

Foxhall, L. 1993. Oil Extraction and Processing Equipment in Classical Greece. In *La production du vin et de l'huile en Méditerranée*. (Actes du symposium international organisé par le Centre Camille Jullian et le Centre Archéologique du Var, Aix-en-Provence et Toulon, 20–22 novembre 1991. *Bulletin de Correspondance Hellénique* Suppl. 26.) Eds. M.-C. Amouretti and J.-P. Brun. 183–200. Athens/Paris.

———. 1997. Appendix I: Ancient Farmsteads, Other Agricultural Sites and Equipment. In *A Rough and Rocky Place: The Landscape and Settlement History of the Methana Peninsula, Greece*. Eds. C. Mee and H. Forbes. 257–268. Liverpool.

———. 2001. Colouring in the Countryside. (Response to Pettegrew, D.K., "Chasing the Classical Farmstead," *JMA* 14.2: 189-209). *JMA* 14.2: 216–222.

———. 2002. Access to Resources in Classical Greece: The Egalitarianism of the Polis in Practice. In *Money, Labour and Land: Approaches to the Economics of Ancient Greece*. Eds. P. Carthedge, E.C. Cohen, and L. Foxhall. 209–220. London.

———. 2003. Cultures, Landscapes, and Identities in the Mediterranean World. *MHR* 18: 75–92.

———. 2007. *Olive Cultivation in Ancient Greece: Seeking the Ancient Economy*. Oxford/New York: Oxford University Press.

Fracchia, H. 1990a. Black Glaze Pottery. In Roccagloriosa I, 231–259.

———. 1990b. Thymiateria. In *Roccagloriosa* I, 261–262 (and 256 fig. 182).

———. 1990c. A Votive Deposit from Roccagloriosa in Western Lucania. In *Italici in Magna Grecia. Lingua, insediamenti e strutture*. Ed. M. Tagliente. 215–232. Venosa.

———. 1993. The Ceramic Evidence from Roccagloriosa: 4th and 3rd centuries. In *Fourth Century BC Magna Graecia: A Case Study*. Ed. M. Gualtieri. 255–280. Jonsered.

———. 2001. The Romanization of the *ager Buxentinus* (Salerno). In *Modalità insediative*, 55–73.

———. 2005. Il paesaggio rurale dell'Alto Bradano fra IV e V secolo d.C. In *Paesaggi e insediamenti rurali*, 133–144.

———. 2008. Rinvenimenti ceramici e trasformazioni dell'assetto insediativo nell'alta valle del Bradano. In *Felicitas Temporum. Dalla terra alle genti: la Basilicata settentrionale tra archeologia e storia*. Eds. A. Russo and H. Di Giuseppe. 289–303. Lavello.

Fracchia, H., and M. Gualtieri. 1999. Roman Lucania and the Upper Bradano Valley. *MAAR* 43: 295–344.

Franciosa, L. 1946. La distribuzione della popolazione nella Lucania in rapporto alle condizioni litologiche del suolo. *BSGRI* Ser. 7, 11.2: 65–78.

Franco, C. 1989. Testimonianze antiche nella Collezione privata Martini Carissimo ad Oria. In *Atti del IX Convegno dei Comuni Messapici, Peceuti e Dauni* (Oria 1984), 233–260. Bari.

Francovich, R., and R. Hodges. 2003. *Villa to Village: The Transformation of the Roman Countryside in Italy, c. 400–1000*. London.

Francovich, R., and G. Vannini. 1977. Saggio di scavo nell'area absidale esterna di S. Miniato al Monte di Firenze. *ArchMed* 4: 213–226.

Frank, S. 1990. *Attische Kelchkratere*. Frankfurt-am-Main/Bern/New York/Paris.

Frankel, R. 1993. The trapetum and the mola olearia. In *La production du vin et de l'huile en Méditerranée*. (Actes du symposium international organisé par le Centre Camille Jullian et le Centre Archéologique du Var, Aix-en-Provence et Toulon, 20–22 novembre 1991. *Bulletin de Correspondance Hellénique* Suppl. 26.) Eds. M.-C. Amouretti and J.-P. Brun. 477–481. Athens/Paris.

Franzoni, C. 2002. Il simposio. In *I Greci* 4.II, 1229–1259.

Frederick, C. 2001. Evaluating Causality of Landscape change: Examples from Alluviation. In *Earth Sciences and Archaeology*. Eds. P. Goldberg, V.T. Holliday, and C.R. Ferring. 55–76. New York.

Fresa, M.P. 1993. Il Melfese. In *Da Leukania a Lucania*, 1–28.

Frey, O.-H. 1998. La Siritide nella prima Età del Ferro. Qualche considerazione. In *Siritide e Metapontino*, 31–43.

Fuchs, F. 1980. Quartäre Küsten-und Flussterrassen in der Umrahmung des Golfs von Tarent (süd-Italien). *Catena* 7: 27–50.

Fusaro, D. 1982. Note di architettura domestica greca nel periodo tardo-geometrico e arcaico. *DialArch* N.S. 4.1: 5–30.

Gabba, E., and M. Pasquinucci. 1979. *Strutture agrarie e allevamento transumante nell'Italia romana (III–I sec. a.C.)*. Pisa.

Gagliardi, V. 2001. Ceramica a vernice nera dal santuario di Punta Stilo: contributi all'analisi delle produzioni. In *Kaulonía, Caulonia, Stilida (e oltre): contributi storici, archeologici e topografici* I. (*ASNSP* Serie IV, Quaderno 11.) Ed. M.C. Parra. 279–318. Pisa.

———. 2004. Il kerameikos di Contrada Lupa: per una revisione dei dati. In *Kaulonía, Caulonia, Stilida (e oltre): contributi storici, archeologici, e topografici* II. (*ASNSP* Serie IV, Quaderno 17.) Ed. M.C. Parra. 493–533. Pisa.

Gais, R.M. 1982. A Crown and Consolation in Rural Southern Italy. *Archaeology* 35.6: 54–56.

Gallant, T.W. 1991. *Risk and Survival in Ancient Greece. Reconstructing the Rural Domestic Economy*. Stanford.

Galli, E. 1926–1928. Metaponto. Esplorazioni archeologiche e sistemazione dell'area del tempio delle Tavole Palatine. *AttiMGrecia* 1926–1928: 63–79.

Garlan, Y. 1988. *Slavery in Ancient Greece*. Trans./rev. J. Lloyd. Ithaca/London.

Garnsey, P. 1988. *Famine and Food Supply in the Greco-Roman World. Responses to Risk and Crisis*. Cambridge.

———. 1999. *Food and Society in Classical Antiquity*. Cambridge.

Gasperetti, G. 1996. Produzione e consumo della ceramica comune da mensa e dispensa nella Campania romana. In *Céramiques communes*, 19–63.

Geer, R.M. 1954. *Diodorus of Sicily* X *(Books XIX.66–110 and XX)*. Cambridge, MA/London: Loeb Classical Library.

Gelichi, S. 1986. Studi sulla ceramica medievale riminese. 2.Il complesso dell'ex Hotel Commercio. *ArchMed* 13: 117–172.

———. 1988. Ceramiche venete importate in Emilia-Romagna tra XIII e XIV secolo. *Padusa* 24: 5–43.

———. 1992. A proposito della ceramica a doppio bagno. *Faenza* 78: 61–68.

———. 1993. La ceramica bizantina in Italia e la ceramica italiana nel Mediterraneo orientale tra XII e XIII secolo: stato degli studi e proposte di ricerca. In *La ceramica nel mondo bizantino tra XI e XV secolo e i suoi rapporti con l'Italia*. Ed. S. Gelichi. 9–46. Florence.

Genovese, G. 1999. *I santuari rurali nella Calabria greca*. Rome.

Gentili, G.V. 1956. Siracusa. Contributo alla topografia dell'antica città. *NSc* Ser. 8, vol. 10: 94–164.

Ghinatti, F. 1961–1962. Ricerca sulla Lega Italiota. *MAP* 74: 117–133.

———. 1973. Aspetti dell'economia agraria della Magna Grecia agli inizi dell'impero. *Critica storica* 10.3: 359–397.

———. 1982. Strutture istituzionali dell'economia arcaica in Magna Grecia. *RIN* 84: 3–94.

———. 1996. *Assemblee greche di Occidente*. Turin.

Ghisetti, F. 1981. Upper Pliocene-Pleistocene Uplift Rates as Indicators of Neotectonic Pattern: An Example from Southern Calabria (Italy). *ZG*, N.F. Suppl. 40: 93–118.

Giacomelli, R. 1988. *Achaea Magno-Graeca. Le iscrizioni arcaiche in alfabeto acheo di Magna Grecia*. Brescia.

Giacometti, D. 1999. Il culto di Artemis a Metaponto. *Ostraka* 8: 407–426.

———. 2005. Metaponto. *Gli dei e gli eroi nella storia di una polis di Magna Grecia*. (*Bibliografia di Studi Relgiosi* 6 / *Studi e Testi* 1.) Cosenza.

Giannelli, C.A. 1969. L'intervento di Archidamo e di Alessandro il Molosso in Magna Grecia. *Critica storica* 8: 1–22.

———. 1974. Gli interventi di Cleonimo e di Agatocle in Magna Grecia. *Critica storica* 11: 353–380.

Giannotta, M.T. 1980. *Metaponto ellenistico-romana*. Galatina.

———. 1983. Un vaso ellenistico rinvenuto a Metaponto. In *Studi in onore di Dinu Adamesteanu*, 49–57.

———. 1994. Tombe messapiche da Vaste e Ortelle. *StAnt* 7: 85–110.

———. 1995. Rinvenimenti tombali da Rocavecchia (1934) al Museo Nazionale di Taranto. *StAnt* 8.2: 39–73.

———. 1996. Rinvenimenti tombali da Rocavecchia (1934) al Museo Provinciale di Lecce: Materiali di corredo e produzioni. *StAnt* 9: 37–98.

———. 1997a. Ricerche sulle necropoli messapiche di età ellenistica. Vaste: vecchi rinvenimenti e nuove problematiche. In *Metodologie di catalogazione dei beni archeologici*. Ed. F. D'Andria. 167–177. Lecce-Bari.

———. 1997b. Una tomba di guerriero del IV sec. a.C. e

la necropoli extra-urbana di Vaste-Fondo Aia. *StAnt* 10: 153–186.
Giardina, A. 1981. Allevamento ed economia della selva in Italia meridionale: trasformazioni e continuità. In *Società romana* I, 87–113.
Giardino, C. 1995. *Il mediterraneo occidentale fra XIV e VIII secolo a.C. Cerchie minerarie e metallurgiche.* (*BAR-IS* 612.) Oxford.
———. 1998. *I metalli nel mondo antico.* Bari.
Giardino, L. 1976a. Il periodo post-annibalico a Heraclea. *Atti Taranto* 15: 549–560.
———. 1976b. L'inizio del periodo romano ad Heraclea. *MGrecia* 11: 19–21.
———. 1978a. Metaponto 1977 – La campagna di scavo nell'area del castrum. *Atti Taranto* 17: 413–429.
———. 1978b. Le ceramiche ellenistiche e romane. In *Leuca*, 121–160.
———. 1980a. Sulla ceramica a pasta grigia di Metaponto e sulla presenza in essa di alcuni bolli iscritti: studio preliminare. *StAnt* 2: 247–287.
———. 1980b. Prime note sull'urbanistica di Grumentum. In *Attività archeologica*, 477–538.
———. 1981. Grumentum: la ricerca archeologica in un centro antico. Mostra documentaria. Galatina.
———. 1982. Metaponto tardo-imperiale e *Turiostu*: proposta di identificazione in margine ad un *miliarium* di Giuliano l'Apostata. *StAnt* 3: 155–173.
———. 1983. Il porto di Metaponto in età imperiale: topografia e materiali ceramici. *StAnt* 4: 5–36.
———. 1990. Heraclea, necropoli meridionale. Le sepolture di II e I sec. a.C. In *Emergenze e problemi*, 73–125.
———. 1991. Grumentum e Metaponto. Due esempi di passaggio dal tardoantico all'alto Medioevo in Basilicata. *MEFRM* 103.2: 827–858.
———. 1996a. Herakleia. In *Arte e artigianato*, 35–43.
———. 1996b. Architettura domestica a Herakleia. Considerazioni preliminari. In *Ricerche sulla casa*, 133–159.
———. 2005. Herakleia e Metaponto: dalla *polis* italiota all'abitato proto-imperiale. *Atti Taranto* 44: 387–432.
Giardino, L., and A. De Siena. 1999a. La costa ionica dall'età del ferro alla fondazione delle colonie: forme e sviluppi insediativi. In *Magna Grecia e Sicilia*, 23–38.
———. 1999b. Metaponto. In *La città greca antica. Istituzioni, società e forme urbane.* Ed. E. Greco. 329–363. Rome.
Giardino, L., E. Lapadula, and R. Auriemma. 1999. La fascia ionica della Basilicata in età tardoantica. Continuità e trasformazioni. *Atti Taranto* 38: 343–368.
Giglio, M. 2005. Unguentari. In *Torre di Satriano* I, 334–341.
Gilotta, F. 1985. *Gutti e askoi a rilievo italioti ed etruschi: teste isolate.* Rome.
Ginouvès, R. 1962. *Balaneutikè. Recherches sur le bain dans l'antiquité grecque.* Paris.
Giorgi, M. 1988a. Ceramica con decorazione a bande di produzione daunia. In *Forentum* I, 172–183.
———. 1988b. Ceramica a vernice nera. In *Forentum* I, 186–205.
Giorgi, M., and S. Martinelli. 1988. Catalogo delle tombe dall'area dell'abitato urbano: A-98. In *Forentum* I, 46–92.
Giraudi, C. 1989. Lake levels and climate for the last 30,000 years in the Fucino area (Abruzzo--Central Italy)--A review. *Palaeogeography, Palaeoclimatology, Palaeoecology* 70 249–260.
Gissing, G. 1996. *By the Ionian Sea. Notes of a Ramble in Southern Italy.* Evanston, IL. (First published 1901.)
Giudice, F. 1997. Le importazioni attiche in Magna Grecia. Appunti per la definizione del quadro di riferimento. *Ostraka* 6: 401–413.
Giudice, G. 2007. *Il tornio, la nave, le terre lontane. Ceramografi attici in Magna Grecia nella seconda metà del V sec. a.C. Rotte e vie di distribuzione.* Rome.
Given, M. 2004. Mapping and Manuring: Can We Compare Sherd Density Figures? In *Side-by-side Survey*, 13–21.
Gladfelter, B.G. 1977. Geoarchaeology: The Geomorphologist and Archaeology. *AmerAnt* 42: 519–538.
Gliozzo, E., C. Fortina, I. Memmi Turbanti, M. Turchiano, and G. Volpe. 2005. Cooking and Painted Ware from San Giusto (Lucera, Foggia): The Production Cycle, from the Supply of Raw Material to the Commercialization of Products. *Archaeometry* 47: 13–29.
Godley, A.D., trans. 1957. *Herodotus with an English Translation* II *(Books III and IV).* Cambridge, MA/London: Loeb Classical Library.
Goldberg, H.J.V., J.A. Weintraub, K.J. Roghmann, and W.S. Cornwell. 1976. Measuring Periodontal Disease in Ancient Populations: Root and Wear Indices in Study of American Indian Skulls. *JP* 47: 348–351.
Goodfield, J., and S. Toulmin. 1965. How Was the Tunnel of Eupalinus Aligned? *Isis* 56: 46–55.
Goodman, A.H., and J.C. Rose. 1990. Assessment of Systemic Physiological Perturbations from Dental Enamel Hypoplasias and Associated Histological Structures. *YPA* 33: 59–110.
Goody, J. 1984. *Cooking, Cuisine and Class: A Study in Comparative Sociology.* Cambridge.
Goudineau, C. 1968. La céramique arétine lisse. (*Fouilles de l'Ecole française de Rome à Bolsena [Poggio Moscini] 1962-1967* vol. 4.) Paris.
Govi, E. 1999. *Le ceramiche attiche a vernice nera di Bologna.* Bologna.
Graepler, D. 1996. La coroplastica funeraria. In *Arte e artigianato*, 228–246.
———. 1997. *Tonfiguren im Grab. Fundkontexte hellenistischer Terrakotten aus der Nekropole von Tarent.* Munich.
Graham, A.J. 1964. *Colony and Mother City in Ancient Greece.* New York.

Grasso, L. 2005. La ceramica miniaturistica votiva di Pompei. In *Depositi votivi e culti dell'Italia antica dall'età arcaica a quella tardo-repubblicana* (Atti del Convegno di Studi, Perugia 1–4 giugno 2000). Eds. A. Comella and S. Mele. 545–554. Città di Castello.

Grave, P.R., E. Robinson, M. Barbetti, Z. Yu, G. Bailey, and R. Bird. 1996–1997. Analysis of South Italian Pottery by PIXE-PIGME. *MedArch* 9/10: 113–125.

Greco, A., and E. Lapadula. 2004. L'insediamento di età tardoromana in località Badia, Cutrofiano (Lecce). Dati preliminari sulla ceramica dipinta. In *Quaderni del Museo della Ceramica di Cutrofiano* 8–9. Ed. S. Matteo. 11–38. Galatina.

Greco, C. 2009. Attività archeologica della Soprintendenza ai beni archeologici della Basilicata. *Atti Taranto* 48: 798–802.

Greco, E. 1979. Ricerche sulla chora poseidoniate: il "paesaggio agrario" dalla fondazione della città alla fine del sec. IV a.C. *DialArch* N.S. 1.2: 7–26.

———. 1981. Dal territorio alla città: lo sviluppo urbano di Taranto. *AION ArchStAnt* 3: 139–157.

———. 1988. La città e il territorio: Problemi di storia topografica. *Atti Taranto* 27: 471–97.

———. 1993. L'impero di Sibari. Bilancio archeologico-topografico. *Atti Taranto* 32: 469–470.

———. 1995. Nella chora di Eraclea di Lucania. I monumenti della Masseria del Concio riconsiderati. In *Sur les pas des Grecs*, 459–465.

———. 2000. L'Ekklesiasterion di Poseidonia-Paestum. In *Rites et espaces en pays celte et méditerranéen: étude comparée à partir du Sanctuaire d'Acy-Romance (Ardennes, France)*. Ed. S. Verger. 337–340. Rome.

———. 2001. Abitare in campagna. *Atti Taranto* 40: 171–201.

Greco, G. 1980. Le fasi cronologiche dell'abitato di Serra di Vaglio. In *Attività archeologica*, 367–404.

———, ed. 1982. *L'evidenza archeologica nel Lagonegrese. Mostra documentaria*. Matera.

———, ed. 1991. *Serra di Vaglio. La "casa dei pithoi."* Modena.

———. 2009. Dalla città greca alla città samnitica: le evidenze dalla Piazza del Foro. *Atti Taranto* 48: 385–444.

Greco, G., and F. Mermati. 2007. Cuma: le indagini nell'area del Foro e la produzione ceramica di età arcaica. In *Ceramiche fini*, 143–162.

Green, J.R. 1972. Oinochoe. *BICS* 19: 1–16.

———. 1976. *Gnathia Pottery in the Akademisches Kunstmuseum in Bonn*. Mainz.

———. 1977. More Gnathia Pottery in Bonn. *AA* 92: 551–563.

———. 1979. Ears of Corn and Other Offerings. In *Studies in Honour of Arthur Dale Trendall*. Ed. A. Cambitoglou. 81–90. Sydney.

———. 1986. Some Gnathia Pottery in the J. Paul Getty Museum. In *Greek Vases in the J.P. Getty Museum*. 116–138. Malibu.

———. 1989. Motif-symbolism and Gnathia Vases. In *Beiträge zur Ikonographie und Hermeneutik. Festschrift für Nikolaus Himmelmann*. Eds. H.U. Cain, H. Gabelmann, and D. Salzmann. 221–226. Bonn.

———. 2001. Gnathia and Other Overpainted Wares of Italy and Sicily: A Survey. In *Céramiques hellénistiques et romaines* III, 57–103.

Guaitoli, M. 2002. Il territorio e le sue dinamiche: osservazioni e spunti di ricerca. *Atti Taranto* 41: 219–252.

Gualandi, M.L., C. Palazzi, and M. Paoletti. 1981. La Lucania orientale. In *Società romana* I, 155–179.

Gualtieri, M. 1990. Louteria. In *Roccagloriosa* I, 305–307.

———. 1999. Nuove forme di uso dell'acqua in età romana. In *Archeologia dell'acqua*, 127–158.

———. 2003. *La Lucania romana. Cultura e società nella documentazione archeologica*. Naples.

———. 2008a. La villa di Masseria Ciccotti di Oppido Lucano: fasi edilizie, architettura, mosaici. In *Felicitas Temporum. Dalla terra alle genti: la Basilicata settentrionale tra archeologia e storia*. Eds. A. Russo and H. Di Giuseppe. 265-287. Lavello.

———. 2008b. La Lucania centro-settentrionale in età romana: la nuova documentazione archeologica. In *Felicitas Temporum. Dalla terra alle genti: la Basilicata settentrionale tra archeologia e storia*. Eds. A. Russo and H. Di Giuseppe. 205-221. Lavello.

———. 2009. Villae e uso del territorio nell'alto Bradano (Regio III) tra tarda Repubblica e primio Impero. In *Agricoltura e scambi*, 341–367.

Guardascione, C. 2002. Le monete. In *Terre di Mare* I, 303–316.

Guarducci, M. 1978. Siris. *RendAccLinc* Ser. 8, Vol. 33: 273–288.

Guerricchio, A., and G. Melidoro. 1988. Urbanizzazione e franosità delle formazioni sabbiose plio-pleistoceniche in Basilicata: il caso di Senise. *MSGLI* 37: 745–774.

Guy, M. 1995. Cadastres en bandes de Métaponte à Agde. Questions et méthodes. In *Sur les pas des Grecs*, 427–444.

Guzzo, P.G. 1978. Importazioni fittili greco-orientali sulla costa jonica d'Italia. *Excursus II*: coppe cosi dette ioniche. In *Céramiques*, 123–128.

———. 1983. Metaponto: dai Neleidi ai Sanniti. *PP* 38: 197–199.

———. 2008 (2009). Oreficerie ornamentali dalla tomba 238 di Crucinia. *BdA* 143: 15–26.

Guzzo, P.G., and E. Greco. 1978. S. Maria del Cedro. Fraz. Marcellina. Loc. S. Bartolo (Cosenza). Scavi di un centro abitato di epoca ellenistica (1973; 1975). *NSc* Ser. 8, Vol. 32: 429–459.

Hall, J. 1998. Grave Goods. In *Necropoleis*, 563–590.

Hall, S.A. 1990. Channel Trenching and Climate Change in the Southern U.S. Great Plains. *Geology* 18: 342–345.
Hänsel, B. 1973. Policoro (Matera). Scavi eseguiti nell'area dell'acropoli di Eraclea negli anni 1965–1967. *NSc* Ser. 8, Vol. 27 400–492.
Hansen, M.H. 1995. Kome. A study in how the Greeks designated and classified settlements which were not poleis. In M.H. Hansen and K. Raaflaub. *Studies in the Ancient Greek Poleis. Historia Einzelschriften* 95. Stuttgart.
———. 1998. *Polis and City-state: An Ancient Concept and Its Modern Equivalent.* (*Acts of the Copenhagen Polis Centre* 5.) Copenhagen.
Hansen, M.H., and T. Fischer-Hansen. 1994. Monumental Political Architecture in Archaic and Classical Greek Poleis. Evidence and Historical Significance. In *From Political Architecture to Stephanus Byzantius: Sources for the Ancient Greek Polis.* (Historia Einzelschriften 87.) Ed. D. Whitehead. 23–90. Stuttgart.
Hansen, P.A. 1985. The Potter Nikomachos and His Dedication (*I.G.* 14.652). *ZPE* 58: 231–233.
Haq, B.U., W.A. Berggren, and J.A. Van Couvering. 1977. Corrected Age of the Pliocene/Pleistocene Boundary. *Nature* 260: 483–488.
Hayes, J.W. 1972. *Late Roman Pottery.* London.
———. 1975. *Roman and Pre-Roman Glass in the Royal Ontario Museum. A Catalogue.* Toronto.
———. 1980. *Ancient Lamps in the Royal Ontario Museum I. Greek and Roman Clay Lamps. A Catalogue.* Toronto.
———. 1984. *Greek and Italian Black-Gloss Wares and Related Wares in the Royal Ontario Museum: A Catalogue.* Toronto.
———. 1985. Sigillate orientali. In *Atlante* II, 1–96.
Hayes, P.P. 1991. Models for the Distribution of Pottery around Former Agricultural Settlements. In *Ploughzone Archaeology*, 81–92.
Hellmann, M.-C. 1987. *Lampes antiques de la Bibliothèque Nationale* II. *Fonds général: lampes pré-romaines et romaines* Paris.
Hempel, K.G. 1996a. Taranto: la ceramica a vernice nera "HFR" e a "pasta grigia." In *Arte e artigianato*, 336–342.
———. 1996b. Metaponto: ceramica a "pasta grigia." In *Arte e artigianato*, 343–345.
———. 2001. *Die Nekropole von Tarent im 2. und 1. Jahrhundert v. Chr. Studien zur materiellen Kultur.* Taranto.
Hempel, K.G., and B. Mattioli. 1994. Contesti tombali di età ellenistica con monete: elementi per una cronologia assoluta. In *Museo Taranto* III.1, 354–390.
Henig, M. 1994. *Classical Gems in the Fitzwilliam Mueum.* Cambridge.
Henneberg, M., and R.J. Henneberg. 1990. Biological Characteristics of the Population in the Chora. In *The Pantanello Necropolis 1982–1989. An Interim Report.* Ed. J.C. Carter. 76–92. Austin.
———. 1994. Treponematosis in an Ancient Greek Colony of Metaponto, Southern Italy, 580–250 BCE. In *The Origin of Syphilis in Europe.* Eds. O. Dutour, G. Palfi, J. Berato, J.-P. Brun. 92–98. Toulon.
———. 1998. Biological Characteristics of the Population Based on Analysis of Skeletal Remains. In *Necropoleis*, 503–556.
———. 2001. Analysis of Human Skeletal and Dental Remains from Metaponto (7th-2nd c. BC). *Atti Taranto* 40: 461–474.
Henneberg, M., R.J. Henneberg, and J.C. Carter. 1992. Health in Colonial Metaponto. *Exploration* 8: 446–459.
Henneberg, M., R.J. Henneberg, and A. De Siena. Forthcoming. Biological Status of the Ancient Urban Population of Metaponto Reconstructed from Skeletal Remains (Necropolis Crucinia, 7th–2nd c. BC, Southern Italy).
Henneberg M., R.J. Henneberg, J.C. Vogel, A. De Siena, and J.C. Carter. 1995. Biological Status of Urban and Rural Populations of the Ancient Greek Colony Metaponto in Italy, 7th–2nd c. BCE. In *1st International Congress on Science and Technology for the Safeguard of Cultural Heritage in the Mediterranean Basin* (November 27–December 2, 1995, Catania, Siracusa, Italy). 397 (extended abstract).
Henneberg, R.J. 1991. Dental Caries. In *Encyclopedia of Human Biology.* Ed. R. Dulbecco. 8 vols. 2.805–815. San Diego.
———. 1998. *Dental Health and Affiliations of Inhabitants of the Ancient Greek Colony in Metaponto, Italy (6th–3rd c. BC).* Doctoral Dissertation, University of the Witwatersrand, Johannesburg, South Africa.
Herdejürgen, H. 1971. *Die tarentinischen Terrakotten des 6. bis 4. Jahrhunderts v. Chr. im Antikenmuseum Basel.* Basel.
Hewlett-Packard Co. 1974. *Hewlett-Packard 55 Statistics Programs.* Cupertino, CA.
Higgins, R.A. 1954. *Catalogue of the Terracottas in the Department of Greek and Roman Antiquities, British Museum* I. London.
———. 1967. *Greek Terracottas.* London.
Hillson, S. 1996. *Dental Anthropology.* Cambridge.
Hodges, R., and H. Patterson. 1986. San Vincenzo al Volturno and the origins of the medieval pottery industry in Italy. In *Ceramica medievale*, 13–26.
Hodkinson, S. and H. Hodkinson. 1981. Mantineia and the Mantinike. *BSA* 76: 240–296.
Hoffmann, A. 2002. *Grabritual und Gesellshaft. Gefässformen, Bildthemen und Funktionen unteritalisch-rotfiguriger Keramik aus der Nekropole von Tarent.* (Internationale Archäologie, Band 76.) Rahden.
Holloway, R.R. 1970. Satrianum. *The Archaeological Investigations Conducted by Brown University in 1966 and 1967.* Providence.
———. 1981. *Italy and the Aegean, 3000-700 BC.* (*Archaeologia Transatlantica* 1.) Louvaine-La-Neuve.
Hooper, W.D. and H.B. Ash, trans. 1979. *Marcus Porcius*

Cato on Agriculture. Marcus Terentius Varro on Agriculture. Cambridge, MA/London: Loeb Classical Library.

Horowitz, A. 1979. *The Quaternary of Israel.* New York.

Hsü, K.J. 1977. Tectonic Evolution of the Mediterranean Basins. In *Ocean Basins and Margins 4A: The Mediterranean.* Eds. A.E.M. Nairn and F.G. Stehli. 29–75. New York.

Hsü, K.J., J. La Brecque, S.F. Percival, R.C. Wright, A.M. Gombos, K. Pisciotto, P. Tucker, N. Peterson, J.A. McKenzie, H. Weissert, A.M. Karpoff, M.F. Carman, Jr., and E. Schreiber. 1984. Numerical Ages of Cenozoic Biostratigraphic Datum Levels. *BGSA* 95.7: 863–876.

Humphreys, S.C. 1982. Fustel de Coulanges and the Greek "genos." *Sociologia del Diritto* 3: 35–44.

Hunt, C., and D.D. Gilbertson. 1995. Human Activity, Landscape Change and Valley Alluviation in the Feccia Valley, Tuscany, Italy. In *Mediterranean Quaternary River Environments*, 167–176.

Hurschmann, R. 1995. Unteritalische Netzlekythoi. *AA*: 667–691.

Iacobone, C. 1988. *Le stipi votive di Taranto (scavi 1885–1934).* Rome.

Idrisi. 1994. *Il libro di Ruggero.* Trans. U. Rizzitano. Palermo.

Imbrie, J., and K.P. Imbrie. 1979. *Ice Ages: Solving the Mystery.* Short Hills, NJ.

Ingravallo, E. 1980. L'insediamento eneolitico di Pizzica Pantanello presso Metaponto (Basilicata). *Atti Società Toscana Scienze Naturali, Memorie A87*: 317–327.

Isager, S., and J.E. Skydsgaard. 1992. *Ancient Greek Agriculture. An Introduction.* London/New York.

Isayev, E. 2001. The Role of the Indigenous Centres in Lucania and Their Collapse in the III c. BC. In *Modalità insediative*, 107–128.

Isler, H.P. 1970. *Acheloos: Ein Monographie.* Bern.

Itzhaki, Y. 1961. Contributions to the Study of the Pleistocene in the Coastal Plain of Israel: Pleistocene Shorelines in the Coastal Plain of Israel. *GSIB* 32: 1–9.

Jaboli, D., and A. Roger. 1954. Esquisse structurale de la Fosse Bradanique (Italie Meridionale). In *Congrès Géologique International.* (Comptes rendus de la XIX session, Alger 1952. Fascicule IX.) 305–324. Algiers.

Jackson, M.M. 2004. Jewellery Evidence and the Lowering of South Italian Ceramic Chronology. *AWE* 3: 283–313.

Jameson, M.H. 1994. Class in the Ancient Greek Countryside. In *Structures rurales*, 55–63.

Jeanmaire, H. 1978. *Dionysos. Histoire du culte de Bacchus: l'orgiasme dans l'antiquité et les temps modernes, origine du théâtre en Grèce, orphisme et mystique dionysiaque, évolution du dionysisme après Alexandre.* Paris.

Jensen, J.R. 1986. *Introductory Digital Image Processing: A Remote Sensing Perspective.* 2nd ed. Upper Saddle River, NJ.

Jircik, N. 1983. A Note on the Red-Figured Pottery from Pizzica. In *Territory of Metaponto*, 32–33.

———. 1991. *The Pisticci and Amykos Painters: the Beginnings of Red-Figured Vase-painting in Ancient Lucania.* Doctoral diss., The University of Texas at Austin.

Johannowsky, W. 1983. *Materiali di età arcaica dalla Campania.* Naples.

Johnson, A. 1983. *Roman Forts of the 1st and 2nd centuries AD in Britain and the German Provinces.* New York.

Johnson, M. 1996. *The Archaeology of Capitalism.* Oxford.

Johnston, A. 1983. The Coins. In *Cozzo Presepe*, 383.

———. 1989. The Bronze Coinage of Metapontum. In Mørkholm Essays: Numismatic Studies in Memory of C.M. Kraay and O. Mørkholm. (*Publications d'histoire de l'art et d'archéologie de l'Université catholique de Louvain* 59. Numismatica lovaniensia 10.) Eds. G. Le Rider, K. Jenkins, N. Waggoner, and U. Westermark. 121–136. Louvain-la-Neuve.

———. 1990. *The Coinage of Metapontum,* Part 3. (*American Numismatic Society Numismatic Notes and Monographs* 164.) New York.

———. 1998. Coins. In *Necropoleis*, 830-833.

Johnston, A. and R.E. Jones. 1978. The "SOS" Amphora. BSA 73: 103–141.

Jones, H.L., trans. 1983. *The Geography of Strabo* III *(Books VI and VII).* Cambridge, MA/London: Loeb Classical Library.

Jones, J.E., A.J. Graham, and L.H. Sackett. 1973. An Attic Country House below the Cave of Pan at Vari. *BSA* 68: 355–452.

Jones, J.E., L.H. Sackett, and A.J. Graham. 1962. The Dema House in Attica. *BSA* 57: 75–114.

Jones, W.H.S. 1907. *Malaria: A Neglected Factor in the History of Greece and Rome.* Cambridge.

———. 1909. *Malaria and Greek History.* Manchester.

Jones, W.H.S. and H.A. Omerod, trans. 1954. *Pausanias. Descriptions of Greece* II *(Books III–V).* Cambridge, MA/London: Loeb Classical Library.

Jost, M. 1999. Les schémes de peuplement de l'Arcadia aux époque archaïque et classique. In *Defining Ancient Arkadia. Symposium April 1–4, 1998. Acts of the Copenhagen Polis Centre,* 6. Eds. T.H. Nielsen and J. Roy. 192–247. Copenhagen.

Judson, S. 1963. Erosion and Deposition of Italian Stream Valleys During Historic Time. *Science* 140: 898–899.

Kahrstedt, U. 1960. *Die wirtschaftliche Lage Grossgriechenlands in der Kaiserzeit.* Historia Einzelschrift 4. Wiesbaden.

Kanowski, M.G. 1984. *Containers of Classical Greece. A Handbook of Shapes.* St. Lucia.

Kavanagh, B.F., and G. Bird. 2000. *Surveying: Principles and Applications.* Upper Saddle River, NJ.

Kayser, B. 1958. *Recherches sur les sols et l'erosion en Italie meridionale: Lucanie.* Paris.

Keay, S.J. 1984. *Late Roman Amphorae in the Western Mediterranean: A Typology and Economic Study. The Catalan*

Evidence. (*BAR-IS* 196.) 2 vols. Oxford.

Keay, S.J., and M. Millett. 1991. Surface Survey and Site Recognition in Spain: The Ager Tarraconensis Survey and Its Background. In *Ploughzone Archaeology*, 129–139.

Keay, S. and D. Williams. 2005. *Roman Amphorae: A Digital Resource*. Archaeology Data Service and the University of Southampton.

Keith, A., and H. Fracchia. 1990a. Coarse Ware. In *Roccagloriosa* I, 262–273.

———. 1990b. Kiln Assemblage. In *Roccagloriosa* I, 273–277.

Kelley, M.A., and C.S. Larsen, eds. 1991. *Advances in Dental Anthropology*. New York.

Kent, J.H. 1948. The Temple Estates of Delos, Rheneia, and Mykonos. *Hesperia* 17.4: 243–338.

Kenyon, F.G., trans. 1897. *The Poems of Bacchylides*. Oxford.

Kingsley, B. 1979. The Reclining Heroes of Taras and Their Cult. *CSCA* 17: 201–220.

Klapisch Zuber, C. 1973. Villaggi abbandonati ed emigrazioni interne. In *Storia d'Italia* V. *I documenti*. 311–364. Turin.

Knox, J.C. 1983. Responses of River Systems to Holocene Climates. In *Late Quaternary Environments of the United States 2: The Holocene*. Eds. H.E. Wright and S.C. Porter. 26–41. Minneapolis.

———. 1984. Fluvial Responses to Small-scale Climatic Changes. In *Developments and Applications of Geomorphology*. Eds. J.E. Costa and P.J. Fleisher. 318–342. Berlin.

Koehler, C.G. 1979. *Corinthian A and B Transport Amphoras*. Doctoral dissertation, Princeton University. Princeton.

———. 1981. Corinthian Developments in the Study of Trade in the Fifth Century. *Hesperia* 50.4: 449–458.

———. 1992. A Brief Typology and Chronology of Corinthian Transport Amphoras / Obshchaia tipologiia I khronologia korinfskikh transportnykh amfor. In *Greek Transport Amphoras / Grecheskie amfory*. Eds. V.I. Kats and S. Iu. Monakhov. 265–279. Saratov.

Koepcke, G. 1964. Goldekorierte attische Schwarzfirniskeramik des vierten Jahrhunderts vor Christus. *AM* 89: 15–71.

Koparal, E., and E. İplikçi. 2004. Archaic Olive Oil Extraction Plant in Klazomenai. In *Klazomenai, Teos and Abdera. Metropoleis and Colony*. (Proceedings of the International Symposium held at the Archaeological Museum of Abdera, 20-21 October 2001.) Eds. A. Moustaka, E. Skarlatidou, M.-C. Tzannes, and Y.E. Ersoy. 221–234. Thessaloniki.

Kraay, C.M. 1966. *Greek Coins*. New York.

Kraft, J.C., S.E. Aschenbrenner, and G. Rapp, Jr. 1977. Paleogeographic Reconstructions of Coastal Aegean Archaeological Sites. *Science* 195: 941–947.

Kraft, J.C., G. Rapp, Jr., and S.E. Aschenbrenner. 1975. Late Holocene Paleogeography of the Coastal Plain of the Gulf of Messina, Greece, and Its Relationship to Archaeological Settings and Coastal Change. *BGSA* 86: 1191–1208.

Krogman, W.M., and M.Y. Iscan. 1986. *Human Skeleton in Forensic Medicine*. Springfield, IL.

Kukla, G.J. 1977. Pleistocene Land-Sea Correlations (1): Europe. *ESR* 13: 307–373.

Kustermann Graf, A. 2002. Selinunte. *Necropoli di Manicalunga. Le tombe della contrada Gaggera*. Catanzaro.

L'Abbate, V. 1979. *Norba e i centri antichi del territorio di Conversano*. Bari.

Labellarte, M. 1988. Via Giuseppe Martino. In *Bari*, 304–339.

———. 1992a. Tomba a fossa di Largo Costantinopoli. In *Principi imperatori vescovi*, 152–155.

———. 1992b. Tomba n. 3 del Regio Tratturo. In *Principi imperatori vescovi*, 506–508.

———. 1995. La ceramica a vernice nera. In *Collezione Chini*, 93–142.

Lacava, M. 1891. *Topografia e storia di Metaponto*. Naples.

Lacey, W.K. 1968. *The Family in Classical Greece*.London.

Lafage, F., M.E. Mariconda, L. Olivieri, and A. Rouveret. 1987. Ceramica di età classica. In *Poseidonia-Paestum* III, 119–138.

Lafond, Y. 1991. Artémis en Achaie. *REG* 104: 410–433.

Laganara Fabiano, C. 1988. Scheda no. 35. *In Il recupero di una città medievale. Fiorentino*. Ed. M.S. Calò Mariani. 86. Bari.

Lamb, H.H. 1966. *The Changing Climate*. London.

———. 1982a. Climate, History, and the Modern World. London.

———. 1982b. Reconstruction of the Course of Post-Glacial Climate over the World. In *Climatic Change in Later Prehistory*. Ed. A.F. Harding. 11–32. Edinburgh.

Lamboglia, N. 1952. Per una classificazione preliminare della ceramica campana. In *Atti del I Congresso Internazionale di Studi Liguri (1950)*. 139–206. Bordighera.

Lamboley, J.-L. 1996. *Recherches sur les Messapiens: IVe-IIe siècle avant J.-C.* (*Bibliothèque des Écoles françaises d'Athènes et de Rome* 292.) Rome.

Lambrugo, C. 2004. Ceramica magnogreca a figure rosse. I vasi miniaturistici. In *Collezione Lagioia*, 239–247.

Langlotz, E., and M. Hirmer. 1965. *The Art of Magna Graecia: Greek Art in South Italy and Sicily*. London.

Lanza, E. 2004. Vasi a reticolo della Collezione Moschini al Museo di Antichità di Torino: riflessioni sulla storia del collezionismo e su una classe ceramica. *QSAP* 20: 21–52.

———. 2005. *Ceramica di Gnathia al Museo di Antichità di Torino*. (*Documenti di archeologia* 41.) Mantova.

Lanza Catti, E. 2008. *La ceramica "di Gnathia" al Museo Nazionale Jatta. Ipotesi di ricontestualizzazione*. (*Antenor Quaderni* 12.) Rome.

La Torre, G.F. 2011. Reflections on Lucanians and Bruttians in Calabria between Hannibal and the principate:

Coloniae civitates foederatae municipia. In *Local Culture of South Italy and Sicily in the Late Republican Period: Between Hellenism and Rome*. Ed. F. Colivicchi. *JRA* Suppl. 83: 145.

Lattanzi, E. 1976. Saggi di scavo a Pomarico. In *Locri Epizefirii, Atti Taranto* 16: 861–863.

———. 1978. Attività archeologica nel Materano. *Atti Taranto* 17: 437–444.

———. 1980. L'attività archeologica in Basilicata nel 1979. In *L'epos Greco in Occidente, Atti Taranto* 19: 399–409.

———. 1983. Un complesso di edifici paleocristiani a Metaponto. In *Lo scavo di S. Giovanni di Ruoti ed il periodo tardoantico in Basilicata*. (Atti della tavola rotonda – Roma, 4 luglio 1981.) 11–19. Bari.

Lattman, L.M. 1958. Technique of Mapping Geologic Fracture Traces and Lineaments on Aerial Photographs. *PE* 24: 568–576.

Leone, D. 2000. Le ceramiche tardoantiche della fattoria di Posta Crusta. In *Ordona* X, 387–436.

———. 2002. Aspetti della circolazione delle merci nell'Apulia tardoantica, tra importazioni e produzioni locali: la ceramica comune dipinta. In *L'Africa romana*, 881–884.

Leone, R. 1998. *Luoghi di culto extraurbani d'età arcaica in Magna Grecia*. Turin.

Lepore, E. 1963. *Ricerche sulla penetrazione romana in Apulia e Lucania fino alla III guerra sannitica*. Bari.

———. 1964. *Il mezzogiorno e l'espansione romana fino alla guerra tarentina*. Bari.

———. 1968. Per una fenomenologia storica del rapporto città-territorio in Magna Grecia. *Atti Taranto* 7: 29–62.

———. 1970. Classi e ordini in Magna Grecia. In *Recherches sur les structures sociales dans l'antiquité classique*. Ed. C. Nicolet. 43–62. Paris.

———. 1973. Problemi dell'organizzazione della chora coloniale. In *Problèmes de la terre*, 15–47.

———. 1974a. Problemi di storia metapontina. *Atti Taranto* 13: 307–325.

———. 1974b. Il dibattito. *Atti Taranto* 13: 335–336.

Letta, C. 1972. *Piccola coroplastica metapontina nel Museo Archeologico Provinciale di Potenza*. Naples.

Leveau, P. 1997. Irrigazione, drenaggio e economia demaniale romana nelle pianure del basso Rodano: L'apporto della ricerca geoarcheologica nella valle dei Baux. In *Uomo, acqua e paesaggio*, 301–310.

Lévêque, P. 1957. *Pyrrhos*. Paris.

———. 1970. Problèmes historiques de l'époque hellénistique en Grande Grèce. *Atti Taranto* 9: 29–70.

Lévêque, P., and P. Vidal-Naquet. 1996. *Cleisthenes the Athenian*. Ed. and trans., D.A. Curtis. Atlantic Highlands, NJ.

Levi, A. 1926. *Le terrecotte figurate del Museo Nazionale di Napoli*. Florence.

Levi, C. 1947. *Cristo si è fermato a Eboli*. 7th ed. Turin.

Lewis, M.J.T. 2001. *Surveying Instruments of Greece and Rome*. Cambridge.

Lezzi-Hafter, A. 1976. *Der Schuwalow-Maler. Eine Kannenwerkstatt der Parthenonzeit*. Mainz.

———. 1988. *Der Eretria-Maler. Werke und Weggefährten* Mainz

Lindholm, R.C. 1987. *A Practical Approach to Sedimentology*. London.

Lippolis, E. 1982. Le testimonianze del culto in Taranto greca. *Taras* 2: 81–135.

———. 1990. La necropoli ellenistica di Taranto. In *Emergenze e problemi*, 15–71.

———. 1994a. Il problema topografico. In *Museo Taranto* III.1, 41–66.

———. 1994b. La necropoli ellenistica. Problemi di classificazione e cronologia dei materiali. In *Museo Taranto* III.1, 238–281.

———. 1995. La documentazione archeologica. In *Culti greci in Occidente. Fonti scritte e documentazione archeologica I*. Eds. E. Lippolis, S. Garraffo, and M. Nafissi. 30–129. Taranto.

———. 1996. La ceramica a figure rosse italiota. In *Arte e artigianato*, 357–362.

———. 2006. La ceramica a figure rosse apula (tarda). In *Museo Taranto* II.2, 431–461.

Lippolis E., and M. Mazzei. 1984. Dall'ellenizzazione all'età tardorepubblicana. In *La Daunia antica. Dalla preistoria all'altomedioevo*. Ed. M. Mazzei. 185–252. Milan.

———. 2005. La ceramica apula a figure rosse: aspetti e problemi. In *Céramique apulienne*, 11–18.

Liseno, M.G. 2004. *Metaponto. Il deposito votivo Favale*. Rome.

Lissi Caronna, E. 1972. Oppido Lucano (Potenza). Rapporto preliminare sulla prima campagna di scavo (1967). *NSc* 26 488–534.

———. 1980. Oppido Lucano (Potenza). Rapporto preliminare sulla seconda campagna di scavo (1968). *NSc* Ser. 8, Vol. 34: 119-297.

———. 1983. Oppido Lucano (Potenza). Rapporto preliminare sulla terza campagna di scavo (1969). *NSc* Ser. 8, Vol. 37 215–352.

———. 1990–1991. Oppido Lucano (Potenza). Rapporto preliminare sulla quarta campagna di scavo (1970). Materiale archeologico rinvenuto nel territorio del comune. *NSc* Ser. 9, Vol. 1–2: 185–488.

———. 1998–1999. Metaponto (Matera). Scavo di parte di una fattoria in contrada Casa Teresa, proprietà Durante. *NSc* Ser. 9, Vol. 9–10: 177–205.

Livingston, R.L. 1971. Intervento. *Atti Taranto* 10: 493–497.

Lloyd, A.H. 1935. The Coin Type of Selinus and the Legend of Empedocles. *Numismatic Chronicle* 15: 73–93.

Lohmann, H. 1979. *Grabmäler auf unteritalischen Vasen.* (*Archäologische Forschungen* 7.) Berlin.
———. 1992. Agriculture and Country Life in Classical Attica. In *Agriculture in Ancient Greece*, 29–60.
———. 1993. *Atene. Forschungen zu Siedlungs- und Wirtschaftsstruktur des klassischen Attika.* 2 vols. Cologne/Weimar/Vienna.
Lombardo, M. 1982. Antileon tirannicida nella tradizione metapontina ed eracleota. *StAnt* 3: 189–205.
———. 1987a. La Magna Grecia dalla fine del V secolo a.C. alla conquista romana. In *Magna Grecia* II, 55–88.
———. 1987b. L'organizazzione militare degli Italioti. In *Magna Grecia* II, 225–280.
———. 1995. Professione medica e magia a Metaponto. *StAnt* 8.2: 95–106.
———. 1998. Siri e Metaponto: esperienze coloniali e storia sociale. In *Siritide e Metapontino*, 45–65.
London, G.A. 1987. Regionalism in Traditional Cypriot Ceramics. *PotTech* 5: 125–136.
Longo, F. 2007. La definizione di un "nuovo" spazio pubblico: l'Agora del Ceramico dalla "nascita" alla spedizione in Sicilia. In *Atene e l'Occidente: i grandi temi. Le premesse, i protagonisti, le forme della comunicazione e dell'interazione, i modi dell'intervento ateniese in Occidente.* (Atti del Convegno Internazionale, Atene, 25-27 maggio 2006.) Eds. E. Greco and M. Lombardo. 117–153. Athens.
Lo Porto, F.G. 1959–1960. Ceramica arcaica delle necropoli di Taranto. *Annuario della Scuola Archeologica ad Atene e delle Missioni Italiane in Oriente* XXI-XXII: 7-230.
———. 1964. Satyrion (Taranto). Scavi e ricerche nel luogo del più antico insediamento laconico in Puglia. *NSc* Ser. 8, Vol. 18: 177–279.
———. 1966. Metaponto. Scavi e ricerche archeologiche. *NSc* Ser. 8, Vol. 20: 136–231.
———. 1969. Metaponto – Tombe a tumulo dell'età del ferro scoperte nel suo entroterra. *NSc* Ser. 8, Vol. 23: 121–170.
———. 1970–1971. Tomba messapica di Ugento. *AttiMGrecia* N.S. 11–12: 99–152.
———. 1973. *Civiltà indigena e penetrazione greca nella Lucania orientale.* (*Monumenti Antichi* 48, Serie Miscellanea I-3.) 149–250. Rome.
———. 1977. Recenti scoperte archeologiche in Puglia. *Atti Taranto* XVI: 725–745.
———. 1977–1979. Una tomba metapontina e l'elmo di Saint Louis nel Missouri. *AttiMGrecia* N.S. 18–20: 171–187.
———. 1980. Medici pitagorici in una *defixio* greca da Metaponto. *PP* 193: 282–288.
———. 1981a. Ricerche e scoperte nell'Heraion di Metaponto. *Xenia* 1: 25–44.
———. 1981b. Metaponto (Matera). Nuovi scavi nella città e nella sua necropoli. *NSc* Ser. 8, Vol. 35: 289–391.
———. 1987. Due iscrizioni votive arcaiche dai dintorni di Taranto. *PP* 232: 39–50.
———. 1988. Testimonianze archeologiche di culti metapontini. *Xenia* 16: 5–28.
———. 1988–1989. Metaponto (Matera). Rinvenimenti nella città antica e nel suo retroterra ellenizzato. *NSc* Ser. 8, Vol. 42–43: 299–441.
———. 1991. *Timmari L'abitato, le necropoli, la stipe votiva.* Rome.
———. 1992. Abitato e necropoli di Toppicelli. In *Principi imperatori vescovi*, 72–102.
———. 2001. Vasi apuli da una tomba da Irsina nel Materano. *MedArch* 4: 91–97.
———. 2002. Scavo stratigrafico sull'acropoli di Satyrion. In *Territori tarantini*, 7–18.
Loprete, T.C. and M.P. Bini. 1989. I dischi fittili di Eraclea. In *Studi su Siris-Eraclea. Archeologia Perusina 8.* 49–74. Rome.
Loth, S.R., and M. Henneberg. 1996a. Mandibular Ramus Flexure: A New Morphologic Indicator of Sexual Dimorphism in the Human Skeleton. *AJPA* 99: 473–485.
———. 1996b. The Taung Child – It's a Boy! Sexually Dimorphic Morphology in the Immature Human Mandible and Its Application to Fossil Hominids. *AJPA* Supplement 22: 152.
———. 2001. Sexually Dimorphic Mandibular Morphology in the First Few Years of Life. *AJPA* 115: 179–186.
Loth, S.R., and M.Y. Iscan. 1994. Morphological Indicators of Skeletal Aging: Implications for Paleodemography and Paleogerontology. In *Biological Anthropology and Aging: Perspectives on Human Variation over the Life Span.* Eds. D.E. Crews and R.M. Garruto. 394–425. New York.
Lowe, J. J., and C. Watson, 1993. Lateglacial and early Holocene pollen stratigraphy of the northern Apennines, Italy. *Quaternary Science Reviews* 12: 727–738.
Lugli, G. 1963. Il sistema stradale della Magna Grecia. *Atti Taranto* 2: 23–37.
Luppino, S., E. Greco, and P.G. Guzzo. 1992. La ceramica a figure rosse. In *Laos* II, 59–78.
Luraghi, N. 1994. *Tirannidi arcaiche in Sicilia e Magna Grecia: da Panezio di Leontini alla caduta dei Dinomenidi.* Florence.
Lynch, K.M. 2001. Pelikai in Use Deposits of Athenian Wells. *Hesperia* 70.2: 171–173.
Maaskant-Kleibrink, M. 1993. Religious Activities on the "Timpone della Motta," Francavilla Marittima, and the Identification of Lagaría. *BABesch* 68: 1–47.
———. 2000. Early Cults in the *Athenaion* at Francavilla Maritima as evidence for a pre-colonial circulation of *nostoi* stories. In *Die Ägäis und das westliche Mittelmeer*, 165–184.
Maccari Poisson, B. 1984. La céramique médiévale. In *Brucato*, 247–450.

Macchioro, S. 1981. Dati storico-archeologici sul territorio di Pomarico in Basilicata durante l'età preromana. *Acme* 34.II: 513–533.

———. 1984. Ceramica apula a figure rosse da Pomarico nel Museo Nazionale Ridola di Matera. *NumAntCl* 13: 49–70.

MacDonald, B.R. 1981. The Emigration of Potters from Athens in the Late Fifth Century B. C. and Its Effect on the Attic Pottery Industry. *AJA* 85, No. 2: 159–168.

Mackenzie, W.S., and A.E. Adams. 1994. *A Colour Atlas of Rocks and Minerals in Thin Section*. London.

Macnamara, E. 1983. Greek Type Cups and Skyphoi. In *Cozzo Presepe*, 321–331.

MacPhail, R.I. 1992. Soil Micromorphological Evidence of Ancient Soil Erosion. In *Past and Present Soil Erosion: Archaeological and Geographical Perspectives*. Eds M. Bell and J. Boardman. 197–215. Oxford

Maddoli, G.F. 1974. I Sanniti a Metaponto. Un capitolo di storia lucana arcaica. *PP* 29: 237–243.

———. 1991. L'Elide in età arcaica: il processo di formazione dell'unità regionale. In *Geografia storica della Grecia antica*. Ed. F. Prontera. 150–173. Rome-Bari.

Maes, K. 1997. Tombe di età ellenistica. In *Ordona* IX, 87–208.

Maetzke, G. 1976. La ceramica di Capaccio. In *Atti del Colloquio Internazionale di Archeologia Medievale* (Palermo-Erice 1974). 535–541. Palermo.

———. 1977. Problemi relativi allo studio della ceramica nell'Italia meridionale nei secc. XI-XIII. In *Ruggero il Gran Conte e l'inizio dello stato normanno*. (*Corpus Membranarum Italicarum «Fonti e Studi»* XII.) 79–100. Rome.

Maffettone, R., et al. 1983. Edificio circolare. In *Poseidonia-Paestum* II, 87–138.

Malatesta, A., and F. Zarlenga. 1988. Evidence of Middle Pleistocene Marine Transgressions along the Mediterranean Coast. *PPP* 68: 311–315.

Mallwitz, A. 1972. *Olympia und seine Bauten*. Munich.

Malone, M.L. 1998. Terracotta Figurines. In *Necropoleis*, 771–785.

Malkin, I. 1987. *Religion and colonization in ancient Greece*. Leiden/New York.

Manacorda, D., L. Paroli, A. Molinari, M. Ricci, and D. Romei. 1986. La ceramica medioevale di Roma nella stratigrafia della Crypta Balbi. In *Ceramica medievale*, 511–544.

Mannino, K. 1996. Gli ateliers attici e la nascita della produzione figurata. In *Arte e artigianato*, 363–370.

———. 2006. *Vasi attici nei contesti della Messapia (480-350 a.C.). Beni Archeologici—Conoscenza e Tecnologie, Quaderno* 5. Bari.

———. 2008. Dalle importazioni attiche alle produzioni italiote: la documentazione nell'area apulo-lucana. *Atti Taranto* 47: 425–443.

Manzo, L. 1989. La ceramica comune. Vasellame da mescita. In *Locri* II, 327–343.

Marabini Moevs, M.T. 1973. *The Roman Thin Walled Pottery from Cosa (1948–1954)*. (*MAAR* 32).

Marchegiani, M.C. 1997. La ceramica tornita *matt-painted*. In *Pomarico Vecchio* I, 1.161–174.

Marchi, M.L. 1991. Ceramica di uso comune. In *Museo Venosa*, 171–175.

———. 2006. *Ager Venusinus*. Ville e villaggi: il paesaggio rurale in età tardoantica. In *Paesaggi e insediamenti rurali*, 173–191.

Marchi, M.L., G. Sabbatini, and M. Salvatore. 1990. Venosa: nuove acquisizioni archeologiche. In *Basilicata/espansionismo romano*, 11–23.

Marinazzo, A. 1978. Carovigno, Via Damiano Chiesa. Tombe messapiche. *RSB* XI: 204–211.

———. 1979. *La necropoli messapica di S. Lorenzo (Francavilla Fontana)*. Mesagne.

Marletta, N. 1997. Le produzioni e i commerci. In *Il Museo Archeologico Nazionale dell'Alta Val d'Agri*. 221–223. Lavello.

Martelli Cristofani, M. 1973. La ceramica greco-orientale. In *CVA Gela* II, II D.

Mårtensson, L., M.-L. Nosch, and E. Andersson Strand. 2009. Shape of things: understanding a loom weight. *Oxford Journal of Archaeology* 28: 373–398.

Martin, A. 1998. La sigillata focese (Phocean Red-Slip/Late Roman C Ware). In *Ceramica in Italia*, 109–122.

Martin, J.M. 1993. *La Pouille du VI° au XII° siècle*. Rome.

Martin, R., and K. Saller. 1957. *Lehrbuch der Anthropologie*. Stuttgart.

Martinelli, S. 1988a. Ceramica acroma. In *Forentum* I, 145–153.

———. 1988b. Decorazione impressa eseguita a stampo. In *Forentum* I, 206-212.

Martorano, F. 2001. Antonio De Lorenzo. Il significato di un riedizione degli studi di topografia reggina antica e medievale. In *Le scoperte archeologiche di Reggio Calabria (1882–1888)*. Ed. F. Martorano. 9–40. Rome.

Maruggi, G.A. 1982. Taranto. Rinvenimenti nella necropoli. *Taras* 2.1–2: 51–80.

———. 1991. Taranto, la necropoli tardo-classica ed ellenistica. In *Vecchi scavi*, 53–83.

———. 1994. La tipologia delle tombe. In *Museo Taranto* III.1, 69–129.

———. 1996a. Crispiano (Taranto), L'Amastuola. In *Ricerche sulla casa*, 197–218.

———. 1996b. Le produzioni ceramiche arcaiche. In *Arte e artigianato*, 247–268.

———. 1996c. Oria (Brindisi). *Taras* 16.1: 67–73

———. 2002. Il territorio a nord di Taranto. In *Territori tarantini*, 43–63.

Masiello, L. 1988. Le collezioni Viola. Le lucerne fittili. In *Museo di Taranto*, 81–112.

———. 1992. Lucerne di età ellenistica in Italia meridionale. *Taras* 12.1: 57–114.

———. 1994. La necropoli ellenistica: le lucerne. In *Museo Taranto* III.1, 337–352.
———. 2006. Lucerne. In *Museo Taranto* II.2, 557–560.
Mastronuzzi, G. 1995. Ricerche archeologiche a Nardò (LE). *StAnt* 8.1: 183–228.
Mastronuzzi, G., V. Melissano, and D. Spano. 2008. Ricerche sulla fase arcaica ad Otranto (LE): il cantiere Mitello *StAnt* 12: 97–149.
Matheson, S. 1995. *Polygnotos and Vase Painting in Classical Athens*. Wisconsin.
Maturo, C. 1997. Varia. In *Pomarico Vecchio* I, 1.249–252.
Mayo, M.E., and K. Hamma. 1982. *The Art of South Italy. Vases from Magna Graecia*. Richmond.
Mazzei, M. 1985. Serpente (San Severo): rinvenimento di una tomba di età ellenistica. *Taras* 5.1: 125–130.
———. 1996. Le armi. In *Arte e artigianato*, 119–134.
Mazzei, M., and L. Pietropaolo. 1995. Ceramica a vernice nera, a pasta grigia, suddipinta e a figure rosse. In *Arpi*, 245–247.
Mazzeo Saracino, L. 1985. Terra Sigillata Nord-Italica. In *Atlante* II, 175–230.
Mazzucato, O. 1985. Primo rapporto sulla ceramica medievale in Albania. *Faenza* 71: 9–19.
McBride, E.F., M.D. Picard, and R.L. Folk. 1994. Oriented Concretions, Ionian Coast, Italy: Evidence of Groundwater Flow Direction. *Journal of Sedimentary Research* A64: 535–540.
McPhee, I. 1997. Stemless Bell-kraters from Ancient Corinth. *Hesperia* 66: 100–145.
McPhee, I., and E. Pemberton. 2004. South Italian and Etruscan Red-figure Pottery from Ancient Corinth. *MedArch* 17: 55–60.
Meiggs, R. 1973. *Roman Ostia*. Oxford.
———. 1982. *Trees and timber in the ancient Mediterranean world*. Oxford.
Meirano V. 1996. Frutti, dolci e focacce in area metapontina: la documentazione coroplastica. *BBasil* 12: 68–101.
Mele, A. 1981. Il pitagorismo e le popolazioni anelleniche d'Italia *AION ArchStAnt* 3: 61–96.
———. 1987. Aristodemo, Cuma e il Lazio. In *Etruria e Lazio arcaico: Atti dell'Incontro di studio, 10–11 novembre 1986*. Ed. M. Cristofani. 155–177. Quaderni del centro di Studio per l'archeologia etrusco-italica 15. Rome.
———. 1996. Culti e miti nella storia di Metaponto. *Hesperia* 7: 9–32.
———. 1998. Culti e miti nella storia di Metaponto. In *Siritide e Metapontino*, 67–90.
———. 2009. Cuma in Opicia tra Grechi e Romani. *Atti Taranto* 48: 77–167.
Melissano, V. 1990. Ricerche archeologiche nel territorio di Cutrofiano (LE). *StAnt* 6: 257–298.
Meloni, P. 1950a. La contesa fra Taranto e Turi per il possesso della Siritide. *RendAccLinc* Ser. VIII, Vol. 5: 574–598.
———. 1950b. L'intervento di Cleonimo in Magna Grecia. *GIF* 3: 103–121.
Melton, F.A. 1956. Problems of the Photogeologist in "Flatland" Regions of Low Dip. *PE* 22.1: 52–63.
Menard, T.D.M 1990. *The Archaic and Classical Pottery from the Sanctuary at Località Santa Venera in Paestum, Italy*. Ann Arbor.
Mendel, G. 1908. *Musées impériaux ottomans. Catalogue des figurines grecques de terre cuite*. Constantinople.
Mercando, L. 1970. Matelica (Macerata). Rinvenimenti di età gallica e di età medievale. *NSc* Ser. 8, Vol. 24: 394–435.
Mercuri, A.M. 1999. Relazione sulle analisi polliniche di 5 campioni provenienti di Pizzica Metaponto. Unpublished report to ICA.
———. 2009. Analisi polliniche su campioni provenienti da Pizzica-Metaponto. Unpublished report submitted to ICA.
Merlo, V. 2003. *Contadini perfetti e cittadini agricoltori nel pensiero antico*. Milan.
Mertens, D. 1974. L'architettura. *Atti Taranto* 13: 187–235.
———. 1980a. Parallelismi strutturali nell'architettura della Magna Grecia e dell'Italia Centrale in età arcaica. In *Attività archeologica*, 37–108.
———. 1980b. Rapporto preliminare sui lavori eseguiti dall' Instituto Archeologico Germanico di Roma nell'area del santuario urbano di Metaponto. In *Metaponto* I, 313–353.
———. 1985. Metapont. Ein neuer Plan des Stadtzentrums. *AA* 1985: 645–671.
———. 1987. Note sull'architettura di Poseidonia-Paestum. Problemi e stato della ricerca. *Atti Taranto* 27: 541–574.
———. 1991. Schnurkonstruktionen. In *Bautechnik der Antike*. (*Diskussionen zur archäologischen Bauforschung* 5.) Eds. A. Hoffman, E.-L. Schwandner, W. Hoepfner, and G. Brands. 155–160. Mainz.
———. 1998. L'architettura e l'urbanistica di Metaponto nel quadro dell'economia locale e dell' evoluzione generale nella Magna Grecia. In *Siritide e Metapontino*, 123–140.
———. 1999. Metaponto: l'evoluzione del centro urbano. In *Storia della Basilicata* 1, 247–294.
———. 2006. *Città e monumenti dei Greci d'Occidente*. Rome.
———. 2010. Tavole Palatine: analisi formale e inquadramento storico-architettonico. In *Il tempio di Hera (Tavole Palatine) di Metaponto: archeologia, archeometria, conservazione*. Ed. L. Lazzarini. 19–36. Pisa-Rome.
Mertens, D., and A. De Siena. 1982. Metaponto: il teatro-ekklesiasterion. *BdA* 67, Ser. 6, Vol. 16: 1–60.
Mertens-Horn, M. 1992. Die archaischen Baufriese aus Metapont. *RM* 99: 1–122.
———. 2001. La scultura di marmo. In *Metaponto/Archeologia*, 71–88.
Miall, A.D. 1978. Lithofacies Types and Vertical Profile Models in Braided River Deposits: A Summary. In *Flu-*

vial Sedimentology. (*Canadian Society of Petroleum Geologists Memoir* 5.) Ed. A.D. Miall. 597–604. Calgary.
———. 1985. Architectural Element Analysis: A New Method of Facies Analysis Applied to Fluvial Deposits. *ESR* 22: 261–308.
Migliorini, C. 1937. Short Account of the Geological Conditions and of the Oil Prospection of a Zone in Southern Italy. *Actes et documents du 2me Congrès Mondial du Pétrole (Paris)*. 1.587–591. Paris.
———. 1952. Lo stato odierno della conoscenza del sottosuolo della Fossa Bradanica. *Atti del VII Convegno Nazionale del Metano e del Petroleo* Vol. 1. (Taormina, 21-24 aprile 1952.) 183–188. Palermo.
Milanese, M. 2001. Geridu. *Archeologia e storia di un villaggio medievale in Sardegna*. Sassari.
———, ed. 2006. *Vita e morte dei villaggi rurali tra medioevo ed età moderna. Dallo scavo della villa de Geriti ad una pianificazione della tutela e della conoscenza dei villaggi abbandonati della Sardegna*. (Atti del Convegno, Sassari-Sorso, 28-29 maggio 2001; *Quaderni del Centro di Documentazione dei Villaggi Abbandonati della Sardegna/QUAVAS* 2.) Florence.
Milanese, M., M. Baldassari, F. Campus, A. Deiana, M. Fiori, G. Gattiglia, E. Marcasciano, G.J. Mullen, A. Panetta, and L. Sanna. 2000. Il villaggio medievale di Geridu, ricerche 1997–1999. In *II Congresso Nazionale di Archeologia Medievale* (Brescia, 28 settembre–1 ottobre 2000; Florence). Ed. G.P. Brogiolo. 254–263.
Miller, M. 1997. *Athens and Persia in the Fifth C. BC*. Cambridge.
Minar, E.L. 1942. *Early Pythagorean Politics in Practice and Theory*. Baltimore.
Ministero dei Lavori Pubblici. 1961. Eventi meteorici del novembre 1959. In *Annali Idrologici* 1959, Part 1. 65–79. Rome.
Miroslav Marin, M. 1982. Scavi del 1929. In *Ceglie Peuceta* I, 77–157.
Moggi, M. 1991. Sinecismi arcaici del Peloponneso. In *La transizione dal Miceneo all'Alto Arcaismo: dal palazzo alla città*. (Atti del convegno internazionale, Roma, 14–19 marzo 1988.) Ed. D. Musti. 155–165. Rome.
Molinari, A. 1997. *Segesta* II. *Il castello e la moschea* (*Scavi 1989-1995*). Palermo.
Mollard, J.D. 1957. Aerial Photographs Aid Petroleum Search. *COGI* 10.7: 91–96.
Mollard-Besques, S. 1954. *Musée national du Louvre. Catalogue raisonné des figurines et reliefs en terre-cuite grecs, étrusques et romains* I. Paris.
———. 1963. *Musée national du Louvre. Catalogue raisonné des figurines et reliefs en terre-cuite grecs et romains* II. Myrina. Paris.
Molnar, S. 1971. Human Tooth Wear, Tooth Function, and Cultural Variability. *AJPA* 34: 175–190.
Montanaro, A.C. 2007. *Ruvo di Puglia e il suo territorio. Le necropoli. I corredi funerari tra la documentazione del XIX secolo e gli scavi moderni*. (*Studia Archaeologica* 160.) Rome
Montepaone, C. 1986. L'apologia di Alexidamos. "L'avventura del cavaliere." *Métis* 1.2: 219–235.
Morano, M. 1994. *Storia di una società rurale. La Basilicata nell'Ottocento*. Bari.
Moore, F.G., trans. 1984. *Livy with an English Translation VI (Books XXIII–XXV)*. Cambridge, MA/London: Loeb Classical Library.
Moore, J.A. 1961. *Selections from the Greek Elegiac, Iambic, and Lyric poets*. Cambridge, MA.
Morard T. 2002. *Les Troyens à Métaponte. Étude d'une nouvelle Ilioupersis de la céramique italiote*. Mainz.
Morel, J.-P. 1965. *Céramique à vernis noir du Forum Romain et du Palatin*. Paris.
——— 1970a. Fouilles à Cozzo Presepe, près de Métaponte. *MEFRA* 82: 73–116.
——— 1970b. Garaguso. *MEFRA* 83: 556–572.
——— 1980. La céramique campanienne: acquis et problèmes. In *Céramiques hellénistiques* I, 85–122.
——— 1981a. *Céramique campanienne: les formes*. (*Bibliotheque des Écoles francaises d'Athenes et de Rome* 244.)
——— 1981b. La produzione della ceramica campana: aspetti economici e sociali. In *Società romana* II, 81–97.
——— 1989. Observations sur l'artisanat dans le Salento à l'époque hellénistique et romaine. In *Salento Porta d'Italia*. (Atti del Convegno Internazionale, Lecce, 27-30 novembre 1986.) 111–119. Galatina.
Morel, J.-P., and M. Picon. 1994. Les céramiques étrusco-campaniennes: recherches en laboratoire. In *Ceramica romana e archeometria*, 23–46.
Moretti J.-C. 2001. *Théatre et souété dans la Grèce antique: une archéologie des pratiques théâtrales*. Paris.
Moreschini, D. 1988. Catalogo delle tombe arcaico-classiche della necropoli di Via Argentina a Taranto. *ASNSP* Ser. 3, Vol. 18.3: 891–936.
Morgan, C. 1993. Review of C. Vatin, *Monuments Votifs de Delphes* (Archaeologia Perusina 10; Archaeologia 100). *CR* N.S. 43: 458.
———. 2003. *Early Greek States*. London.
Morgan, C., and J.J. Coulton. 1997. The Polis as a Political Entity. In *The Polis as an Urban Center and as a Political Community*. (*Acts of the Copenhagen Polis Centre* 4.) Ed. M.H. Hansen. 87–144. Copenhagen.
Morgan, C., and J.M. Hall. 1996. Achaian Poleis and Achaian Colonisation. In *Introduction to an Inventory of Poleis*. (*Acts of the Copenhagen Polis Centre* 3.) Ed. M.H. Hansen. 164–232. Copenhagen.
Morgan, C., and T. Whitelaw. 1991. Pots and Politics: Ceramic Evidence for the Rise of the Argive State. *AJA* 95.1: 79–108.

Moritz, L.A. 1958. *Grain-mills and Flour in Classical Antiquity.* Oxford.

Mörner, N.-A. 1979–1980. The Northwestern European "Sea-Level Laboratory" and Regional Holocene Eustasy. *PPP* 29: 281–300.

Morrone M. 1998. I materiali. In *Il castello di Santa Severina. Ricerche archeologiche.* Ed. R. Spadea. 124–127. Catanzaro.

Morter, J., and J. Hall. 1998. Dating of Tombs. In *Necropoleis*, 449–454.

Morter, J., and J.R. Leonard. 1998. Storage Amphorae. In *Necropoleis*, 731–747.

Munzi, P. 1999. Laos: aspetti di vita quotidiana attraverso lo studio del materiale ceramico. In *Terra degli Enotri*, 91–98.

———. 2007. Un contesto arcaico da Cuma: le ceramiche decorate, non figurate, di produzione locale. In *Ceramiche fini*, 109–130.

Musti, D. 1983. Metaponto: note sulla tradizione storica. *RivFil* 111: 265–291.

———. 1988. *Strabone e la Magna Grecia. Città e popoli dell'Italia antica.* Padua.

Myrto, H. 1995. Un luogo di culto con altare a gradini a Durazzo. *StAnt* 8.1: 259–274.

Nafissi, M. 1989. Distribution and trade. In C.M. Stibbe. *Laconian Mixing Bowls. A history of the kraters Lakonikos from the seventh to the fifth century BC.* 68–88. Amsterdam.

National Imagery and Mapping Agency. 1989. *The Universal Grids: Universal Transverse Mercator (UTM) and Universal Polar Stereographic (UPS).* (TM8358.2.) Bethesda.

———. 1990. *Datums, Ellipsoids, Grids, and Grid Reference Systems.* (TM8358.1.) Bethesda.

———. 1997. *Department of Defense World Geodetic System 1984: Its Definition and Relationships with Local Geodetic Systems.* (TR8350.2.) 3rd ed. Bethesda.

Naumann, R. 1958. *Palinuro. Ergebnisse der Ausgrabungen* I. *Topographie und Architektur.* Heidelberg.

Naumann, R., and B. Neutsch. 1960. *Palinuro. Ergebnisse der Ausgrabungen* II. *Nekropole: Terrassenzone und Einzelfunde.* Heidelberg.

Nava, M.L. 1997. Attività archeologica in Basilicata. *Atti Taranto* 36: 481–483.

———. 1999. L'attività archeologica in Basilicata nel 1998. *Atti Taranto* 38: 689–732.

———. 2000. L'attività archeologica in Basilicata nel 1999. *Atti Taranto* 39: 675–726.

———. 2001. L'attività archeologica in Basilicata nel 2000. *Atti Taranto* 40: 939–980.

———. 2002. L'attività archeologica in Basilicata nel 2001. *Atti Taranto* 41: 719–765.

———. 2003a. L'attività archeologica in Basilicata nel 2002. *Atti Taranto* 42: 653–717.

———. 2003b. Metodologie integrate di tutela archeologica. Il caso della Basilicata. In *Lo sguardo di Icaro*, 129–141.

———. 2004. L'attività archeologica in Basilicata nel 2003. *Atti Taranto* 43: 933–1000.

———. 2005. L'attività archeologica in Basilicata nel 2004. *Atti Taranto* 44: 313–386.

Neboit, R. 1975. *Plateaux ed collines de Lucanie orientale et des Pouilles: étude morphologique.* Doctoral dissertation, University of Paris. Paris.

———. 1977. Un example de morphogenese accélérée dans l'antiquite: les vallées du Basento et du Cavone en Lucanie (Italie). *Méditerranée* 31.4: 39–50.

———. 1980. Morphogenese et occupation humaine dans l'antiquite. *BAGF* 466: 21–27.

———. 1983. *L'homme et l'erosion.* Clermont-Ferrand.

———. 1984. Érosion des sols et colonisation grecque en Sicile et en Grande Grèce. *BAGF* 499: 5–13.

Neils, J. 1997. "Lost" and Found: Adam Buck's Wedding of Dionysos. In *Athenian Potters and Painters*, 231–240.

Neutsch, B. 1967. *Archäologische Forschungen in Lukanien* II: *Herakleiastudien.* (*Mitteilungen des Deutschen Archäologischen Instituts, Römische Abteilung.* Ergänzungsheft 11.) Heidelberg.

———. 1968. Neue archäologische Entdeckungen in Siris und Herakleia. *AA* 1968: 753–796.

———. 1981. Documenti artistici del santuario di Demetra a Policoro. *Atti Taranto* 20: 149–173.

Nevett, L.C. 1999. *House and Society in the Ancient Greek World.* Cambridge.

Nicholls, R.V. 1952. Type, Group, and Series: A Reconsideration of Some Coroplastic Fundamentals. *BSA* 47: 217–226.

Nikiforova, K.V. 1978. Status of the Boundary between Pliocene and Pleistocene. In *Contributions to the Geologic Time Scale.* (*American Association of Petroleum Geologists Studies in Geology* 6.) Eds. G.V. Cohee, M.F. Glaessner, and H.D. Hedberg. 171–178. Tulsa.

Nir, D. 1983. *Man, A Geomorphological Agent.* Jerusalem.

Nitti, A. 1983. *Il Museo Archeologico di Mesagne. Catalogo.* (Quaderni del Museo Archeologico di Mesagne.) Mesagne.

Noe, S.P., and A. Johnston. 1984. *The Coinage of Metapontum*, Parts 1 and 2. (*American Numismatic Society Numismatic Notes and Monographs* 32 and 47.) New York.

Novak, S.A. 2000. Battle-related Trauma. In *Blood Red Roses: The Archaeology of a Mass Grave from the Battle of Towton, AD 1461.* Eds. V. Fiorato, A. Boylston, and C. Knüsel. 90–102. Oxford.

Novellis, D. 2003. Santa Maria del Castello (Castrovillari – CS): un nuovo santuario rurale ai confini della chora di Sibari? *Polis* I.1: 11–63.

Nowicka, M. 1975. *Les maisons à tour dans le monde grec.* (Academia Scientiarum Polona, Bibliotheca Antiqua 15.) Warsaw.

Noyé, G. 1976. La céramique médiévale de Pietra San Giovanni (Basilicata). *MEFRM* 88: 693–743.
———. 1984. Les enseignements géographique des fouilles du chateau de Metaponte. *BAGF* 499: 21–23.
———. 1987. Quelques données sur les technique de construction en Italie centro-meridionale (X-XII siècle). In *Artistes, artisans et production artistiques au Moyen Age*. Ed. X. Barral y Altet. 275–306. Paris.
———. 1988. Quelques observations sur l'évolution de l'habitat en Calabre du V° au XI° siècles. *StBiz* 25: 57–138.
Nuzzi, D. 2008. *Rituale funerario*. Tesi di specializzazione, Università degli Studi di Bari.
Oakley, J.H. 1988. Attic red-figured skyphoi of Corinthian shape. *Hesperia* 57: 165–191.
———. 1997. *The Achilles Painter*. Mainz-am-Rhein.
Ogniben, L. 1969. Schema introduttivo alla geologia del confine calabro-lucano. *MSGLI* 8: 453–763.
Olbrich, G. 1976. Ein Heiligtum der Artemis Metapontina? Zur Ikonographie der Terrakotta-Figuren von S. Biagio bei Metapont. *PP* 31: 376–408.
———. 1979. *Archaische Statuetten eines metapontiner Heiligtums*. (*Studia Archaeologica* 23.) Rome.
———. 1986. Friese und Pinakes aus Magna Grecia. *PP* 41: 122–152.
———. 1996. Verwandtes un Benachbartes zu den archaischen Terrakotten. In *Quellheiligtum der Demeter*, 181–190.
Olcese, G. 2004. Anfore greco-italiche antiche: alcune osservazioni sull'origine e sulla circolazione alla luce di recenti ricerche archeologiche ed archeometriche. In *Metodi e approcci archeologici: l'industria e il commercio nell'Italia antica / Archeological Methods and Approaches: Industry and Commerce in Ancient Italy*. (*BAR-IS* 1262.) Eds. E. De Sena and H. Dessales. 173–192. Oxford.
Oleson, J.P. 1974. Underwater Survey and Excavation in the Port of Pyrgi (Santa Severa). *JFA* 4, 3: 297–308.
Olivero Ferrero, G. 1989. La ceramica a vernice nera. Il vasellame di IV e III strato. In *Locri* II, 71–106.
Oldfather, C.H., trans. 1956. *Diodorus of Sicily IV (Books IX–XII)*. Cambridge, MA/London: Loeb Classical Library.
Olson, G.W., compiler. 1976. *Criteria for Making and Interpreting a Soil Profile Description: A Compilation of the Official USDA Procedure and Nomenclature for Describing Soils*. Lawrence, KA.
O'Neill, K., W. Yielding, J. Near, J.E. Coleman, P.S. Wren, and K.M. Quinn. 1999. Halai: The 1992–1994 Field Seasons. *Hesperia* 68.3: 285–341.
Orlandini, P. 1966. Lo scavo del Thesmophorion di Bitalemi e il culto delle divinità ctonie a Gela. *Kokalos* 12: 8–35.
———. 1972. Aspetti dell'arte indigena in Magna Grecia. *Atti Taranto* 11: 273–308.
———. 1976. Scavi archeologici in località Incoronata presso Metaponto. *Acme* 29.1: 29–39.
———. 1986. Incoronata—Scavi dell'Università Statale di Milano. In *Greci sul Basento*, 29–39.
———. 1999. La colonizzazione ionica della Siritide. In *Storia della Basilicata* 1, 197–210.
Orsi, P. 1916. Caulonia. Campagne archeologiche del 1912, 1913 e 1915. *MonAnt* 23: 685–947.
Orton, C., P. Tyers, and A. Vince. 1993. *Pottery in Archaeology*. Cambridge/New York.
Osanna, M. 1988. Ceramica di produzione coloniale arcaica a bande e a vernice rossa. In *Forentum* I, 154–157.
———. 1992. *Chorai coloniali da Taranto a Locri. Documentazione archeologica e ricostruzione storica*. Rome.
———. 1996. Metaponto. In *Arte e artigianato*, 44–49.
———. 1999. Territorio coloniale e frontiera. La documentazione archeologica. *Atti Taranto* 37: 273–292.
———. 2001. Fattorie e villaggi in Magna Grecia. *Atti Taranto* 40: 203–220.
———. 2008. L'attività archeologica in Basilicata nel 2007. *Atti Taranto* 47: 911–944.
Osborne, R. 1985a. Buildings and Residence on the Land in Classical and Hellenistic Greece: The Contribution of Epigraphy. *BSA* 80: 119–128.
———. 1985b. *Demos. The Discovery of Classical Attika*. Cambridge.
———. 1986. Island Towers: The Case of Thasos. *BSA* 81: 166–178.
———. 1987. *Classical Landscape with Figures: The Ancient Greek City and Its Countryside*. London: G. Philip.
———. 1988. Social and Economic Implications of the Leasing of Land and Property in Classical and Hellenistic Greece. *Chiron* 18: 279–323.
———. 1992. "Is It a Farm?" The Definition of Agricultural Sites and Settlements in Ancient Greece. In *Agriculture in Ancient Greece*, 21–27.
———. 1996. *Greece in the Making, 1200–479 BC*. London/New York: Routledge.
———. 1998. Early Greek colonization? The nature of Greek settlement in the West. In *Archaic Greece: New Approaches and New Evidence*. Eds. N. Fisher and H. Van Wees. 251–269. London.
Osmaston, H. 1980. Patterns in Trees, Rivers, and Rocks in the Mulu Park, Sarawak. *GeoJ* 146: 33–49.
O'Sullivan, D., and D. Unwin. 2003. *Geographic Information Analysis*. Hoboken.
Otto, B. 1996. Die grossen Göttinnen des Quellheiligtums von Herakleia. In *Quellheiligtum der Demeter*, 141–151.
———. 2005. Il santuario sorgivo di Siris-Herakleia nell'odierno comune di Policoro. In *Spazio del rito*, 5–18.
Otvos, E.G. 1981. Tectonic Lineaments of Pliocene and Quaternary Shorelines. *NGCG* 9: 398–404.
Palazzo, P. 1990. Brindisi. Località "La Rosa": i reperti

anforam. *Taras* 10.1: 141–156.

———. 1992. Nuovi rinvenimenti in località "La Rosa" – Brindisi. *Taras* 12.1: 115–130.

———. 1993. Anse con bollo rinvenute in loc. Rosa (Brindisi). *Taras* 13.1–2: 227–236.

Palmentola, F. 2006. Un gruppo di kylikes a figure e a decorazione nera. *Archeologia Classica* 57: 321–336.

Palumbo, M.R. 1986. *Le terrecotte figurate di tipo greco in Daunia, Peucezia e Messapia*. Lecce.

Pancrazzi, O. 1979. Passim in *Cavallino*, 51-215.

Panella, C. 1993. Merci e scambi nel Mediterraneo tardoantico. In *Storia di Roma* 3.2. *I luoghi e le culture*. Eds. A. Carandini, L. Cracco Ruggini, and A. Giardina. 613–697. Turin.

———. 1996. Lo studio delle ceramiche comuni di età romana: qualche riflessione. In *Céramiques communes*, 9–15.

Pannuzi, S., and A.R. Staffa. 1997. Le più antiche produzioni di maiolica in Abruzzo (secc. XIII-XIV). In *La protomaiolica*, 97–119.

Papanastasiou, A. 2004. *Relations Between Red-Figured and Black-Glazed Vases in Athens of the 4th Century BC.* (BAR International Series 1297.) London.

Parente, A.R. 1999. Rinvenimenti tombali di età preromana in Lucania. *In Trouvailles monétaires de tombes.* (Actes du deuxième colloque international du Groupe suisse pour l'étude des trouvailles monétaires, Neuchâtel, 3-4 mars 1995.) Eds. O.F. Dubuis, S. Frey-Kupper, and G. Perret. 141–149. Lausanne.

———. 2009. Per un'economia del territorio in Lucania di IV e III sec. a.C.: la documentazione numismatica. In *Verso la città. Forme insediative in Lucania nel mondo italico fra IV e III sec. a.C.* (Atti delle giornate di studio, Venosa, 13-14 maggio 2006.) Ed. M. Osanna. 45–61. Venosa.

Parker, R. 1987. Festivals of the Attic Demes. In *Gifts to the Gods: Proceedings of the Uppsala Symposium 1985*. Eds. T. Linders and G. Nordquist. 137–147. Uppsala.

———. 1996 *Athenian Religion: A History*. Oxford.

Patera, A. 1986. Tomba a camera tarantina. *Taras* 6.1–2: 31–64.

Patitucci Uggeri, S. 1976. Saggio stratigrafico nell'area di S. Pietro degli Schiavoni a Brindisi: relazione preliminare 1975–76. *RS* 9: 133–200.

———. 1977. *La ceramica medievale pugliese alla luce degli scavi di Mesagne*. Mesagne.

———. 1979. Protomaiolica brindisina Gruppo I. *Faenza* 65.6: 241–253.

———. 1984. Per una revisione della protomaiolica, il contributo degli scavi di Brindisi. In *San Lorenzo Maggiore*, 391–414.

———. 1995. La nuova ceramica dell'età federiciana: la protomaiolica. In *Federico II e l'Italia. Percorsi, luoghi, segni e strumenti*. 111–123. Rome.

Paton, W.R., trans. 1916–1918. *The Greek Anthology with an English Trnaslation*. Loeb Classical Library. Cambridge, MA: Harvard University Press.

Patterson, H. 2001. The Pottery. In *San Vincenzo al Volturno* 3. *The Finds from the 1980-86 Excavations*. Eds. J. Mitchell and I.L. Hansen. 297–322. Spoleto.

Patterson, H., and D. Whitehouse. 1992. Medieval Domestic Pottery. In *Excavations at Otranto* II. *The Finds*. Eds. F. D'Andria and D. Whitehouse. 87–195. Galatina.

Pavolini, C. 1981. Le lucerne nell'Italia romana. In *Società romana* II, 139–184 and 278–288.

———. 1982. Ambiente e illuminazione. Grecia e Italia tra il VII e il III sec. a.C. *Opus* I, 2: 291–313.

———. 1988. Le lucerne romane fra il III sec. a.C. e il III sec. d.C. In *Céramiques hellénistiques* II, 139–166.

———. 1994. Lucerne italiche tardo-repubblicane. In *Ancient and Traditional Ceramics*. (Seminar held at the European University Centre for Cultural Heritage, Ravello, March 19-24, 1990.) Ed. T. Hackens. 77–88. Rixensart.

———. 2000. Illuminazione, archeologia della. In *Dizionario di archeologia*. Eds. R. Francovich and D. Manacorda. 168–172. Bari.

Payne, H. 1931. *Necrocorinthia. A Study of Corinthian Art in the Archaic Period*. Oxford.

Peacock, D.P.S. 1982. *Pottery in the Roman World: An Ethno-archaeological Approach*. London/New York.

Peacock, D.P.S., and M.G. Fulford. 1984. *Excavations at Carthage: The British Mission I.2. The Avenue du President Habib Bourguiba, Salammbo: The Pottery and Other Ceramic Objects from the Site*. London.

Peacock, D.P.S., and D.F. Williams. 1986. *Amphorae and the Roman Economy. An Introductory Guide*. London/New York.

Pečírka, J. 1973. Homestead Farms in Classical and Hellenistic Hellas. In *Problèmes de la terre*, 113–147.

Pedio, T. 1990. *Centri scomparsi in Basilicata*. Venosa.

Pedley, J., and J. Sevink. 1985. Excavations at Paestum 1984. *AJA* 89.1: 53–60.

Pedley, J. and M. Torelli. 1984. Excavations at Paestum 1983. *AJA* 88: 367–376.

Peduto, P. 1979. Archeologia medievale in Campania. *La voce della Campania* 7: 247–262.

———. 1980. La ceramica di un sepolcreto d'età longobarda in Altavilla Silentina, provincia di Salerno. *Faenza* 66: 229–232.

———. 1994. La Campania. In *Storia dell'alto Medioevo*, 279–297.

Pelagatti, P. 1970. Akrai (Siracusa). Ricerche nel territorio. *NSc* Ser. 8, Vol. 24: 436–499.

———. 1980–1981. L'attività della Soprintendenza alle Antichità della Sicilia Orientale. *Kokalos* 26–27: 694–736.

Pelosi, A. 1991. Dinamiche territoriali del VII secolo a.C. nell'area sirite-metapontina. *DialArch* Ser. 3, Vol. 9.1–2: 49–74.

Pelosi, A. and R. Cantilena. 1989. Area Galiano B. In *Laos* I, 97–100.

Peluso, P. 1992. L'opus doliare. In *Locri* IV, 257–262.

Pérez Ballester, J. 2002. *Vasos sobrepintados italiotas del Museo Arqueológico Nacional de Madrid*. Murcia.

Peroni, R. and F. Trucco, eds. 1994. *Enotri e Micenei nella Sibaritide*. Taranto.

Perrone, M.L. 2003. Il deposito votivo del tempio di Giunone Lucina a Norba. In *Santuari e luoghi di culto nell'Italia antica*. (*Atlante Tematico di Topografia Antica* 12.) Eds. L. Quilici and S. Quilici Gigli. 353–384. Rome.

Pesce, G. 1936. Metaponto. Ritrovamenti varii. *NSc* Ser. 6, vol. 12.10–12: 439–449.

Petitot, B.R. 2000. *Catalogue des lampes grecques et romaines. Les collections du Musée de l'Arles Antiques*. Arles.

Petsas, P. 1966. Ο τάφοζ των Λευκαδίων. Athens.

Pettegrew, D.K. 2001. Chasing the Classical Farmstead: Assessing the Formation and Signature of Rural Settlement in Greek Landscape Archaeology. *JMA* 14.2: 189–209.

Pettijohn, F.J. 1963. *Chemical Composition of Sandstones – Excluding Carbonate and Volcanic Sands*. (*USGS Professional Paper* 440–S.) Washington, D.C.

Petitti, P.M. 1997. Le lucerne. In *Pomarico Vecchio* I, 1: 147–154.

Pianu, G. 1990. *La necropoli meridionale di Eraclea 1. Le tombe di secolo IV e III a.C.* Rome.

Piccarreta, F. 2003. Fotografia aerea archeologica: istituzioni e storia. In *Lo sguardo di Icaro*, 67–74.

Pieri, M. 1966. Tentativo di ricostruzione paleogeografico-strutturale dell'Italia centro-meridionale. *GeolRom* 5: 407–424.

Pieri, P., L. Sabato, and M. Tropeano. 1996. Significato geodinamico dei caratteri deposizionali e strutturali della Fossa bradanica nel Pleistocene. *MSGLI* 51: 501–515.

Pierobon, R. 1980. Il materiale. *NSc* Ser. 8, Vol. 34: 353–388.

Pierro, E. 1984. *Ceramica "ionica" non figurata e coppe attiche a figure nere*. Rome.

Piggott, S. 1965. *Ancient Europe from the Beginnings of Agriculture to Classical Antiquity*. Chicago.

———. 1983. *The Earliest Wheeled Transport: From the Atlantic Coast to the Caspian Sea*. London.

Pimpl, H. 1997. *Perirrhanteria und Louteria. Entwicklung und Verwendung großer Marmor- und Kalksteinbecken auf figürlichem und säulenartigem Untersatz in Greiechenland*. Berlin.

Pinto-Cisternas, J., J. Moggi-Cecchi, and E. Pacciani. 1995. A Morphological Variant of the Permanent Upper Lateral Incisor in Two Tuscan Samples from Different Periods. In *Aspects of Dental Biology: Palaeontology, Anthropology and Evolution*. Ed. J. Moggi-Cecchi. 333–339. Florence.

Pirazzoli, P.A. 1976. Sea Level Variations in the Northwest Mediterranean during Roman Times. *Science* 194: 519–521.

———. 1991. *World Atlas of Holocene Sea-Level Changes*. Amsterdam.

Pirzio Biroli Stefanelli, L. 1977. Tabelle fittili tarantine relative al culto dei Dioscuri. *ArchCl* 29.2: 310–398.

Pizzo, M. 1995a. Bacini fittili. In *Incoronata* 3, 93–95.

———. 1995b. Ceramica comune. In *Incoronata* 3, 97–100.

———. 1998–1999. Vassallaggi (S. Cataldo, Caltanissetta). La necropoli meridionale, scavi 1956. *NSc* Ser. 9, Vol. 9–10: 207–388.

Plana-Mallart, R. 1994. *La chora d'Emporion. Paysage et structures agraires dans le nord-est catalan à la période pré-romaine*. (*Annales Littéraires de l'Université de Besançon* 544. *Centre de Recherches d'Histoire Ancienne* 137. *Espaces et paysages* 2.) Paris.

———. 2001. *D'emporion à Emporion: la colonie et son territoire*. *Atti Taranto* 40: 545–566.

Pomar, L., and M. Tropeano. 2001. The Calcarenite di Gravina Formation in Matera (Southern Italy): New Insights for Coarse-grained, Large-scale, Cross-bedded Bodies Encased in Offshore Deposits. *AAPG* 85.4: 661–689.

Pontrandolfi, A. 1999. *Storia della Bonifica Metapontina*. Matera and Rome.

Pontrandolfo, A. 1996. Per un'archeologia dei Lucani. In *Greci, enotri e lucani*, 171–181.

———. 2000. La ceramica attica di IV secolo in area tirrenica. In *La céramique attique du IVe siècle en Méditerranée occidentale*. (Actes du colloque international organisé par le Centre Camille Jullien. Arles, 7–9 décembre 1995.) 121–130. Naples.

———. 2004. Il mondo "indigeno." *Atti Taranto* 44: 83–109.

Pontrandolfo, G.A. and Rouveret, A. 1992. *Le tombe dipinte di Paestum*. Modena.

Pope, K.O., and T.H. Van Andel. 1984. Late Quaternary Alluviation and Soil Formation in the Southern Argolid: Its History, Causes and Archaeological Implications. *JAS* 11: 281–306.

Poso, C.D. 1988. *Il Salento normanno*. Galatina.

Postrioti, G. 1996. *La stipe votiva del Tempio "E" di Metaponto*. (*Corpus delle stipi votive in Italia* IX. Regio III.2. *Archaeologica* 117.) Rome.

Potter, T.W. 1987. *Roman Italy*. London

Pouzadoux, C. 2002. Tra mondo greco e mondo lucano. Immagine e mito nella ceramografia del IV secolo a.C. In *Immagine e mito*, 121–128.

———. 2009. La céramique à figures rouges. In *Civita di Tricarico* I. *Le quartier del la Maison du Monolithe et l'enceinte intermédiare*. Ed. O. De Cazanove. 350–371. Rome.

Prag, A.J.N.W. 1983a. Black-glazed Ware. In *Cozzo Presepe*, 351–364.

———. 1983b. The Terracottas. In *Cozzo Presepe*, 379–381.

———. 1992a. Attic Black Glaze. In *Gravina*, II.66–67.

———. 1992b. Black Glaze Ware. In *Gravina*, II.67–131.

———. 1992c. Grey Glaze Ware. In *Gravina*, II.134–155.

———. 1992d. The Lamps. In *Gravina*, II.209–218.

Prag, A.J.N.W., and M. Broughton. 1992. Unguentaria. In *Gravina*, II.155–156.

Prag, A.J.N.W., F. Schwizer, and J.L.W Williams. 1974. Hellenistic Glazed Wares from Athens and Southern Italy: Analytical Techniques and Implications. *Archaeometry* 16: 153–187.

Praturlon, A. 1980. Cadre geologique general de l'Italie su place dans le cadre européen et mediterranéen. In *Introduction a la geologie generale d'Italie.* (*Guide a l'excursion* 122°. XXVI Congrès Géologique International, Paris.) 3–14. Paris.

Prazzoli S. 1996. La ceramica a vernice nera. Taranto: la documentazione del pozzo di via C. Nitti. In *Arte e artigianato*, 333–336.

Preacco, M.C. 1989. La ceramica a vernice nera. Il vasellame del I strato. In *Locri* II, 195–255.

Preacco Ancona, M.C. 1992. La ceramica fine della "Casa dei Leoni." In *Locri* IV, 113–171,

———. 1996. La ceramica a vernice nera: La Calabria. In *Arte e artigianato*, 346–353.

———. 1997. La ceramica sovraddipinta. In *Pomarico Vecchio* I, 1.127–137.

———. 1999a. La ceramica a vernice nera. In *Oppido Mamertina*, 270–289.

———. 1999b. La ceramica a pasta grigia. In *Oppido Mamertina*, 291–292.

———. 1999c. Le lucerne. In *Oppido Mamertina*, 297–298.

Price, M.J. 1968. Early Greek Bronze Coinage. In *Essays in Greek Coinage Presented to Stanley Robinson*. Eds. C.M. Kraay and G.K. Jenkins. 90–104. Oxford.

Prieto, A. 2000. Field Survey 2000, Metaponto. In *Ancient Territories 2000*, 15–22.

———. 2005. *Landscape Organization in Magna Graecia.* Doctoral dissertation, The University of Texas at Austin. Austin

Prieto, A., and W. Polleichtner. 2008. Erodoto e Siri. *Atti Taranto* 47: 181–195.

Pritchett, W.K. 1953. The Attic Stelai, Part I. *Hesperia* 22.3: 225–299.

———. 1956. The Attic Stelai, Part II. *Hesperia* 25.3: 210–317.

Prohászka, M. 1998. Metal Objects and Coins. In *Necropoleis*, 787–834.

Pucci, G. Terra Sigillata Italica. In *Atlante* II, 359–406.

Pugliese Carratelli, G. 1988. L'orfismo in Magna Grecia. In *Magna Grecia* III. *Vita religiosa e cultura letteraria, filosofica e scientifica*. Ed. G. Pugliese Carratelli. 159–170. Milan.

———. 1999. "Siris" e Metaponto. In *Storia della Basilicata* 1, 183–195.

Purcell, N. 1994. South Italy in the Fourth Century BC. In *The Cambridge Ancient History VI. The Fourth Century BC.* Eds. D.M. Lewis, J. Boardman, S. Hornblower, and M. Ostwald. 381–403. Cambridge.

Quercia, A. 2004. Appendix 3: la ceramica da fuoco dello scarico 1 dal *kerameikos* di Metaponto. In *I sostegni di fornace dal kerameikos di Metaponto.* (*Beni Archeologici – Conoscenza e Tecnologie* Quaderno 3.) Ed. V. Cracolici. 175–199. Bari.

———. 2008. I residui organici nella ceramica. Stato degli studi e prospettive di ricerca. In *Uomini piante e animali nella dimensione del sacro*. Eds. F. D'Andria, J. De Grossi Mazzorin, and G. Fiorentino. 209–213. Bari.

Quilici, L. 1967. *Siris-Heraclea* (*Forma Italiae*, Regio III, Vol. I). Rome.

Quilici, L., and S. Quilici Gigli, eds. 2002. *Carta archeologica della valle del Sinni 2. Da Valsinni a San Giorgio Lucano e Cersosimo.* (*Atlante Tematico di Topografia Antica* Suppl. X/2000-2002.) Rome.

———, eds. 2003. *Carta archeologica della valle del Sinni 7. Da Episcopia a Latronico a Seluci e Monte Sirino.* (*Atlante Tematico di Topografia Antica* Suppl. X/2000-2002.) Rome.

Quilici Gigli, S. 1997. L'irreggimentazione delle acque nella trasformazione del paesaggio agrario dell'Italia centro-tirrenica. In *Uomo, acqua e paesaggio*, 193–212.

Rabinowitz, A. 2004. *Symposium, Community, and Cultural Exchange in Archaic Sicily and South Italy.* Doctoral dissertation, University of Michigan. Ann Arbor.

———. 2009. Drinking from the same cup: Sparta and late Archaic commensality. In *Sparta: Comparative Approaches*. Ed. S. Hodkinson. 113–192. Swansea.

Race, W.H., trans. 1997. *Pindar, Olympian Odes, and Pythian Odes*. Loeb Classical Library.

Rackham, H. 1932. *Aristotle: Poetics*. Loeb Classical Library. Cambridge, Mass./London.

Radi, G. 1999. Il neolitico. In *Storia della Basilicata* 1, 31–65.

Raffi, S. 1986. The Significance of Marine Boleal Molluscs in the Early Pleistocene Faunas of the Mediterranean Area. *PPP* 52: 267–289.

Raffy, J., B. Dumas, P. Gueremy, and R. Lhenaff. 1981. Uplift and Quaternary Marine Terraces to the East of Villa San Giovanni (Calabria, Italy). *ZG*, N.F. Suppl. 40: 119–125.

Rathbone, D.W. 2003. The Control and Exploitation of the *Ager* publicus in Italy under the Roman Republic. In *Tâches publiques et entreprise privée dans le monde romain.* Ed. J.-J. Aubert. (*Actes du Diplôme d'Etudes Avancées Universités de Neuchâtel et de Lausanne.*) 135–178.

Raviola, F. 1993. Fra continuità e cambiamento: Atene, Reggio e Leontini. In *Hesperìa. Studi sulla grecità di Occidente* 3. Ed. L. Braccasi. 85–97. Rome.

Rendell, H. 1982. Clay Hillslope Erosion Rates in the Basento Valley, S. Italy. *GeogAnn* 64A: 141–147.

———. 1986. Soil Erosion and Land Degradation in Southern Italy. In *Desertification in Europe*. (Proceedings of the Informational Symposium in the EEC Programme on Climatology, held in Mytilene, Greece, 15-18 April 1984.) Eds. R. Fantechi and N.S. Margaris. 184–193. Dordrecht.

Rheo Bumbalova, M. 1979. *La collezione Meo-Evoli. Ceramica italiota a figure rosse*. Fasano.

Riavez, P. 2000. Atlit-ceramica RMR. In *Ceramica invetriata*, 207–222.

Riccardi, A. 1989a. Fase IIb. L'edificio tardo-arcaico (seconda metà VI-metà IV sec. a.C.). In *Monte Sannace*, 132–154.

———. 1989b. Scavo G: Settore 3. In *Monte Sannace*, 65–78.

———. 2002–2003. Bitonto (Bari). *Taras* 23, 1-2: 65-78.

———. 2003. *Gli antichi Peucezi a Bitonto. Documenti ed immagini dalla necropoli di via Traiana*. Bari: Edipuglia.

———. 2006. Ceramica a fasce e di stile misto. In *Museo Taranto* II.2, 351-385.

———. 2008. *Donne e guerrieri da Ruvo e Bitonto. Le scoperte del III millennio. Catalogo del Museo archeologico della fondazione De Palo-Ungaro* 2. Bari.

Ricchetti, E. 1992a. Ceramica a fasce. In *Principi imperatori vescovi*, 240–242.

———. 1992b. Ceramica a vernice nera. In *Principi imperatori vescovi*, 249-256.

———. 1992c. Ceramica nello stile di Gnathia. In *Principi imperatori vescovi*, 257–258.

Ricchetti, G. 1967. Lineamenti geologici e morfologici della media valle del fiume Bradano. *BSGLI* 86: 607–622.

Ricci, A. 1985. Ceramica a pareti sottili. In *Atlante* II, 231–357.

Ricci, M. 1998. La ceramica comune dal contesto di VII secolo della Crypta Balbi. In *Ceramica in Italia*, 351–382.

Rice, P.M. 1984. Change and Conservatism in Pottery-producing Systems. In *The Many Dimensions of Pottery*. Ed. S.E. Van der Leeuw. 231–294. Amsterdam.

Richter, G.M.A. 1968. *Engraved Gems of the Greeks, Etruscans, and Romans* I. London.

Rinaldi, M. 2005. Ceramica comune. In *Torre di Satriano* I, 222–239.

Rizza, G. 1955. Leontini. Campagne di scavo 1950–51 e 1951–52: la necropoli della Valle San Mauro, le fortificazioni meridionali della città e la porta di Siracusa. *NSc*: 281–376.

Rizzitelli, C. 2000. Ceramiche sigillate italiche ed africane dall'area della domus A. In *Ordona* X, 267–284.

Rizzo, M.A. 1990. *Le anfore da trasporto e il commercio etrusco arcaico. Complessi tombali dall'Etruria meridionale*. Roma.

Robinson, E.G.D. 1990. Between Greek and Native: The Xenon Group. In *Greek Colonists and Native Populations*, 251–266.

———. 1996. La ceramica sovraddipinta monocroma: vasi del Gruppi di Xenon e del Cigno Rosso. In *Arte e artigianato*, 447–452.

———. 2004. Reception of Comic Theatre amongst the Indigenous South Italians. *MedArch* 17: 193–212.

———. Forthcoming. Archaeometric Analysis of Ceramics. In *Contestualizzare la ceramica greca*. (Atti della Summer School, Cavallino 2001.) Eds. F. D'Andria and G. Semeraro. Bari.

Robinson, E.S.G. 1964. Carthaginian and Other South Italian Coinages of the Second Punic War. *Numismatic Chronicle* Ser. 7, Vol. 4: 37–64.

Robinson, E.W. 1997. *The First Democracies: Early Popular Government Outside Athens*. (*Historia* Einzelschriften 107.) Stuttgart.

Rocchietti, D. 2002. *Aree sepolcrali a Metaponto. Corredi e ideologia funeraria fra VI e III secolo a.C*. Potenza.

———. 2003. *Reperti vascolari della necropoli di Metaponto già conservati al Museo di Reggio Calabria*. Tesi di specializzazione, Università degli Studi di Torino.

Roda, C. 1967. Diario dei lavori (riunione di lavoro sul Plio-Pleistocene). *AAGSNC* Ser. 6, 18: 13–25.

Rolley, C. 1981. Siris. Le problème artistique. *Atti Taranto* 20: 175–195.

Romei, D. 1986. La ceramica medioevale di Roma nella stratigrafia della Crypta Balbi: ceramica acroma depurata. In *Ceramica medievale*, 523–529.

———. 1992. Ceramica invetriata monocroma verde. In A. De Crescenzo, I. Pastore, and D. Romei, *Ceramiche invetriate e smaltate del Castello di Salerno dal XII al XV secolo*. 15–37. Naples.

———. 2001. Il deposito di X secolo nell'esedra della Crypta Balbi. Ceramica a vetrina pesante e ceramica a vetrina sparsa. In *Roma dall'antichità al Medioevo. Archeologia e storia*. Eds. M.S. Arena, P. Delogu, L. Paroli, M. Ricci, L. Saguì, and L. Vendittelli. 524–527. Milan.

Rossi, F. 1989. Fase III. Dall'ellenismo alla romanizzazione (seconda metà del IV secolo a.C. – I secolo d.C.). In *Monte Sannace*, 155–210.

———. 1992. Ipogei e tombe di Via Lavello. In *Principi imperatori vescovi*, 482–490.

Rossi, L. 1992. Tomba a cassa di Via Cerignola. In *Principi imperatori vescovi*, 156–160.

Rossi, L. and van der Wielen-van Ommeren, F. 1983. *Canosa II. Ritrovamenti sulla strada Canosa-Ofanto. Un corredo canosino*. Bari.

Rossi, S., C. Auroux, and J. Mascle. 1983. The Gulf of

Taranto (Southern Italy): Seismic Stratigraphy and Shallow Structure. *MarGeol* 51: 327–346.
Rossignani, M.P. 1973. Ceramica comune. In *Scavi di Luni*. 410–418. Rome.
Rossiter J. 1977. The Lamps. In *Monte Irsi*, 199–202.
Röthlisberger, F. 1976. Gletscher- und Klimaschwankungen im Raum Zermatt, Ferpècle und Arolla. *Les Alpes* 52 année 3–4: 59–152.
Roubis, D. 1996. Ricerche archeologiche nell'abitato indigeno di Difesa San Biagio (Montescaglioso). In *Ricerche sulla casa*, 235–253.
Rouveret, A. 1978. Le classi dei materiali. In *Leuca*, 91–115.
———. 1983. Plateia An. In *Poseidonia-Paestum* II, 146–151.
Rubinich, M. 1992. La ceramica fine della frequentazione arcaica e del sacello. In *Locri* IV, 85–111.
Rüdiger U. 1967. Santa Maria d'Anglona (Com. Tursi, Prov. Matera). Rapporto preliminare sulle due campagne di scavi negli anni 1965 e 1966. *NSc* Ser. 8, Vol. 21: 331–353.
Rupp, D.W., T. Gregory, K.W. Schaar, J. Possey, J. Moody, D. Sullivan, C. Runnels, and J. Wickers. 1983. Archaeological Survey: Definitions. In *Archaeological Survey*, 17–30.
Rupp, D.W., and D.R. Keller. 1983. Introduction. In *Archaeological Survey*, 4–6.
Russo, A. 1988. Ceramica a vernice rossa e bruna parziale. In *Forentum* I, 213–220.
———. 1993a. Mancamasone - complesso rurale. In *Da Leukania a Lucania*, 30–32.
———. 1993b. Moltone di Tolve – complesso residenziale. In *Da Leukania a Lucania*, 39–42.
———. 1999. Il ruolo dell'acqua nei luoghi sacri della Basilicata antica. In *Archeologia dell'acqua*, 127–158.
Russo, A. and H. Di Giuseppe, eds. 2008. *Felicitas Temporum. Dalla terra alle genti: la Basilicata settentrionale tra a archeologia e storia*. Lavello.
Russo, G. 1996. La ceramica a vernice nera. Taranto. Catalogo. In *Arte e artigianato*, 329–332.
Russo Tagliente, A. 1992. Edilizia domestica. In *Apulia e Lucania. Ellenizzazione e società nella tipologia abitativa indigena tra VIII e III secolo a.C.* Galatina.
———. 1995. *Armento. Archeologia di un centro indigeno. BdArch* 35–36.
Rutter, N.K., ed. 2001. *Historia Nummorum*. Italy. London.
Sabato, L. 1996. Delta calcareo terrazzato nella Calcarenite di Gravina (Pleistocene inferiore). Minervino, Murge nord-occidentali. *MSGLI* 51: 517–526.
Sabbione, C. 1982. Le aree di colonizzazione di Crotone e Locri Epizefiri nell'VIII e VII sec. a.C. *ASAA* 60 (N.S. 46): 251–300.
———. 1984. L'artigianato artistico. *Atti Taranto* 23: 245–301.
Sacchi, C. 1990. Problemi storico-archeologici della Siritide e del Metapontino tra VIII e VII secolo. *PP* 251: 135–160.
Saint-Non, J.C.R. 1781–1782. *Voyage pittoresque ou description des Royaumes de Naples et de Sicilie*. Paris.
Sallares, R. 2002. *Malaria and Rome: A History of Malaria in Ancient Italy*. Oxford.
Salmon, E.T. 1970. *Roman Colonization under the Republic*. Ithaca.
Salvatore, M.R. 1980. Rinvenimenti ceramici sotto la Cattedrale di Bari. In *Atti del X Convegno Internazionale della Ceramica*. 153–172. Albisola.
———. 1982a. La ceramica altomedievale nell'Italia meridionale: stato e prospettive della ricerca. *ArchMed* 9: 47–66.
———. 1982b. Ceramiche medievali dal castello di Bari. In *Atti del XI Convengno Internazionale della Ceramica*. 81–93. Albisola.
———. 1984. Ceramica medievale da Policoro (Basilicata). In *San Lorenzo Maggiore*, 429–449.
———. 1985. Il Medioevo. In *Museo Policoro*, 141–145.
Sammis, J.W. 1968. *The Gnathian Pelike: Local Centers of Production*. Doctoral dissertation, University of Pennsylvania. Philadelphia.
Sapelli, M. 1979. Lucerne fittili delle civiche raccolte archeologiche. Milan.
Saprykin, S. 1994. *Ancient Farms and Land-plots on the Khora of Khersonesos Taurike*. Amsterdam.
Sartori, F. 1953. *Problemi di storia costituzionale italiota*. Rome.
———. 1967. Eraclea di Lucania: profilo storico. In *Herakleiastudien*, 16–95.
Saunders, S.L., and J. du Plat Taylor. 1992. Wheel-made Painted Ware. In *Gravina*, II.13–41.
Savelli, S. 2006. Le anfore da trasporto. In *Cuma*, 103–126.
Scarano, G. 1992. Le necropoli di Metaponto in età ellenistica. *BBasil* 8: 17–23.
Scarfi, B.M. 1961. Gioia del Colle (Bari). Scavi nella zona di Monte Sannace. Le tombe rinvenute nel 1957. *Monumenti di Antichità* 45: 145–332.
Schauenburg, K. 1995. Zur Chevrongruppe. *RM* 102: 63–69.
———. 2000a. Some Remarks on the Underworld Painter. In *Periplous. Papers on Classical Art and Archaeology Presented to Sir John Boardman*. Eds. G.R. Tsetskhladze, A.J.N.W. Prag, and A.M. Snodgrass. 253–262. New York.
———. 2000b. *Studien zur unteritalischen Vasenmalerei* II. Kiel.
———. 2005. *Studien zur unteritalischen Vasenmalerei*, Band VII-VIII. Kiel.
Scheibler, I. 1964. Exaleiptra. *JdI* 79: 72–108.

Schlager, H., and U. Rudiger. 1967. S. Maria d'Anglona (Com. Tursi, Prov. Matera). *NSc* Ser. 8, Vol. 21: 347–348.

Schlotzhauer, U. 2000. Die südionischen Knickrandschalen: Formen und Entwicklung der sog. Ionischen Schalen in archaischer Zeit. In *Die Ägäis und das westliche Mittelmeer*, 407–416.

Schmidt, E. 1994. *Katalog der antiken Terrakotten, Martin-von-Wagner-Museum der Universität Würzburg* I. Mainz.

Schmiedt, G., and R. Chevallier. 1959a. Caulonia e Metaponto: applicazioni della fotografia aerea in ricerche di topografia antica nella Magna Grecia. *L'Universo* 39.2: 349–370 and 39.5: 993–1032.

———. 1959b. *Caulonia e Metaponto: applicazioni della fotografia aerea in ricerche di topografia antica nella Magna Grecia*. Firenze.

Schnapp, A., A. Pelosi, and R. Cantilena. 1989. Area Galiano B. In *Laos* I, 97–100.

Schneider, R. 1985. Analyse palinologique dand l'Aspromonte en Calabre (Italie meridionale). *Cahiers Ligures de Prehistoire et de Protohistoire* 2: 278–288.

Schneider-Herrmann, G. 1971. Gruppe früh-klassischer tarantiner Terrakottaköpfe. *BABesch* 46: 138–143.

———. 1977. *Apulian Red-figured Paterae with Flat or Knobbed Handles*. (*BICS* 34.) London.

Schofield, A.J. 1991. Interpreting Artefact Scatters: An Introduction. In *Ploughzone Archaeology*, 3–8.

Schojer, T. 2001. Ginosa Marina (Taranto). *Taras* 21.1: 124–126.

———. 2002. Il N.W. tarantino. In *Territori tarantini*, 65–86.

Schon, R. 2000. On a Site and Out of Sight: Where Have Our Data Gone? (Response to Bintliff, J.L., P. Howard, and A.M. Snodgrass, "The Hidden Landscape of Prehistoric Greece," *JMA* 12.2: 139–168.) *JMA* 13.1: 107–111.

Schultz, C.B. 1941. The Pipy Concretions of the Arikaree. *NUB* 2.8: 69–82.

Schumm, S.A. 1956. Evolution of Drainage Systems and Slopes in Badlands at Perth Amboy, New Jersey. *BGSA* 67: 597–646.

———. 1977. *The Fluvial System*. New York.

Schürmann, W. 1989. *Katalog der antiken Terrakotten in Badischen Landesmuseum Karlsruhe*. (*SIMA* 84.) Göteborg.

Scionti, R., and P. Tarentini. 1990. Manduria. Emergenze archeologiche tra preistoria e medioevo. In *Emergenze e problemi*, 127–292.

Sconfienza, R. 1996. Sistemi idraulici in Magna Grecia: classificazione preliminare e proposte interpretative. *Bollettino Storico della Basilicata* 12: 25–66.

———. 2005. Fortificazioni tardo classiche ed ellenistiche in Magna Grecia. *BAR Int. Series 1341*: 33–40.

Scott, E.C. 1979. Dental Wear Scoring Technique. *AJPA* 51: 213–218.

Segre, A.G. 1978. Il pleistocene ed il paleolitico della Basilicata. In *Atti della XX Riunione Scientifica dell'Istituto Italiano di Preistoria e Protostoria in Basilicata*. 15–39. Florence.

Selby, M.J. 1993. *Hillslope Materials and Processes*. 2nd ed. Oxford.

Selli, R. 1962. Le Quaternaire marin du versant adriaticionien de la péninsule italienne. *Quaternaria* 6: 391–411.

———. 1977. The Neogene/Quaternary Boundary in the Italian Marine Formations. *GGeol* 41: 81–105.

Semeraro, G. 1983. Otranto dal VI sec. a.C. all'età ellenistica (Scavi 1977-1979). *StAnt* 4: 125–212.

———. 1996. Ceramica geometrica e orientalizzante. In *Arte e artigianato*, 269–280.

———. 1997. Ἐν νηυσί. *Ceramica greca e società nel Salento arcaico*. Bari.

Sestieri, P.C. 1940. Metaponto. Campagna di scavi (marzo– aprile 1939). *NSc* Ser. 7, Vol. 1: 51–122.

———. 1940–1941. Note sulla plastica di Metaponto. *Le Arti* 3: 92–100.

Shear, Jr., T.L. 1978. Tyrants and Buildings in Archaic Athens. In *Athens Comes of Age: From Solon to Salamis*. 1–19. Princeton.

Shepherd, G. 2005. Dead Men Tell No Tales: Ethnic Diversity in Sicilian Colonies and the Evidence of the Cemeteries. *OJA* 24.2.: 115–136.

Sichtermann, H. 1966. *Griechische Vasen in Unteritalien*. Tübingen.

Siciliano, A., and G. Sarcinelli. 2004. Metapontum-Siris/Herakleia. La documentazione numismatica. In *Presenza e funzioni della moneta nelle chorai delle colonie greche dall'Iberia al Mar Nero*. (Atti del XII convegno organizzato dall'Università "Federico II" e dal centro internazionale di studi numismatici, Napoli, 16–17 giugno 2000.) 247–304. Rome.

Signore, M.G. 1996. Rilievi fittili con recumbente dal ceramico di Metaponto. *StAnt* 9: 299–359.

Silvestrelli, F. 2004. L'archeologia della produzione in Italia meridionale. Il caso di Metaponto. In *Metodi e pratica della cultura materiale. Produzione e consumo dei manufatti*. Ed. R. Giannichedda. 107–116. Bordighera.

———. 2005. Le fasi iniziali della ceramica a figure rosse nel kerameikos di Metaponto. In *Céramique apulienne*, 113–123.

———. 2008. La distribuzione della ceramica a figure rosse dei Pittori di Creusa, di Dolone e dell'*anabates*. In *La collana e le perle. Studi offerti a Mario Torelli per il suo settantesimo compleanno*. 271–292. Venosa.

———. Forthcoming a. Il kerameikos di Metaponto. In *Contestualizzare la ceramica greca*. (Atti della Summer School, Cavallino 2001.) Eds. F. D'Andria and G. Semeraro. Bari.

———. Forthcoming b. *Ricerche sul kerameikos di Metaponto. Impianti artigianali e modi di produzione dall'età arcaica all'ellenismo*. Bari.

———. Forthcoming c. Figured Pottery. In *The Study of Ancient Territories. Chersonesos and Metaponto. 2008 Annual Report, Institute of Classical Archaeology*. Ed. J.C. Carter. Austin.
Siviero, G.B. 1972. I bacini delle chiese ferraresi. In *Atti del V Convegno Internazionale della Ceramica*. 243–259. Albisola.
Skibo, J.M. 1992. *Pottery Function. A Use-alteration Perspective*. New York/London.
Slapsak, B., and B. Kirigin. 2001. Pharos and Its Chora. *Atti Taranto* 40: 567–591.
Small, A.M. 1977a. Site B. In *Monte Irsi*, 11–56.
———. 1977b. Wheel Made Painted Wares. In *Monte Irsi*, 114-117.
———. 1977c. Black Glaze. In *Monte Irsi*, 117-132.
———. 1977d. Cooking Pots. In *Monte Irsi*, 133-134.
———. 1977e. Plain Wares. In *Monte Irsi*, 134-136.
———. 1991. Late Roman Rural Settlement in Basilicata and Western Apulia. In *Roman Landscapes. Archeological Survey in the Mediterranean Region*. (*Archaeological Monographs of the British School at Rome* 2.) Eds. G. Barker and J. Lloyd. 204–222. London.
———. 1992a. Botromagno: An Introduction. In *Gravina*, I.3-18.
———. 1992b. The Field Survey of 1965. In *Gravina*, I.25–27.
———. 1999. L'occupazione del territorio in età romana. In *Storia della Basilicata* 1, 559–600.
———. 2006. Le analisi al radiocarbonio e la fine della villa di S. Giovanni di Ruoti. In *Paesaggi e insediamenti rurali*, 127–131.
Small, A.M., B. Roe, J.W. Hayes, C.J. Simpson, G. Guzzetta, M. Mackinnon, and S.G. Monckton. 1994. A Pit Group of c. 80-70 BC from Gravina di Puglia. *PBSR* 49: 197–260.
Smith, A.G., and N.H. Woodcock. 1982. Tectonic Syntheses of the Alpine-Mediterranean Region: A Review. In *Alpine-Mediterranean Geodynamics*. (Geodynamics Series No. 7.) Eds. H. Berckhemer and K. Hsü. 15–38. Washington, D.C.
Smith, C.F., trans. 1958. *Thucydides with an English Translation IV (Books VII and VIII)*. Cambridge, MA/London: Loeb Classical Library.
Snodgrass, A.M. 1998. Rural Burial in the World of Cities. In *Et pouvoir: idéologies, pratiques et interprétations*. (Actes du Colloque "Theories de la Nécropole Antique," Lyon 21–25 janvier 1995.) Eds. S. Marchegay, M.-T. Le Dinahet, and J.-F. Salles. 37–41. Lyons.
Snyder, J.P. 1987. *Map Projections: A Working Manual*. (*USGS Professional Publication 1835*.) Washington, DC.
Soil Survey Staff, 1999. *Soil Taxonomy: A Basic System of Soil Classification for Making and Interpreting Soil Surveys*, Second Edition. Agriculture Handbook Number 436, Natural Resources Conservation Service, United States Department of Agriculture.
Söldner, M. 2007. ΒΙΟΣ ΕΥΔΑΙΜΩΝ. *Zur Ikonographie des Menschen in der rotfigurigen Unteritaliens. Die Bilder aus Lukanien*. Möhnsee.
Solovyov, S.L. 1999. *Ancient Berezan: The Architecture, History, and Culture of the First Greek Colony in the Northern Black Sea*. Eds. J. Boardman and G. Tsetskhladze. *Colloquia Pontica* 4. Leiden.
Sorrenti, G. 2001. Battisteri paleocristiani della Basilicata: il caso di Metaponto. In *L'edificio battesimale in Italia. Aspetti e problemi*. (Atti dell'VIII Congresso Nazionale di Archeologia Cristiana, Genova, Sarzana, Albenga, Finale Ligure, Ventimiglia, 21-26 settembre 1998.) 1131–1148. Bordighera.
Spadea, R. 1984. La Topografia. *Atti Taranto* 23: 119–165.
Stanley Price, N., et al. 1983. Archaeological Survey: Theory and Methodology. In *Archaeological Survey*, 31–44.
Stazio, A. 1974a. Osservazioni sulla monetazione di Metaponto. *Atti Taranto* 13: 67–106.
———. 1974b. Il dibattito. *Atti Taranto* 13: 333–334.
———. 1999. Le emissioni monetarie dei centri greci. In *Storia della Basilicata* 1, 455–469.
Stea, G. 1991. La ceramica grigia di VII sec. a.C. dall'Incoronata di Metaponto. *MEFRA* 103.2: 405–442.
———. 1999. Forme della presenza greca sull'arco ionico della Basilicata: tra emporìa e apoikìai. In Κοινὰ, 49–71.
Stevenson, R.B K. 1947. The Neolithic Cultures of South-east Italy. *PPS*, N.S. 13: 85–100.
Stewart. A. 1990. *Greek Sculpture*. New Haven and London.
Sthamer, E. 1995. *L'amministrazione dei castelli nel regno di Sicilia sotto Federico II e Carlo I d'Angiò*. Bari.
Stibbe, C.M. 1986. Il cratere laconico. In *Studi sulla ceramica laconica* (Atti del Seminario. Perugia, 23–24 febbraio 1981). 75–89.
———. 1989. *Laconian Mixing Bowls. A history of the kraters Lakonikos from the seventh to the fifth century BC* Amsterdam.
Stillwell MacKay, T. 1967. More Byzantine and Frankish Pottery from Corinth. *Hesperia* 36.3: 249–320.
Stone, S.C., III. 1982. *Roman Pottery from Morgantina in Sicily*. Doctoral dissertation, University of Michigan. Ann Arbor.
Stoop, J.M.W. 1974–1976. Francavilla Marittima: acropoli sulla Motta. *AttiMGrecia* 15–17: 107–168.
Strzheletskiy, S.F. 1961. *Klery Khersonesa Tavricheskogo: Khersonesski Sbornik* 6. Simferopol.
Sullivan, D. 1983. Preliminary Results of Pollen Analysis, Pizzica-Pantanello, 1982–1983. In *Territory of Metaponto*, 35.
Surra, A., and M.C. Preacco Ancona. 1997. La ceramica a vernice nera: forme aperte. In *Pomarico Vecchio* I, 1.65–82.

Tagliente, M. 1982. Lo scavo 1967: le classi di materiali dagli ambienti. *StAnt* 3: 13–45.

———. 1986. Policoro: nuovi scavi nell'area di Siris. In *Siris-Polieion*, 129–133; Pls. 25–30.

———. 2006. L'attività archeologica in Basilicata nel 2005. *Atti Taranto* 45: 725–753.

Tagliente, M., and M. Lombardo. 1985. Nuovi documenti su Pisticci in età arcaica. *PP* 222: 284–307.

Tagliente, P. 1996–1997. *La ceramica medievale invetriata dallo scavo dell'abitato di Torre di Mare (Metaponto). Saggi I e II. 1995.* Tesi di Specializzazione, Università degli Studi della Basilicata.

———. 2000a. La ceramica invetriata policroma nel basso Salento. In *Ceramica invetriata*, 167–182.

———. 2000b. Ceramica bicroma e a doppio bagno in Terra d'Otranto. In *Atti del XXXIII Convegno Internazionale della Ceramica*. 383–390. Albisola.

———. 2001. La Terra d'Otranto e la Grecia. Contatti culturali attraverso le fonti archeologiche dell'età medievale: la ceramica invetriata policroma e la protomaiolica. In *Archeologia e tradizioni popolari*. 461–484. Lecce.

———. 2002. Lecce: uno scarico di fornaci della fine del Quattrocento. Primi dati. *ArchMed* 29: 543–555.

———. 2003a. La ceramica da cucina. In *Medioevo e Rinascimento al Castello Carlo V di Lecce*. Eds. P. Arthur, A. Bramato, P. Tagliente, and B. Vetere. 41. Lavello.

———. 2003b. La ceramica dell'età di transizione nella Puglia meridionale: la fine delle «RMR» e l'inizio delle produzioni graffite. In *Atti del III Convegno Nazionale di Archeologia Medievale*. Ed. S. Gelichi. 306–311. Florence.

———. 2004. Il casale medievale di Sombrino attraverso l'evidenza della ceramica invetriata. In *Supersano. Un paesaggio antico del basso Salento*. Eds. P. Arthur and V. Melissano. 85–99. Galatina.

Taliercio Mensitieri, M. 1989. La riduzione ponderale in Magna Grecia. *DialArch* Ser. 7, Vol. 3: 31–52.

———. 1999. Le emissioni monetarie dei Lucani. In *Storia della Basilicata* I, 471–485.

Tamime, A.Y., D.G. Dalgleish, and W. Banks. 1991. Introduction. In *Feta and Related Cheeses*. Eds. R.K. Robinson and A.Y. Tamime. 11–48. West Sussex.

Tassinari, S. 1993. *Il vasellame bronzeo di Pompei*. Rome.

Taylor, D.M. 1957. Cosa: Black-glaze Pottery. *MAAR* 25: 65–193.

Taylor, J. 2000. Cultural Depositional Processes and Post-depositional Problems. In *Ploughsoil Assemblages*, 16–26.

Tempesta, A.L. 1995. Antefisse fittili da Metaponto. *BBasil* 11: 41–69.

———. 2001. Commercio e artigianato. In *Metaponto/Archeologia*, 89–102.

———. 2002. La casa. In *I Greci* 4.II, 1125–1227.

———. 2005. Attestazioni del mito di Io e del culto di Dioniso a Metaponto: alcune considerazioni. In *Spazio del rito*, 83–91.

Terrenato, N. 2000. The Visibility of Sites and the Interpretation of Field Survey Results: Towards an Analysis of Incomplete Distributions. In *Ploughsoil Assemblages*, 60–71.

Theodorescu, D. 1988. Elements d'urbanisme et de topographie. État actuel et perspectives. *Atti Taranto* 27: 501–538.

Theseider, Dupré, L. 1992. Le lucerne. In *Locri* IV, 186–203.

Thompson, D.B. 1963. *Troy. Supplementary Monograph 3. The Terracotta Figurines of the Hellenistic Period.* Princeton.

Thompson, H.A. 1934. Two Centuries of Hellenistic Pottery. *Hesperia* 3.4: 311–480.

———. 1988. Building for a More Democratic Society. The Athenian Agora after Ephialtes. In Πρακτικά του 12ου Διεθνούς Συνεδρίου Κλασικής Αρχαιολογίας (Αθήνα, 4-11 Σεπτεμβρίου 1983). 4 vols. 4.198–204. Athens.

Thompson, S.M. 2000. The Still Hidden Landscape. (Response to Bintliff, J.L., P. Howard, and A.M. Snodgrass, "The Hidden Landscape of Prehistoric Greece," *JMA* 12.2: 139–168.) *JMA* 13.1: 111–115.

———. 2001. The Metaponto Archaeological Survey 2001. In *Ancient Territories 2001*, 72–80.

———. 2004. Side-by-side and Back-to-front: Exploring Intra-regional Latitudinal and Longitudinal Comparability in Survey Data. Three Case Studies from Metaponto, Southern Italy. In *Side-by-side Survey*, 65–85.

Threipland, L.M., and M. Torelli. 1970. A Semi-subterranean Etruscan Building in the Casale Pian Roseto (Veii) Area. *PBSR* 38: 62–121.

Thulin, C., ed. 1971. *Corpus agrimensorum romanorum*. Stuttgart.

Tigano, G. 2000. Isolato S. Via Industriale. Lo scavo e i primi dati sui materiali. In *Da Zancle a Messina. Un percorso archeologico attraverso gli scavi*. 123–155. Messina.

Tinè Bertocchi, F. 1985. *Le necropoli daunie di Ascoli Satriano e di Arpi*. Genova.

Tinelli, M. 2000–2001. *Frammenti di vita quotidiana dal casale medievale di Quattro Macine (Giuggianello, LE).* Tesi di laurea, Università degli Studi di Lecce.

Tocco, G. 1991. La villa ellenistica di Monte Moltone. *BdArch* 9: 89–94.

Tocco Sciarelli, G. 1981. L'espansione di Siris tra l'Agri e il Sinni. *Atti Taranto* 20: 223–235.

Todd, J.E. 1896. Log-like Concretions and Fossil Shores. *AmerGeol* 17: 347–349.

Tomay, L., P. Munzi, and M. Gentile. 1996. Ceramiche di produzione locale. In *Santuari della Magna Grecia in Calabria*. Ed. E. Lattanzi. 213–220. Naples.

Torelli, M. 1977. Greci e indigeni in Magna Grecia: ideologia religiosa e rapporti di classe. *StStor* 18: 45–61.

Toxey, A. 1998. Black-glazed Tomb Markers. In *Necropoleis*, 692–693.

Toxey, A.P., and J.C. Carter. 1998. Unglazed Pottery. In *Necropoleis*, 719–730.

Toynbee, A. 1965. *Hannibal's Legacy: The Hannibalic War's Effects on Roman Life*. London.

Tramutoli, M., T. Pescatore, M.R. Senatore, and L. Mirabile. 1984. Interpretation of Reflection High-Resolution Seismic Profiles Through the Gulf of Taranto. *BOTA* 2: 33–52.

Trendall, A.D. 1960. Il Pittore del Ciclope. *AttiMGrecia* N.S.3: 85–92.

———. 1989. *Red-figure Vases of South Italy and Sicily: A Handbook*. London.

———. 1990. On the Divergence of South Italian from Attic Red-figure Vase-painting. In *Greek Colonists and Native Populations*, 217–230.

Tricart, J., and A. Cailleux. 1972. *Introduction to Climatic Geomorphology*. Trans. C.J. Kieweit de Jonge. New York.

Trotter, M., and G.C. Gleser. 1952. Estimation of Stature from Long Bones of American Whites and Negroes. *AJPA* 10: 463–514.

———. 1977. Corrigenda: Estimation of Stature from Long Limb Bones of American Whites and Negroes. *AJPA* 47: 355–356.

Trump, D. 1966. *Central and Southern Italy Before Rome*. New York.

Turchiano, M. 2000. La cisterna e il suo contesto. Materiali tardoantichi dalla domus B. In *Ordona* X, 343–386.

———. 2002. Aspetti della circolazione delle merci nell'*Apulia* tardoantica, tra importazioni e produzioni locali: la ceramica sigillata africana. In *L'Africa romana*, 858–870.

Turner, II, C.G., C.R. Nichol, and G.R. Scott. 1991. Scoring Procedures for Key Morphological Traits of the Permanent Dentition: The Arizona State University Dental Anthropology System. In *Advances in Dental Anthropology*. Eds. M.A. Kelley and C.S. Larsen. 13–31. New York.

Uggeri, G. 1969. Κλῆροι arcaici e bonifica classica nella χώρα di Metaponto. *PP* 124: 51–71.

Ugolini, D. 1983. Tra perirrhanteria, louteria e thymiateria. *MEFRA* 95: 449–472.

Uguzzoni, A., and F. Ghinatti. 1968. *Le tavole greche di Eraclea*. Rome.

Vagnetti, L., ed. 1982. *Magna Grecia e mondo miceneo: Nuovi documenti*. Taranto.

Vail, P.R., R.M. Mitchum, Jr., and S. Thompson III. 1977. Seismic Stratigraphy and Global Changes in Sea Level, Part 4: Global Cycles of Relative Changes in Sea Level. In *Seismic Stratigraphy: Applications to Hydrocarbon Exploration*. (*Memoirs of the American Association of Petroleum Geologists* 26.) Ed. C.E. Payton. 83–98. Tulsa, OK.

Valduga, A. 1973. Fossa Bradanica. In *Geologia dell'Italia*. Ed. A. Desio. 692–695. Turin.

Valentini, M. 2004. *L'insediamento altomedievale nelle campagne toscane*. Florence.

Valentini, G., C. Cherubini, and F.M. Guadagno. 1979. Caratteristiche geotecniche dei sedimenti argillosi pleistocenici tra Pisticci ed il mare. In *Giornate di studio*, 569–599.

Vallet, G. 1968. La cité e son territoire dans le colonies grecques d'Occident. *Atti Taranto* 7: 67–142.

———. 1974. Le "stenopos" des Muses à Métaponte. In *Mélanges de philosophie, de littérature et d'histoire ancienne offerts à Pierre Boyancé*. (*Collection de l'École française de Rome* 22.) 749–759. Rome.

Vallet, G., and F. Villard. 1955. Mégara Hyblaea V. Lampes du VIIe siècle et chronologie des coupes ioniennes. *MEFRA* 67: 7–34.

Van Andel, T.H., T.W. Jacobsen, J.B. Jolly, and N. Lianos. 1980. Late Quaternary History of the Coastal Zone near Franchthi Cave, Southern Argolid, Greece. *JFA* 7: 389–402.

Van Andel, T.H., and E. Zangger. 1990. Landscape Stability and Destabilisation in the Prehistory of Greece. In *Man's Role*, 139–157.

Van Andel, T.H., E. Zangger, and A. Demitrack. 1990. Land Use and Soil Erosion in Prehistoric and Historical Greece. *JFA* 17.4: 379–396.

Van Bemmelen, R.W. 1975. Driving Forces of Mediterranean Orogeny (Tyrrhenian Test-Case). In *Geology of Italy*, I.263–292.

Van Compernolle, T. 1996. Coppe di tipo ionico. In *Arte e artigianato*, 299–305.

———. 2000. Les céramiques ioniennes en Méditerranée centrale. In *Ceràmiques jònies d'època arcaica: centres de producció i comercialització al Mediterrani occidental*. (Actes de la taula rodona celebrada a Empùries, 26-28 de maig 1999; *Monografies Emporitanes* 11.) Eds. P. Cabrera Bonet and M. Santos Retolaza. 89–100. Barcelona.

Vanderhoeven, M. 1976. La terre sigillée lisse. Campagnes de fouilles de 1963 à 1974. In *Ordona* V, 79–182.

———. 1988. La terre sigillée. Campagnes de fouilles de 1976 à 1986. In *Ordona* VIII, 85–218.

Vandermersch, C. 1994. *Vins et amphores de Grande Grèce et de Sicile, IVe-IIIe s. avant J.-C.* (*Études du Centre Jean Bérard* 1.) Naples.

Vanderpool, E., J.R. McCredie, and A. Steinberg. 1962. Koroni: A Ptolemaic Camp on the East Coast of Attica. *Hesperia* 31.1: 26–61.

Vandiver, P.B., and C.G. Koehler. 1986. Structure, Processing, Properties, and Style of Corinthian Transport Amphoras. In *Ceramics and Civilization* 2. *Technology and Style*. Ed. W.D. Kingery. 173–215. Columbus.

Vatin, C. 1991. La moisson d'or de Métaponte. In *Monuments votifs de Delphes*. Ed. C. Vatin. 53–67. Rome.

Vegas, M. 1973. *Cerámica común romana*. Barcelona.

Vera, D. 1994. Il sistema agrario tardoantico: un modello. In *Storia dell'alto Medioevo*, 136–138.

Verbicaro, G. 2006. Uno scarico di materiali nell'area dell'edificio termale. In *Ricerche nel santuario di Hera Lacinia*. Ed. R. Spadea. 81–91. Rome.

Verhoogen, V. 1956. *Terres cuites grecques aux Musées Royaux d'Art et d'Histoire*. Bruxelles.

Vernon-Harcourt, L.F. 1896. *Rivers and Canals: The Flow, Control, and Improvement of Rivers and the Design, Construction, and Development of Canals Both for Navigation and Irrigation, with Statistics of the Traffic on Inland Waterways.* 2nd ed. Oxford: Clarendon Press.

Vezzani, L. 1967. I depositi plio-pleistocenici del litorale ionico della Lucania. *AAGSNC* Ser. 6, 18: 159–180.

———. 1975. Lithostratigraphic Complexes and Evidence for Tectonic Phases in the Molise-Puglia-Lucania Apennines. In *Structural Model of Italy*, 329–363.

Villa, L. 1994. Le anfore tra tardoantico e medioevo. In *Ad mensam. Manufatti d'uso da contesti archeologici fra tarda antichità e medioevo*. Ed. S. Lusuardi Siena. 335–431. Udine.

Vita-Finzi, C. 1969. The Mediterranean Valleys: Geological Changes in Historical Times. Cambridge.

———. 1975. Late Quaternary Alluvial Deposits in Italy. In *Geology of Italy*, II.329–340.

Vittoria, E. 2000. *Materiale inedito proveniente dalla necropoli di Metaponto in località Pizzica, proprietà D'Onofrio*. Tesi di laurea, Università degli Studi della Tuscia (Viterbo).

Volpe, G. 1990. *La Daunia nell'età della romanizzazione. Paesaggio agrario, produzione, scambi*. Bari.

———. 1996. *Contadini, pastori e mercanti nell'Apulia tardoantica*. Bari.

———. 1999. Paesaggi della Puglia tardoantica. *Atti Taranto* 38: 267–329.

Von Falkenhausen, V. 1978. *La dominazione bizantina nell'Italia meridionale dal IX al XI secolo*. Bari.

Von Fritz, K. 1940. *Pythagorean Politics in Southern Italy. An Analysis of the Sources*. New York.

Waagé, F.O. 1934. Preliminary Report on the Medieval Pottery from Corinth, I: The Prototype of the Archaic Italian Maiolica. *Hesperia* 3.2: 129–139.

Wade, M. 2002. Black Gloss Pottery from the Pantanello Sanctuary: A Progress Report. In *The Study of Ancient Territories. Chersonesos and Metaponto. 2002 Annual Report, Institute of Classical Archaeology*. Ed. J.C. Carter. 67–68. Austin.

Walters, H.B. 1926. *Catalogue of the Engraved Gems and Cameos, Greek, Etruscan and Roman in the British Museum*. London.

Ward-Perkins, J.B., M.A. Cotton, H.B. van der Poel, E. Macnamara, J. du Plat Taylor, and A. Carter. 1969. Excavations at Botromagno, Gravina di Puglia: Second Interim Report, 1967-1968. *PBSR* 37: 100–157.

Waters, M.R. 1992. *Geoarchaeology: A North American Perspective*. Tucson.

Watson, J.S. 1853. *Justinus, Marcus Junianus; Nepos, Cornelius; Eutropus*. London.

Webb, W.P. 1951. *The Great Frontier*. Austin.

Webster, T.B.L. 1968. Towards a Classification of Apulian Gnathia. *BICS* 15: 1–33.

Wells, S.G., L.D. McFadden, and J.C. Dohrenwend. 1987. Influence of Late Quaternary Climatic Changes on Geomorphic and Pedogenic Processes on a Desert Piedmont, Eastern Mojave Desert, California. *QR* 27: 130–146.

West, M.L. 1971. *Iambi et elegi graeci ante Alexandrum cantati* I, *Archilochus, Hipponax, Theognidea*. Oxford.

Wheatley, D., and M. Gillings. 2002. *Spatial Technology and Archaeology: The Archaeological Applications of GIS*. London/New York.

Whitbread, I.K. 1995. *Greek Transport Amphorae: A Petrological and Archaeological Study*. (*Fitch Laboratory Occasional Papers* 4.) Athens.

White, H., trans. 1912. *Appian's Roman History with an English Translation* I *(Books I–VIII)*. Cambridge, MA/London: Loeb Classical Library.

Whitehouse, D. 1966a. Medieval Painted Pottery in South and Central Italy. *MedievArch* 10: 30-44.

———. 1966b. Ceramiche e vetri medioevali provenienti dal castello di Lucera. *BdA* 3–4: 171–178.

———. 1967. The Pottery. In Mallet, M., and D. Whitehouse, "Castel Porciano: An Abandoned Medieval Village of the Roman Campagna." *PBSR* 35: 113–146.

———. 1969. Italy. In "Red-painted and Glazed Pottery in Western Europe from the Eighth to the Twelfth Century." Ed. J.C. Hurst. *MedievArch* 13: 137–143.

———. 1970. Excavations at Satriano: A Deserted Medieval Settlement in Basilicata. *PBSR* 38: 188–219.

———. 1988. The Medieval Pottery. In *Ordona* VIII, 296–321.

Whitehouse, D., and R. Whitehouse. 1969. Excavations at Anglona. *PBSR* 37: 34–75.

Whitehouse, R.D. 1968. The Early Neolithic of Southern Italy. *Antiquity* 42: 188–193.

———. 1981. Prehistoric Settlement Patterns in Southeast Italy. In *Papers in Archaeology* II. *Archaeology and Italian Society: Prehistoric, Roman and Medieval Studies*. (*BAR-IS* 102.) Eds. G. Barker and R. Hodges. 157–165. Oxford.

Whitehouse, R., and J. Wilkins. 1989. Greeks and Natives in South-east Italy. In *Centre and Periphery: Comparative Studies in Archaeology*. Ed. T. Champion. 102–126. London.

Whitelaw, T.M. 1994. An Ethnoarchaeological Study of Rural Land-use in North-west Keos: Insights and Implications for the Study of Past Aegean Landscapes. In *Structures rurales*, 163–176.

Wikander, O., and C. Wikander. 2000. The Greek World. In *Handbook of Water Technology*. Ed. O. Wikander. 557–702. Leiden.

Will, E.L. 1982. Greco-Italic Amphoras. *Hesperia* 51.3: 338–356.

Wilson, A. 2000. Land Drainage. In *Handbook of Ancient Water Technology*. Ed. Ö. Wikander. 303–317. Leiden.

Winter, F. 1903. *Die Typen der figürlichen Terrakotten* I. Berlin/Stuttgart.

Wise, D.U., R. Funiciello, M. Parotto, and F. Salvini. 1981. Topographic Lineament Swarms: Clues to Their Origin from Fracture Domain Analysis of Italy CABS (abstr.). In *Program of the Annual Meeting of the Geological Society of America*. 584. Cincinnati.

———. 1985. Topographic Lineament Swarms: Clues to Their Origin from Domain Analysis of Italy. *BGSA* 96: 952–967.

Wiseman, T.P. 1964. Via Anniae. *PBSR* 32: 21–37.

Wright, J.C., J.F. Cherry, J.L. Davis, E. Mantzourani, S.B. Sutton, R.F. Sutton, Jr. 1990. The Nemea Valley Archaeological Project: A Preliminary Report. *Hesperia* 59.4: 579–659.

Wright, K.I. 2000. The Social Origins of Cooking and Dining in Early Villages of Western Asia. *PPS* 66: 89–121.

Wuilleumier, P. 1939. *Tarante des origins à la conquete romaine*. Paris.

Yntema, D. 1990a. *The Matt-painted Pottery of Southern Italy. A General Survey of the Matt-painted Pottery Styles of Southern Italy During the Final Bronze Age and the Iron Age*. Galatina.

———. 1990b. A Specific Group of Black-Gloss Ware Excavated at Valesio: The HFR Group and Its Connections. *StAnt* 6: 167–186.

———. 1993. *In Search of an Ancient Countryside. The Amsterdam Free University Field Survey at Oria Province of Brindisi, South Italy (1981–1983)*. Amsterdam.

———. 1995a. Salento and the Eastern Mediterranean in the Middle and Late Hellenistic Period: Some "Eastern" Ceramic Evidence (Fine Wares) from Valesio, Province of Brindisi. *StAnt* 8.2: 387–404.

———. 1995b. The Artefacts of the Roman Period. In *Mutatio Valentia. The Late Roman Baths at Valesio, Salento*. Ed. J.S. Boersma. 265–294. Amsterdam.

———. 2001. *Pre-Roman Valesio. Excavations of the Amsterdam Free University at Valesio, Province of Brindisi, southern Italy. I. The Pottery*. Amsterdam.

———. 2005. *Conspectus Formarum of Apulian Grey Gloss Wares* (ceramica a pasta grigia): *Inventory of Forms of the Apulian Grey Gloss Wares*. Amsterdam.

Yorston, R.M., V.L. Gaffney, and P.J. Reynolds. 1990. Simulation of Artefact Movement Due to Cultivation. *JAS* 17: 67–83.

Young, J.H. 1956. Studies in South Attica: Country Estates at Sounion. *Hesperia* 25: 122–146.

Zaccagnino, C. 1998. *Il thymiaterion nel mondo greco. Analisi delle fonti, tipologia, impieghi*. Rome.

Zaccaria Ruggiu, A. 1980. *Le lucerne fittili del Museo Civico di Treviso*. Rome.

Zamani, A., and H. Maroukian. 1981. A Morphotectonic Investigation in Northwestern Crete: The Peninsula of Akrotiri. *ZG*, N.F. Suppl. 40: 151–164.

Zancani Montuoro, P. 1974–1976. La leggenda di Epeio. *AttiMGrecia* 15–17: 93–106.

———. 1980–1982. Francavilla Marittima, necropoli e ceramico a Macchiabate. Zona T. *AttiMGrecia* 21–23: 7–129.

Index

Italic page numbers indicate material in tables or figures; **bold** page numbers indicate material in catalogs.

Illustration Credits

ALP	Ann Patterson
ALC	Aldo La Capra
AP	Alberto Prieto
BB	Barbara Bini
BP	Bruno Perini
CDA	Cesare D'Annibale
CP	Claes Petersson
CR	Cesare Raho Chris Reneau
CW	Chris Williams
DB	Deena Berg
DD	Dan Davis
EH	Esteban Hinojosa
ELC	Erminia Lapadula
EL	Elisa Lanza Catti
JCC	Joseph Coleman Carter
JT	Jessica Trelogan
JTA	James T. Abbott
ML	Max Limoncelli
MLG	Maria Luisa Giulianni
PHD	Peter H. Dana
PO	P. Orlandini
PD	Pam Dyson
RLF	Robert L. Folk
SMT	Steve M. Thompson
TH	Tom Hale
VM	Vera Massaro